# Metallurgy for Engineers

TO
MY WIFE

# Metallurgy for Engineers

E. C. ROLLASON, MSc, PhD, FIM.
Formerly Chairman of School of Metallurgy and
Hanson Professor and Head of Industrial Metallurgy, University of Birmingham.
Previously Henry Bell Wortley Professor of Metallurgy, University of Liverpool.
Delegate Director and Research Manager, Murex Welding Processes Ltd.

EDWARD ARNOLD

© E. C. Rollason 1973

FIRST PUBLISHED 1939
by Edward Arnold (Publishers) Ltd
41 Bedford Square, London WC1B 3DQ

REPRINTED 1940, 1941, 1942, 1943 (twice), 1945, 1947
SECOND EDITION 1949
REPRINTED 1951, 1954, 1955, 1957, 1958, 1959
THIRD EDITION 1961
REPRINTED 1963, 1964, 1965, 1968, 1970
FOURTH EDITION 1973
REPRINTED 1975, 1977, 1978, 1980, 1982, 1984

ISBN 0 7131 3282 5


All Rights Reserved. No part of this publication may be reproduced, stored in a retrieval system, or transmitted in any form or by any means, electronic, mechanical, photocopying, recording or otherwise, without the prior permission of Edward Arnold (Publishers) Ltd.

Printed in Great Britain by
Richard Clay (The Chaucer Press) Ltd,
Bungay, Suffolk

# Preface to the Fourth Edition

In this fourth edition the overall format of previous editions has been retained but the electrical and magnetic alloys have been collected together, with additions, into a separate chapter and a new chapter added which deals briefly with the characteristic structures and properties of polymers which are now used extensively by engineers.

The opportunity has been taken of changing to SI units where possible but at the time of writing many British Standards have not been converted and some indecision exists as to the actual units to be used finally. For example, notch tough steels are being graded on their strengths in hectobars, but the trend is now towards $N/mm^2$. British Standards have changed greatly in their designations of alloys and this is particularly so in the revised BS 970 in which the well known En Nos. have been changed. These changes have been incorporated in the test but a limitation has been imposed by the fact that only the 'imperial unit' specification is available.

The chapter on light alloys has been largely re-written and additions have been made to most chapters. Details of many new alloys have been added together with new strengthening methods such as thermomechanical treatments, fibre reinforcement, etc. New topics such as fracture mechanics, superplasticity, Pourbaix diagrams, new steel retreatment processes, selection of alloys and machining are dealt with briefly.

Acknowledgments are made to many friends who have given me helpful comments, especially Dr. W. M. Doyle with regard to aluminium alloys.

# Preface to the First Edition

The need of the engineer for metallurgical training has been recognised in University and Technical College courses of education and by the Institutions of Mechanical, Production and Civil Engineers. The object of this book is to provide a condensed description of metallurgical theory and practice, both ferrous and non-ferrous, now required by engineering students. It should also cover much of the physical metallurgy necessary in the examinations of the City and Guilds of London Institute, and should therefore be useful to young metallurgists and welders. In its condensed form the book should prove of the greatest assistance when used in conjunction with a teacher.

Practical details of specimen preparation are given which should enable the reader to commence the macro- and micro-examination of metals. Chaps. 4 to 6 deal with general metallurgical principles common to all metals, ferrous and non-ferrous.

Certain theories are highly controversial, and in a small general book it is impossible to discuss fully the pros and cons. In most cases one explanation of the phenomenon has been given, as in the author's opinion a working theory helps the student to remember the practical facts. Reference to 'amorphous metal' has been avoided.

Only sufficient production metallurgy has been given to show how certain properties of a metal depend on its process of manufacture. It is hoped that the book will carry the student on to such a point that he can read with profit the numerous specialised publications now available.

Except where otherwise stated, the photographs have been taken by the author in the Metallurgy Department of the University or the County Technical College, Wednesbury.

The author is indebted to Messrs T. G. Bamford, MSc, AIC, W. Chapman, MSc, J. W. Jones, MSc, F. Johnson, DSc, and T. Wright, PhD, for helpful criticisms; and for the loan of blocks to Messrs Murex Welding Processes, Ltd, *Metallurgia, Sheet Metal Industries* and R. & J. Beck, Ltd.

# Contents

*Handwritten annotations:*
96-101, 160-165, (192-194), 200-203, 212-215, (224-233)
(270-278), 280-283, 245-248, 256-7, 408-9, 59-70
Ch7, Ch19, Ch20, Ch1, Ch8,

| | | |
|---|---|---:|
| | Preface to the Fourth Edition | v |
| | Preface to the First Edition | vi |
| | Abbreviations used in Text | viii |
| 1 | Mechanical and Non-destructive Tests | 1 |
| 2 | The Macro-examination of Metals | 39 |
| 3 | Micro-examination | 45 |
| 4 | Solidification of Metals | 59 |
| 5 | Equilibrium Diagrams and their Interpretation | 71 |
| 6 | Hardening of Metals | 93 |
| 7 | Deformation and Annealing of Metals | 104 |
| 8 | Corrosion and Oxidation Phenomena | 138 |
| 9 | Irons and Carbon Steels | 152 |
| 10 | Heat-treatment of Steels | 175 |
| 11 | Alloy Steels | 212 |
| 12 | Stainless, Creep and Heat-resisting Steels | 245 |
| 13 | Cast Irons | 270 |
| 14 | Electrical and Magnetic Alloys | 286 |
| 15 | Copper and its Alloys | 300 |
| 16 | Aluminium, Magnesium and Light Alloys | 327 |
| 17 | Titanium and other New Metals | 351 |
| 18 | Miscellaneous Non-ferrous Metals and Alloys | 368 |
| 19 | Polymers | 378 |
| 20 | Metallurgical Aspects of Metal-joining | 393 |
| 21 | The Measurement of Temperature | 415 |
| | Selected Bibliography | 429 |
| | Appendix I Periodic Table of Elements | 431 |
| | II Metal Compatibility | 432 |
| | III Conversion Data | 433 |
| | IV Comparison of En Nos. and Equivalent Values in BS 970 | 435 |
| | Index | 436 |

# Abbreviations used in Text

*TS = Ultimate tensile strength.
*YP = Yield point.
*LP = Limit of proportionality.
*PS = Proof stress (0·1%).
BS = British Standard Specification.
BH = Brinell hardness.
El = Elongation per cent on 5·65 $\sqrt{A}$ gauge length.
VPN = Vickers pyramid number (Hardness).
RA = Reduction of area, per cent.
$\mu$m = micron = one-thousandth mm.
Å = ångström unit = one-ten thousandth micron.
MeV = million electron volts.

*These values are given in newtons per square millimetre and also in tons per square inch where appropriate.

| | | | | | |
|---|---|---|---|---|---|
| Al | = Aluminium | Au | = Gold | Rh | = Rhodium |
| Sb | = Antimony | Hf | = Hafnium | Se | = Selenium |
| A | = Argon | H | = Hydrogen | Si | = Silicon |
| As | = Arsenic | In | = Indium | Ag | = Silver |
| Ba | = Barium | Ir | = Iridium | S | = Sulphur |
| Be | = Beryllium | Fe | = Iron | Ta | = Tantalum |
| Bi | = Bismuth | Kr | = Krypton | Te | = Tellurium |
| B | = Boron | Pb | = Lead | Th | = Thorium |
| Cd | = Cadmium | Mg | = Magnesium | Sn | = Tin |
| Ca | = Calcium | Mn | = Maganese | Ti | = Titanium |
| C | = Carbon | Mo | = Molybdenum | W | = Tungsten |
| Ce | = Cerium | Ni | = Nickel | U | = Uranium |
| Cr | = Chromium | Nb | = Niobium | Zn | = Zinc |
| Co | = Cobalt | P | = Phosphorus | Zr | = Zirconium |
| Cu | = Copper | Pu | = Plutonium | | |
| Ge | = Germanium | Ra | = Radium | | |

## Greek Letters

| | | | | | |
|---|---|---|---|---|---|
| $\alpha$ | alpha | $\eta$ | eta | $\pi$ | pi |
| $\beta$ | beta | $\theta$ | theta | $\rho$ | rho |
| $\gamma$ | gamma | $\iota$ | iota | $\sigma$ | sigma |
| $\Delta$, $\delta$ | delta | $\varkappa$ | kappa | $\tau$ | tau |
| | | $\lambda$ | lambda | $\phi$ | phi |
| $\varepsilon$ | epsilon | $\mu$ | mu | $\omega$ | omega |
| $\zeta$ | zeta | $\nu$ | nu | | |

# 1 Mechanical and Non-destructive Tests

*Introduction*

During the last fifty years vast strides have been made in the advance of metallurgy. As a result, a wide range of materials is now available for the engineer. Among them are alloys showing a great increase in strength and in resistance to fatigue which have proved useful in saving weight in aeroplanes and automobiles; alloys having a high resistance to corrosion and others possessing considerable strength at elevated temperatures are invaluable for chemical plant and for boilers working at high temperatures and pressures; new tool steels and alloys have increased production rates; and finally, a number of metals which a few years ago were little more than laboratory curiosities.

The use of such new alloys, however, often presents problems of new methods in the forge, heat-treatment or machine shop. If the full advantages are to be obtained it is essential that the engineer be equipped with the fullest possible information as to the peculiar characteristics of the materials he is using.

Too frequently the engineer is only concerned with the mechanical properties of the metals he uses, but composition, casting, shaping and heat-treatment have a tremendous effect on the ultimate behaviour of the new alloys. Consequently, some knowledge of metallurgical theory and practice is becoming more and more essential to the modern engineer, and this involves such things as microscopic examination of metals and the interpretation of the structures revealed. There exists a close relationship between structure and properties of metals (p. 52) and the structure is established by the past history of the metal. For example in

(*a*) MANUFACTURE, where imperfections are grown into the metal (p. 93)
(*b*) FABRICATION

 (i) by *alloying* (Chap. 5);
 (ii) by *grain size control*, by casting (p. 61), by heat-treatment (p. 175), by cold working and annealing (p. 121) or hot working (p. 127);
 (iii) by *mechanical working*, producing work hardening and directional properties (p. 117);
 (iv) by *heat-treatment* to harden by allotropic transformation or by precipitation of particles.

(*c*) SERVICE, by corrosion (p. 139), creep (p. 21) and nuclear irradiation (p. 361).

At his first contact with physical metallurgy the engineer finds himself confronted with peculiar equilibrium diagrams, and micro-constituents which are denoted by letters of the Greek alphabet. At first it is discouraging, but a little study will soon produce familiarity with the essential jargon of the metallurgist and will enable the structures of metals and alloys of industrial importance to be considered in the light of their constitution.

The subject of mechanical testing of metals is an important one to both the engineer and the metallurgist. During the seventeenth century Robert Hooke established the famous law named after him, which forms the basis of many mechanical tests, but it was the industrial revolution that provided the impetus to the study of mechanical testing. The development of modern transport, particularly the aeroplane, is associated with a vast increase in varieties of alloys and the necessity for making full use of their properties. More recently, the development of nuclear energy has stimulated the need for entirely new metals and alloys. Today, more attention is being given to the interpretation of the test results in terms of service performance, but this is hampered by the fact that some of the tests are of a purely empirical nature. Nevertheless, many of these have been proved to give reliable indications of the ability of the material to perform certain types of duty.

Mechanical tests are employed, too, in *investigational* work in order to obtain data for use in design; also for *acceptance* work, the main purpose of which is to check whether the material meets the specification. For this latter purpose the tests should yield the information accurately, rapidly and economically.

In the U.K. the British Standards Institution (BSI) provides facilities for the drafting and publication of specifications, which frequently contain clauses relating to method of manufacture, chemical composition, heat-treatment, the selection of test pieces and mechanical tests. The Air Ministry also issues DTD specifications of aircraft materials not dealt with by the BSI. In America the American Society for Testing Materials (ASTM) publish yearly proceedings which contain research papers and tentative standards. The ASTM standards are published every three years and contain standard specifications and methods of testing adopted by the Society. The New International Association for Testing Materials provides means for discussing tests and issuing papers in English, French and German, but it does not put out definite specifications.

## Hardness

The property of hardness largely determines the resistance to scratching, wear, penetration, machinability and the ability to cut. Some thirty methods have been used for measuring hardness, but only four important tests can be considered here.

# HARDNESS

(*a*) *The Brinell test* consists of indenting the surface of the metal by a hardened steel ball under a load, and measuring the average diameter of the impression with a low-power portable microscope fitted with a scale. The spherical area is calculated from the diameter of the impression. The Brinell number, H, is:

$$H = \frac{\text{load}}{\text{area}} = \frac{P}{\frac{\pi D}{2}[D - \sqrt{(D^2 - d^2)}]} \text{ kg/mm}^2$$

where P = load (kilogrammes)
D = diameter of ball (millimetres)
d = diameter of impression (millimetres).

To obviate calculation, the Brinell numbers are usually obtained from tables giving values of impression diameters and corresponding Brinell numbers. Typical Brinell numbers are given for a few materials in Table 1, compared with hardness values obtained by other tests.

The diameter of the ball is usually 10 mm and the load 3000 kg for steel, 1000 kg for copper and 500 kg for aluminium. If other sizes of ball are used the load is varied according to the relation: $\frac{P}{D^2}$ = constant. The constant is 30 for steel, 10 for copper and 5 for aluminium.

The *time* of loading is 15 seconds and the thickness of the specimen should not be less than 10 times the depth of the impression.

The *piling up* of metal round the edge of the impression indicates a low rate of hardening by deformation; *sinking* denotes the ability for work-hardening (Fig. 1).

The approximate tensile strength (N/mm$^2$) of steels can be obtained by multiplying the Brinell number by 3·54 for annealed condition and by 3·24 for quenched and tempered steels.

TABLE 1   COMPARISON OF HARDNESS VALUES

| Material | Brinell | | Vickers Pyramid Number | Rockwell Scale | | Shore |
|---|---|---|---|---|---|---|
| | Impression dia, mm | Number | | C | B | |
| Soft brass | — | 60 | 61 | — | — | — |
| Mild steel | 5·20 | 131 | 131 | — | 74 | 20 |
| Soft chisel steel | 3·95 | 235 | 235 | 22 | 99 | 34 |
| White cast iron | 3·00 | 415 | 437 | 44 | 114 | 57 |
| Nitrided surface | 2·25 | 745 | 1050 | 68 | — | 100 |

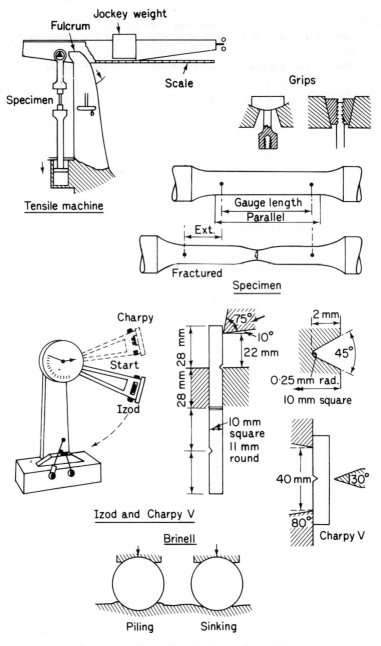

FIG. 1 Tensile and Izod Charpy V machines

Errors arise when the Brinell test is used on very hard materials, resulting in low values owing to: (a) the spherical shape of indentor, (b) flattening of the ball.

(b) *The Vickers machine.* The errors mentioned above are eliminated in this machine by using a diamond square-based pyramid which does not readily deform and which gives geometrically similar impressions under different loads. The angularity of the pyramid is 136° and loads ranging from 5 to 120 kg can be used. The rate and duration of loading are controlled by a piston and a dashpot of oil. A microscope can be swung over the square impression, the diagonals of which are measured between knife edges instead of hair wires or scale, and the reading is taken from a digit counter. The load divided by the contact area of the impression gives the Vickers pyramid number (VPN).

Since the impressions are small, the machine is very suitable for testing polished and hardened material.

The Brinell and Vickers hardness values are practically identical up to a hardness of 300. The Brinell number is not reliable above 600, as shown in Table 1.

(c) *Rockwell test.* This uses either a steel ball 1·58 mm diameter loaded with 100 kg (Scale B) or 150 kg on a diamond cone having a 120° angle (Scale C). The penetrator is first loaded with a minor load of 10 kg to take out any slack in the machine and the indicator, for measuring the depths of the impression, set to zero. The above major load is applied and after its removal the dial gauge records the depth of the impression in terms of Rockwell numbers. The Rockwell test is particularly useful for rapid routine tests on finished products.

(d) *The Shore scleroscope* consists of a small diamond-pointed hammer weighing 2·5 g, which is allowed to fall freely from a height of 250 mm down a glass tube graduated into 140 equal parts. The height of the first rebound is taken as the index of hardness. This test is used for testing rolls, dies and gears and leaves no visible impression.

## The tensile test

This is a commonly used test for indicating the strength and ductility of a metal. *Ductility* may be defined as the property which enables metals to be drawn into wire, involving the use of a tensile force. *Malleability* is the property of being permanently extended in all directions when subjected to compressive forces. The tensile testing machine consists of two essential parts: (a) the *straining* device, such as a screw and nut, driven electrically, or a plunger subjected to oil pressure; (b) the *measurement of the load* on the specimen, determined from the oil pressure, care having been taken to obviate errors due to friction, or from the position of a jockey-weight on a beam balanced upon a knife edge and also attached to one end of the specimen. During a test the weight is slid along the beam in order to keep it floating as the load on the specimen increases (Fig. 1).

The specimen may be round or rectangular in cross-section, depending on the nature of the product being tested. The centre is usually reduced in section to form the gauge length. A sharp change in section from the parallel portion to the shoulders should be avoided, otherwise concentrations of stress occur and a brittle material would fracture at an apparently low stress. With ductile materials deformation occurs at such places, distributing the concentration of stress over a wider area. A similar effect occurs when the line of action of the force is eccentric to the axis of the specimen. This effect can be obviated by the use of self-aligning grips fitted with spherical seats in the testing machine. The specimen may be held by flanges, pins, screw thread or serrated wedges (Fig. 1).

Before placing the specimen in the machine its diameter is measured with a micrometer, and two small 'pop' marks are made at a distance apart corresponding to the gauge length on which the extension is to be measured.

*Stress-strain curve*

The relation of extension to the applied stress is shown in Fig. 2A for mild steel. Up to E the extension of the specimen is very small and necessitates the use of magnifying devices—called *extensometers*—for its measurement. On removing the load the specimen returns to its previous dimensions and exhibits elasticity. The stress at E is called the *elastic limit* and in some metals point E does not coincide with P, which denotes the point at which the curve deviates from a straight line; the stress corresponding to P is called the *limit of proportionality*. Up to P, therefore, the stress is proportional to the strain and can be calculated:

$$\text{Stress} = \frac{\text{Load}}{\text{Cross-sectional area}} \text{ e.g., N/mm}^2$$

$$\text{Strain} = \frac{\text{Extension of gauge length}}{\text{Original gauge length}}$$

Therefore by Hooke's Law

$$\frac{\text{Stress}}{\text{Strain}} = \frac{\frac{\text{Load}}{\text{Cross-sectional area}}}{\frac{\text{Extension of gauge length}}{\text{Original gauge length}}}$$

$$= E, \text{ a constant known as Young's Modulus.}$$

Young's Modulus is fixed by the nature of the material and for ordinary steel the value of about 200 kN/mm$^2$ is not much affected by composition

# THE TENSILE TEST

or heat-treatment, but decreases with temperature. The value for steels decreases 25% between 15 °C and 600 °C. The springiness of a material is indicated by its Young's Modulus and when stiffness is required by design, the Modulus has to be considered when changing from mild steel to high tensile steels with similar values or to non-ferrous metals with low values of Young's Modulus.

Recent investigations have shown, however, that the limit of proportionality becomes lower as more sensitive instruments are used to measure the extension. Some of the high values now in use are due to the inability of the extensometer to detect small amounts of permanent extension.

As the load on the test piece is increased beyond the elastic limit there comes a point at which there is a sudden extension, indicated by the drop of the beam and continued extension with a lower load. If the load is removed, the specimen does not recover its original dimensions and it is said to have undergone plastic deformation or plastic flow. A coating of brittle lacquer is useful for indicating plastic yielding.

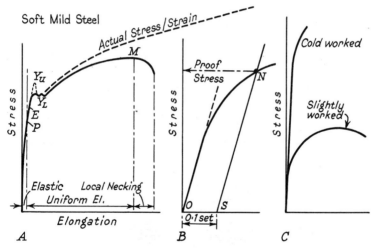

FIG. 2 Stress-elongation curves

In Fig. 2A the *upper yield point* is denoted by $Y_U$, the highest stress before sudden extension occurs, and its value is affected by surface finish, shape of test piece and rate of loading. The *lower yield point*, which is normally measured in commercial testing, is denoted by $Y_L$, the lowest stress producing the large elongation (p. 113).

This large elongation of the metal only occurs in a very few materials such as wrought iron and mild steels. 'Stretcher strains' (p. 119) occur in the latter owing to this pronounced yield point, but can be prevented by temper rolling the steel, with the result that the stress-strain curve is smoothed out (Fig. 2c). *Heavy* cold working raises the ultimate tensile stress, yield point and limit of proportionality, but reduces the elongation.

Much engineering design has been based on tensile strength rather than yield strength, which is the primary criterion of load-carrying ability. With the older materials of engineering the ratio of yield to tensile strength is fairly constant, but the newer materials, especially alloy steels, have much higher ratios and design should be based on yield strength or proof stress, unless fatigue (p. 24) is an important criterion.

As the load is increased the specimen continues to extend (i.e. plastic deformation) until a constriction occurs in the gauge length and the beam drops. This corresponds to the *maximum load* (M, Fig. 2A) on the specimen. The load now acts on a diminishing area and produces a stress sufficient to break the specimen. The *actual breaking stress*, which is never used in practice, is obtained by dividing the breaking load by the actual sectional area of the fracture. The use of such actual stresses would give the broken curve in Fig. 2A, the elongation being calculated from the diameter of the restricted region.

Much fundamental significance is being attached to the true stress-strain curve. The slope of the straight-line portion from maximum load to fracture seems to afford a measure of work-hardening capacity of the metal; while the area under the curve indicates degree of toughness.

*Commercial results*

In a commercial test the two broken ends of the specimen are fitted together and the distance between the gauge marks and also the smallest diameter of the local neck are measured. The results are calculated as follows:

$$\text{Yield point} = \frac{\text{Yield load}}{\text{Original cross-sectional area}}$$

$$\text{Tensile stress (TS)} = \frac{\text{Maximum load}}{\text{Original cross-sectional area}}$$

$$\text{Elongation, \% on gauge length} = \frac{\text{Extension}}{\text{Original gauge length}} \times 100$$

$$\text{Reduction of area, \%} = \left(\frac{\text{Original area} - \text{Final area}}{\text{Original area}}\right) \times 100.$$

*Proof stress*

For the harder steels and non-ferrous metals there is no sharply defined yielding of the material, and a curve is obtained as shown in Fig. 2B. For aircraft materials it has become essential to specify a stress which corresponds to a definite amount of permanent extension, and it is called *proof stress*, commonly obtained as illustrated in Fig. 2B. A line, SN, is drawn from S, so that OS represents 0·1%, or some other permanent set. SN is drawn parallel to the rectilinear portion of the curve and cutting it at N. The stress corresponding to N is the 0·1% *proof stress*. The material fulfils the specification if, after the proof stress is applied for 15 seconds and removed, the specimen has not permanently extended more than 0·1% of the gauge length.

*Elongation*

The total elongation of a test piece is made up of two separate parts:
(*a*) *Uniform extension*, which occurs up to the maximum load (M) and is proportional to the gauge length.
(*b*) *Local extension*, due to the necking. This is independent of the gauge length, but varies with the cross-sectional area of the specimen. As the gauge length is increased, the effect of the necking on the value of the total elongation decreases, as shown in Fig. 3. It is essential, therefore, that the

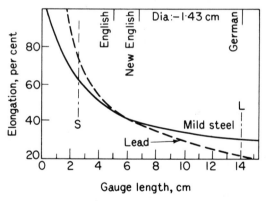

Fig. 3 Effect of gauge length on percentage elongation

gauge length should be stated when an elongation figure is mentioned. In practice, it is found that geometrically similar test bars of a given material deform similarly (Barba's Law), giving the same percentage elongation if:

$$\frac{\text{Gauge length}}{\sqrt{(\text{Cross-sectional area})}} = \text{Constant},$$

which allows for variation of the gauge length with the cross-sectional area of a specimen. Unfortunately the same constant is not used in various parts of the world and consequently values of elongation cannot be directly compared with each other without the use of formulae. The gauge length used in England was $4\sqrt{(\text{Area})}$, but has now been changed to $5\cdot65\sqrt{(\text{Area})}$ i.e. 5 dia for round specimens (ISO) in Germany, $11\cdot3\sqrt{(\text{Area})}$; and in America, $4\cdot47\sqrt{(\text{Area})}$.

Values of elongation obtained on such arbitrary gauge lengths do not afford a guide as to the relative manner in which two different materials will behave in service. From Fig. 3 it will be seen that if measurement is made on the gauge length S, the steel appears to be better, while if the gauge length L is used the reverse is the case. For steel plates a fixed gauge length of 200 mm is used with a width varying with the thickness of the plate.

The values of elongation and of reduction of area indicate the ductility of a metal. A high reduction of area indicates that the material has a low rate of hardening by deformation. 18/8 stainless steel has a high elongation but a low reduction of area, and this indicates that the alloy is tough but hardens extremely rapidly when deformed.

## Speed of loading

It is necessary to take into account the wide variation of actual strain rates in a tensile test which may affect the results obtained. For measurement of yield stresses a rate of strain and of stress not exceeding 0·15 min and 0·3 N/mm$^2$/s respectively is recommended. Speed is particularly important in the case of soft metals and also of hard metals tested at high temperatures. At high testing speeds the tensile stress increases and the elongation decreases. At sufficiently high speeds there may not be time to conduct away the heat produced by plastic working and the temperature may rise sufficiently to soften the deforming material. Plastic flow may thus be concentrated in zones which eventually rupture. Outside the zone of failure little deformation is evident and an apparent brittle failure is indicated.

On the other hand, the embrittling effect of hydrogen on high tensile steel becomes more effective on slow or sustained dead loading (p. 232).

## Effect of previous deformation

With steels deformation only slightly in excess of the yield point has the effect of lowering the limit of proportionality. After a rest at room temperature or a short time at 100–300 °C, the proportional limit for tension stresses is raised, accompanied by a reduction in the corresponding value for compression stresses. (See also pp. 113, 171.)

|                     | Yield Point N/mm² | Proportional Limit N/mm² |
|---------------------|-------------------|--------------------------|
| First loading       | 256               | 238                      |
| Overstrained        | 266               | 164                      |
| Strained and rested | 357               | 337                      |

*Heavy* deformation affects the stress-strain curve as shown in Fig. 2c.

## Fracture

*Ductile fracture.* A pure and inclusion free metal can elongate under tension to give approx 100% RA and a point fracture, Fig. 4. Most alloys contain second phases which lose cohesion with the matrix or fracture and the voids so formed grow as dislocations flow into them. Coalescence of the voids forms a continuous fracture surface followed by failure of the remaining annulus of material usually on plane at 45° to the tension axis. The central fracture surface consists of numerous cup-like depressions generally called dimples. The shape of the dimples is strongly influenced by the direction of major stresses—circular in pure tension and parabolic under shear. Dimple size depends largely on the number of inclusion sites. Fig. 5(*a*) shows typical dimples.

Some important features of ductile fracture can be summarised as follows:

(1) Pure metals and solid solutions that are relatively free from second phase particles (including impurity particles) are usually more ductile than strong two phase alloys.
(2) The local stress required for hole *nucleation* at particles depends on their resistance to cracking and the strength of their bond with the matrix.
(3) The local stress generated at the particles depends on the flow strength of the alloy, the applied strain and the shape and size of the particles.
(4) *Growth* of the holes, so that they coalesce to form a macroscopic fracture, depends on the applied stresses being *tensile*. Much higher ductilities are achieved in *compressive* straining.

In *cleavage* fracture the material fails along well defined crystallographic planes within the grain but the crack path is affected by grain boundaries and inclusions. Basically a cleavage fracture surface contains large smooth areas separated by cleavage steps and feathers, river markings and cleavage tongues which are the direct result of crack path disturbances—Fig. 5(*b*).

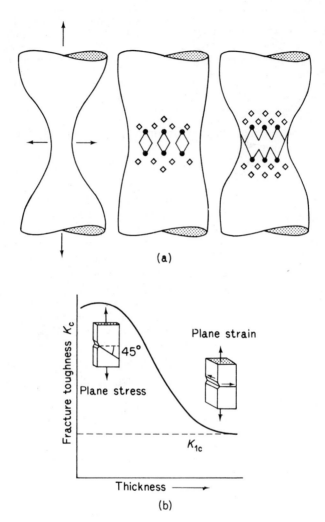

FIG. 4 (*a*) Stages in ductile fracture from inclusions
(*b*) Fracture toughness *v* thickness

*Intercrystalline* fracture is characterised by separation of the grains to reveal a surface composed of grain boundary facets, Fig. 5(*c*). This type of fracture is found in stress-corrosion (p. 120) creep hot tearing (p. 23) and hydrogen embrittlement (p. 131).

*Fatigue fractures* are characterised by striations (Fig. 5(*d*)) representing the extent of crack propagation under each cycle of loading.

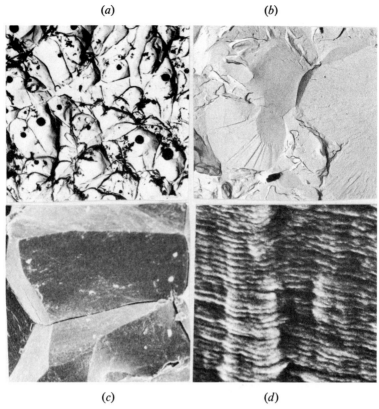

FIG. 5 (a) Dimples in a ductile fracture of mild steel (×5000)
(b) Cleavage patterns in HS steel fracture (×12 000)
(c) Intergranular fracture in low alloy steel (×1500)
(d) Fatigue striations in Nimonic 80A (×7000) (*A. Strang*)
(Scanning electron micrographs)

*Compound stresses and brittle fracture*

The failure of some American all-welded ships during the Second World War has stimulated much work on the causes of brittle fracture in steel.

In the tensile test plastic deformation involves shearing slip along crystal planes within the crystals (p. 104), but in the presence of tension of equal magnitude in each principal direction, shearing stresses are absent, plastic deformation is prevented and a brittle fracture occurs as soon as the cohesive strength of the material is exceeded. Equal triaxial tension stresses do not arise frequently in practice, but it is common to find a triaxial tension superimposed on a unidirectional tension, and if the margin between cohesive strength and plastic yield strength is small, a brittle fracture may occur in a material ordinarily considered highly ductile. Compound stresses arise in a weld in very thick plate and in a tube under internal pressure and an axial tension.

The actual stress-strain curve in Fig. 2 may be called the *plastic yield stress*-strain curve. This is shown in Fig. 6 with *cohesive stress*-strain curves, B, N, F. If the two curves intersect at Y, brittle fracture occurs preceded by plastic deformation, which decreases as the cohesive strength curve becomes lower with respect to the yield stress-strain curve. Orowan has shown that if the yield stress is denoted by Y, the strength for brittle

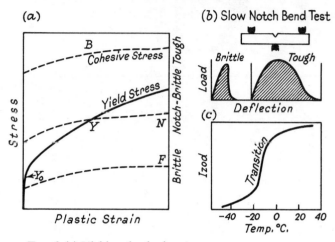

FIG. 6 (*a*) Yield and cohesive stress curves
(*b*) Slow notch bend test
(*c*) Effect of temperature on the Izod value of mild steel

fracture by B (both Y and B depend on the plastic strain), and the initial value of Y (for strain = 0) by $Y_0$, we have the following relation:

The material is brittle if $B < Y_0$;
The material is ductile but notch-brittle if $Y_0 < B < 3Y$;
The material is not notch-brittle if $3Y < B$.

The factor 3 takes into account the stress increase at a notch.

Whether the material is notch-brittle or not depends on the very small margin between B and 3Y.

Carbon steel is an exceptional material, because 3Y and B are so close, and this is the reason why the results of Izod tests seem to be so erratic, and why notch brittleness is so sensitive to slight variations of composition, previous treatment and temperature.

Brittle fracture is characterised by the very small amount of work absorbed and by a crystalline appearance of the surfaces of fracture, often with a chevron pattern pointing to the origin of fracture, due to the formation of discontinuous cleavage cracks which join up (Fig. 7). It can occur at a low stress of 75–120 N/mm² with great suddenness; the velocity of crack propagation is probably not far from that of sound in the material. In this type of fracture plastic deformation is very small, and the crack need not open up considerably in order to propagate, as is necessary with a ductile failure.

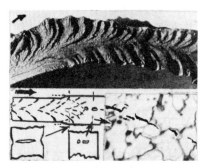

FIG. 7 Steel brittle fracture surface with chevron markings. Micrograph shows discontinuous cracks ahead of main crack

The work required to propagate a crack is given by Griffith's formula:

$$\sigma \simeq \sqrt{E\gamma/c}$$

where $\sigma$ = tensile stress required to propagate a crack of length $c$
$\gamma$ = surface energy of fracture faces
E = Young's modulus

Orowan modified the Griffith theory to include a plastic strain energy factor, $p$, since some plastic flow is always found near the fracture surface:

$$\sigma \simeq \sqrt{E(\gamma + p)/c}$$

When the temperature is above the brittle–ductile transition temperature, $p$ is large and the stress, $\sigma$, required to make the crack grow will also be large. Below the transition temperature the metal is brittle and $p$ will be smaller. The stress necessary to cause crack growth, therefore, will be reduced. The reason for the increasing speed of crack propagation, once a crack has started, is clear from both Griffith's and Orowan's equations: as the crack grows in length, the stress required for propagation continually decreases.

**Notched bar tests**

Notches produce triaxial tension stresses which increase the ratio of tensile to shear stresses and the use of an Izod test is an attempt to subject metals to a controllable triaxial tension, and not to measure the resistance of the material to impact, since the velocity of testing is too low. Test pieces, either 10 mm$^2$ section or 11 mm diameter (Fig. 1), are gripped in a vice, with the root of the notch level with the top of the vice. A weighted pendulum, swinging on ball bearings, is raised to a standard height and allowed to strike the specimen on the same side as the notch. The striking energy of 163 J is partially absorbed in fracturing the specimen, and this amount is indicated by the pointer. The sharpness of the notch largely controls the test, but geometrically similar notches do not produce the same results on large parts as they do on small test pieces. Consequently true behaviour of a material under the unmeasurable triaxial tension in service can only be obtained by testing full-sized parts as they will be stressed in service. The Izod figures, therefore, have *no design value*. A material of 40 J is not 'twice as strong as one of 20 J' and with equal room temperature Izod values, two materials may respond quite differently to temperature (see p. 18), size of specimen and dimensions of notch.

The Izod test is accurate for its particular geometry and stress conditions and is useful to the metallurgist in detecting differences due to casting, mechanical and heat treatment, not indicated by the tensile test. For material susceptible to notch-brittle fracture it gives a guide to the resistance against failure at a discontinuity or change in section and also indicates the resistance of a material to the *spread* of a crack after it has formed. Notched-bar test results form no basis for approval or condemnation of a metal for services under conditions of notch-fatigue (p. 28).

In assessing the notch ductility of steel the important factors are state of stress (notch effect), speed of deformation and temperature. Earlier, the effect of temperature was ignored and steels were classified as tough or brittle by the results of an Izod test at room temperature. Because it is so much easier to test specimens over a range of temperature the Charpy V-notch test has displaced the Izod test in recent years. In the Charpy machine the specimen, 10 mm$^2$ section is supported as a beam, 40 mm span using the standard Izod notch instead of the Charpy keyhole notch. The Charpy machine has a lighter tup but a higher striking velocity, giving a striking energy of 280 J. In spite of the same notch geometry the results of the two tests differ; in general the Izod test gives higher energy values and lower crystalline areas of fracture than the Charpy V-notch test. Some authorities consider that the area of brittle fracture is a better criterion than an energy value and this is easier to estimate on large specimens. Many tests employing the full plate thickness have been introduced. A slow V-notch bend test 70 × 225 mm is used by Van der Veen, while the Tipper test employs a notched tensile specimen. These various tests do not indicate the same transition temperature for a steel but they usually rate steels in the same order of merit. Of much more use to the engineer is the Robertson test. In this a specimen is subjected to a uniform applied stress and also to a temperature gradient. At the cold end is a sharp defect—a saw cut—and a crack is initiated from this defect by means of a blow. The crack propagates under these conditions and the temperature at the point where the failure changes from brittle to ductile is known as the crack arrest temperature. From a series of such tests carried out over a range of applied stresses, Robertson finds a temperature at which the stress required to propagate a brittle fracture increases rapidly.*

*Fracture mechanics and fracture toughness*

Although the Charpy test has been used as a guide and a quality control test for toughness in the past its limitations, from the design point of view, are becoming increasingly apparent especially with the development of new materials, for which previous experience is lacking. The use of linear fracture mechanics is now being applied to many design problems to give a fracture-safe structure containing a given defect size.

Essentially, the technique involves a study of the stresses and strains at the tips of sharp cracks or defects and relates them to the applied loading. Under *elastic* conditions this stress field can be described in terms of a single parameter known as the *stress intensity factor K* which should not be confused with stress concentration factor, which describes the ratio between actual and nominal stress at a discontinuity.

The critical value of $K$, necessary for crack growth under a static load, is called $K_c$—the *fracture toughness*—which as plate thickness increases, approaches a minimum value $K_{1c}$ under plane strain conditions (i.e. high triaxiality and restraint in thick sections†) and becomes a material property in the same sense as a proof stress. $K_{1c}$ is called the critical plane strain stress intensity factor, Fig. 4. Thus given $K_{1c}$ it is possible to specify acceptance limits for sizes and shape of defect so that fracture will not occur at the design stress. This method is very suitable for high strength materials, but in lower strength alloys used in conventional engineering structures fracture is often associated with local plasticity which invalidates elastic analysis. However, the occurrence of this yielding brings with it the separation of the crack surface, even at the crack tip just before crack propagation. This critical *crack opening displacement* (COD = $\delta$) can be regarded as a parameter for measuring fracture toughness. For unstable fracture initiation in relatively ductile materials $\delta$ is found to reach a critical value $\delta_c$ depending on material and testing conditions‡.

For determining $K_{1c}$ or $\delta_c$ various tension and bend tests are used with

*Met. Review*, 1957, 2, 195.

†Thickness $B$ should be $\geqslant 2 \cdot 5 \dfrac{(K_{1c})^2}{YS}$  1 represents the mode of stressing YS = Yield

‡COD $\delta = \dfrac{4G}{\pi YS}$  Where $G$ is crack extension force which is related to the surface energy term in Griffith formula.

central, single or double edge notches, but in all of them it is essential to use sharp natural cracks, which are usually obtained by producing a small fatigue crack extension at the tip of a machine notch. The opening of the notch is recorded against the applied force, and crack growth is represented by an incremental increase of opening without an increase in force. The object is to determine the force at which a given amount of crack extension has taken place, i.e. in terms of a given deviation from linearity.

$K_{1c}$ increases greatly with control of impurities in steel and with hot working, but it decreases with increase of yield strength. With high strength steel with medium carbon content and aluminium alloys the permissible design stress could be *reduced* by the use of higher strength material if the size of defects remain constant. However, acceptable fracture toughness combined with high strength ($>$ 1500 N/mm$^2$) is available in low carbon maraging and ausformed steels p. 232. Fracture mechanics permits assessment in situations involving very brittle material, such as cobalt-iron alloys, used for magnetic rather than mechanical properties, yet the designer may wish to be assured that fast fracture will not occur. It may also be used to establish quantitatively the effect of fabricating procedures, such as welding.

The Pellini diagram, Fig. 8, illustrates in a simple manner many of the important factors concerned with fast fracture, i.e. the relation of service stress, flaw size and crack arrest temperature. The crack arrest temperature (CAT) curve is defined by Robertson type tests and is shifted to the right with thick sections as shown. The diagram defines crack arrest or instability initiation conditions but in practice it is usually used to define permitted service temperature at a given stress in terms of the crack arrest line.

## Properties at low temperatures

Aircraft and chemical processing equipment are now required to work at subzero temperatures and the behaviour of metals at temperatures down to $-150\,°C$ needs consideration,§ especially from the point of view of welded design where changes in section and undercutting at welds may occur.

An increase in tensile and yield strength at low temperature is characteristic of metals and alloys in general. Copper, nickel, aluminium and austenitic alloys retain much or all of their tensile ductility and resistance to shock at low temperatures in spite of the increase in strength.

In the case of unnotched mild steel, the elongation and reduction of area is satisfactory down to $-130\,°C$ and then falls off seriously. It is found almost exclusively in ferritic steels, however, that a sharp drop in Izod value occurs at temperatures around $0\,°C$ (see Figs. 6 and 9). The transition temperature at which brittle fracture occurs is lowered by:

§See H. W. Gillett. Project 13 of the Joint ASME–ASTM Res. Comm. on Effect of temperature on the properties of metals.

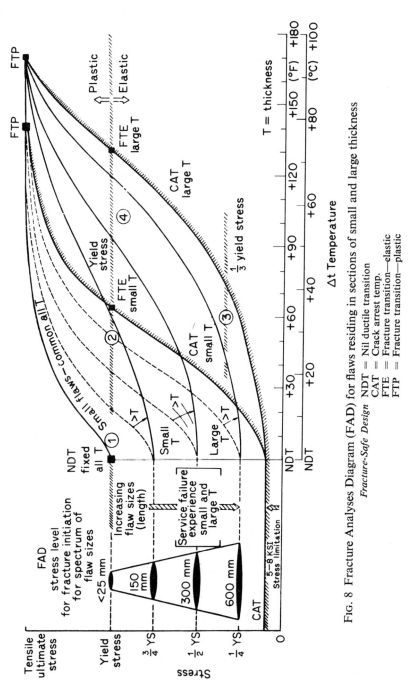

Fig. 8 Fracture Analyses Diagram (FAD) for flaws residing in sections of small and large thickness

*Fracture-Safe Design*   NDT = Nil ductile transition
CAT = Crack arrest temp.
FTE = Fracture transition—elastic
FTP = Fracture transition—plastic

(a) a decrease in carbon content, less than 0·15% is desirable.
(b) a decrease in velocity of deformation.
(c) a decrease in depth of 'notch'.
(d) an increase in radius of 'notch', e.g. 6 mm minimum.
(e) an increase in nickel content, e.g. 9% (see p. 21).
(f) a decrease in grain size. It is desirable, therefore, to use steel deoxidised with aluminium (i.e. fine grained, p. 167) normalised to give fine pearlitic structure (p. 176) and to avoid the presence of bainite (p. 194) even if tempered subsequently.
(g) an increase in manganese content. Mn/C ratio should be greater than $2\frac{1}{2}$, preferably 8.

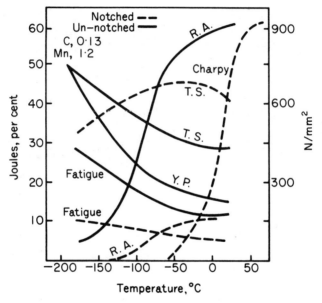

FIG. 9 Effect of low temperatures on the mechanical properties of steel in plain and notched conditions

Surface grinding with grit coarser than 180 and shot-blasting causes embrittlement at −100 °C due to surface work-hardening, which, however, is corrected by annealing at 650–700 °C for 1 h.

This heat-treatment also provides a safeguard against initiation of brittle fracture of welded structures by removing residual stresses.

Where temperatures lower than −100 °C or where notch-impact stresses are involved in equipment operating below zero, it is preferable to use an 18/8 austenitic or a non-ferrous metal. The 9% Ni steel provides an attractive combination of properties at a moderate price. Its excellent toughness is due to a fine grained structure of tough nickel-ferrite devoid of embrittling carbide networks which are taken into solution during tempering at 570°C to form stable austenite islands. This tempering is particularly important because of the low ferrite-austenite transformation temperatures (see Fig. 147). A 4% Mn Ni rest iron is suitable for castings for use down to −196°C. Care should be taken to select plates without surface defects and to ensure freedom from notches in design and fabrication.

Fig. 10 shows tensile and impact strengths for various alloys.

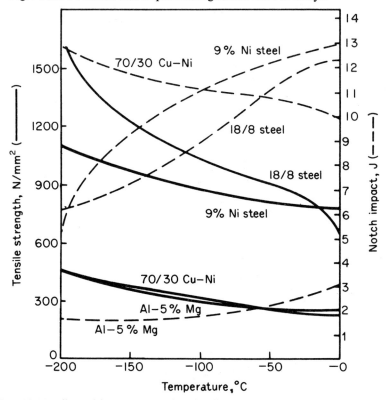

Fig. 10 Tensile and impact strengths of various alloys at subzero temperatures

## Properties at high temperatures

Creep is the slow plastic deformation of metals under a constant stress, which becomes important in:

(1) The soft metals used at about room temperature, such as lead pipes and white metal bearings.
(2) Steam and chemical plant operating at 450–550°C.
(3) Gas turbines working at high temperatures.

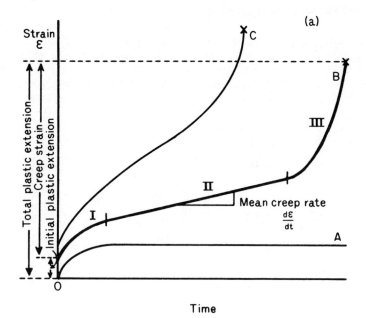
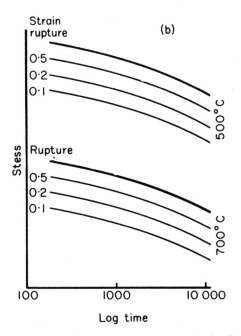

FIG. 11(*a*) Family of creep curves at stresses increasing from A to C
(*b*) Stress-time curves at different creep strain and rupture

## PROPERTIES AT HIGH TEMPERATURES

Creep can take place and lead to fracture at static stresses much smaller than those which will break the specimen when loaded quickly in the temperature range 0·5–0·7 of the melting point Tm K.

The variation with time of the extension of a metal under different stresses is shown in Fig. 11(a). Three conditions can be recognised:

(a) The *primary* stage, when relatively rapid extension takes place but at a decreasing rate. This is of interest to a designer since it forms part of the total extension reached in a given time, and may affect clearances.
(b) The *secondary* period during which creep occurs at a more or less constant rate; sometimes referred to as the minimum creep rate. This is the important part of the curve for most applications.
(c) The *tertiary* creep stage when the rate of extension accelerates and finally leads to rupture. The use of alloys in this stage should be avoided; but the change from the secondary to the tertiary stage is not always easy to determine from creep curves for some materials.

The limited nature of the information available from the creep curve is clearer when a family of curves is considered covering a range of operating stresses.

As the applied stress decreases the primary stage decreases and the secondary stage is extended and the extension during the tertiary stage tends to decrease. Modifying the temperature of the test has a somewhat similar effect on the shape of the curves.

Design data are usually given as series of curves for constant creep strain (0·01, 0·03%, etc.), relating stress and time for a given temperature. It is important to know whether the data used are for the secondary stage only or whether it also includes the primary stage, Fig. 11(b).

In designing plant work at temperatures well above atmospheric temperatures, the designer must consider carefully what possible maximum strains he can allow and what the final life of the plant is likely to be. The permissible amounts of creep depend largely on the article and service conditions. Examples for steel are:

|  | Rate of Creep mm/min | Time, h | Maximum Permissible Strain, mm |
|---|---|---|---|
| Turbine rotor wheels, shrunk on shafts | $10^{-11}$ | 100 000 | 0·0025 |
| Steam piping, welded joints, boiler tubes | $10^{-9}$ | 100 000 | 0·075 |
| Superheater tubes | $10^{-8}$ | 20 000 | 0·5 |

In designing missiles data are needed at higher temperatures and stresses and shorter time (5–60 min) than are determined for creep tests. This data is often plotted as isochronous stress-strain curves (see also polymers, p. 378).

*Creep tests*

For long-time applications it is necessary to carry out lengthy tests to get design data because it is dangerous to extrapolate from short time tests which may not produce all the structural changes, e.g. spheroidation of carbide (p. 176; Fig. 49(*e*)). For alloy development and production control short time tests are used.

*Long time creep tests*

A uniaxial tensile stress is applied by means of a lever system to a specimen (similar to that used in tensile testing) situated in a tubular furnace and the temperature is very accurately controlled. A very sensitive mirror extensometer (of Martens type) is used to measure creep rate of $1 \times 10^{-8}$ strain/h. From a series of tests at a single temperature a limiting creep stress is estimated for a certain arbitrary small rate of creep, and a factor of safety is used in design.

*Short time tests*

The *rupture test* is used to determine time-to-rupture under specified conditions of temperature and stress with only approximate measurement of strain by dial gauge during the course of the experiments because total strain may be around 50%. It is a useful test for sorting out new alloys and has direct application to design where creep deformation can be tolerated but fracture must be prevented.

**Fatigue**

The term 'fatigue' is used to describe the failure of a material under a repeatedly applied stress. The stress required to cause failure, if it is applied a large number of times, is much less than that necessary to break the material with a single pull.

Fatigue causes over 80% of the operating failures of machine elements, but in many of these the stress cycles may be very complex with occasional high peaks, for example the gust loading of aircraft wings. Experience has shown that for satisfactory correlation with service behaviour full-size or large-scale specimens must be tested under conditions as close as possible to those existing in service. Nevertheless some basic information about fatigue properties of materials can be obtained from simplified laboratory tests.

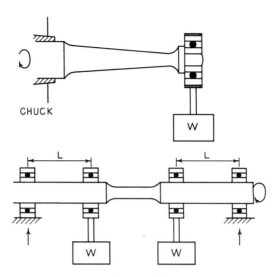

Fig. 12 Fatigue testing machine

A *Wöhler-type machine*, commonly used, is shown in Fig. 12. The specimen, in the form of a cantilever, forms the extension of a shaft, driven by an electric motor. Dead loading is applied to the specimen through a ball-bearing. As the specimen rotates there is a sinusoidal variation of stress which is greatest at the surface and zero at the centre. The dimensions of the specimen vary according to the make of machine. In some cases the specimen may be tapered (Amsler) or equal two-point loading employed in order to obtain a uniform surface stressing over a considerable length of specimen. Other machines use tensile (with or without compression), impact, torsional and combined bending and torsional stresses.

Several specimens are subjected to reversals of stresses, each at a different intensity, until failure occurs or until 10 million cycles have been endured. From the results a stress-cycle (S/N or S/log N) curve is plotted, as in Fig. 13(*b*). The rate of reversal of stress has a negligible effect on the result at normal temperatures.

For the ferritic steels below 200 °C titanium and its alloys, magnesium and some aluminium–magnesium alloys at room temperature the S/N curve tends to become horizontal after about 10 million cycles at a stress which is called the *fatigue limit*. (This may be due to strain-ageing in these materials.)

With most non-ferrous metals and austenitic steels no definite fatigue limit is found even after 50–100 million cycles of stress.

The *endurance limit* at N cycles of a specimen is the stress which just produces fracture after N stress cycles.

Recent studies have shown that fatigue is a statistical phenomena and that therefore some failures may occur prematurely and a scatter band is shown in Fig. 13.

The *fractured surface* in a fatigue failure consists of two areas. One portion is smooth, discoloured and has rippled markings like a mussel shell indicating the gradual creeping of the crack from one or more centres; while the remainder of the surface has either a crystalline or a fibrous appearance which indicates the final tearing, which occurs when the area can no longer sustain the load. Figs 13(a), 14 and 5(d) show these effects.

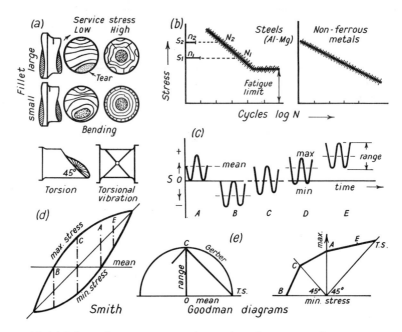

Fig. 13 (a) Effect of stress concentration and service stress on fracture
(b) S/Log N curves  (c) Cyclic variations
(d) Smith diagram  (e) Goodman diagrams

Fig. 13(a) also shows how the fracture surface is related both to stress concentration and service stresses.

Fatigue arises from a complex of engineering and metallurgical factors some of which are considered below.

(1) The simpler stress conditions used in rotary bending machines applies only to a few engineering applications such as shafts, but in the majority of cases the mean stress is not zero. Empirical rules have been suggested to enable the engineer to estimate the range of stress for some other mean

stress values. Some typical pulsating stress cycles are shown in Fig. 13(c), and Fig. 13(e) shows a representation of the Gerber and Goodman relations between range of stress to mean stress. As the mean stress increases the stress range ($\pm$), i.e. the difference between upper and lower stress limits, which can be safely superimposed on the mean stress decreases and becomes zero at TS. This fact is also expressed in the form of Goodman and Smith diagrams Fig. 13(e) and (d). In the Goodman diagram maximum stress is plotted against minimum stress while in the Smith diagram both maximum and minimum stress values are plotted against mean stress. The letters correspond to the forms of cyclic variation in Fig. 13(c). These diagrams are mentioned because most Continental specifications for welded structures subjected to fatigue loading specify the permissible stresses in this form.

(2) *Cumulative damage.* Service conditions often involve a load spectrum of different stress ranges and Miner's hypothesis enables the groups of amplitude to be replaced by an equivalent number of cycles of a single magnitude. Referring to Fig. 13(b), if the material is subjected to $n_1$ stress cycles at stress amplitude $S_1$ at which the expected life is $N_1$, then the fraction of life used up is $\frac{n_1}{N_1}$. If stress is now raised to $S_2$ for $n_2$ cycles, then fraction of life used is $\frac{n_2}{N_2}$. For any number of stress levels failure will occur when

$$\frac{n_1}{N_1} + \frac{n_2}{N_2} + \frac{n_3}{N_3} + \ldots = 1, \text{ i.e. } \sum \frac{n}{N} = 1$$

For safety the empirical rule is often taken as $\sum \frac{n}{N} = 0.6$.

(3) *Overstressing and understressing.* Specimens subjected to alternating cycles of stress greater than the endurance limit for short periods (e.g. $S_1$, $S_2$) exhibit a lowering of the endurance limit and this is known as overstressing. This damage is greater with high tensile steels than with mild steels. In specimens subjected to a large number of cycles of stress slightly *less* than the fatigue limit the resultant fatigue limit is raised. This is known as understressing or coaxing. When the amplitude is increased in a series of steps it can improve but not always make good the damage of overstressing. This effect may be connected with the strengthening effect of strain-ageing (p. 170) in steel and it may also explain why some fatigue cracks form so far and then fail to propagate into material which is harder than it was originally.

(4) Failure of connecting rod bolts is often due to looseness, while some shaft failures arise from torsional vibration, the fracture occurring in two directions at 45° to the axis, whereas torsion causes failure in one direction (Fig. 13).

The failures are more often due to bad design or assembly or to poor machining, than to the quality of the metal.

The initial crack usually starts from a point of high stress concentration, such as a keyway, toolmark, oil-hole, sharp fillet, slag inclusion or other 'stress raiser'.

Stress concentration factors (or multiples) can be obtained from formula or from photo-elastic studies of two dimensional plastic models.

A very sharp fillet may reduce the fatigue limit to half the value obtained on a well-prepared specimen for mild steel and to a quarter for high tensile alloy steels. Under dynamic stresses, fillet welds should be replaced by butt welds.

Normal cast iron is not so sensitive to the notch effect because the graphite flakes act as notches even in a smooth specimen. A notched bar of 18/8 stainless steel, which work hardens very readily, shows a higher endurance limit than a polished unnotched bar. Notches, however, drastically reduce the effective fatigue strength of high tensile steels and unless care is taken in design and manufacture they are little better than mild steels under dynamic loading.

(5) There is no direct relation between the Izod value and the fatigue limit.

(6) The fatigue strength of steels decreases slightly with increasing size of test piece, the reduction being greater in material of high tensile strength and in specimens having discontinuities of section.

(7) The fatigue limit of steel is usually increased slightly by increasing the speed of testing from 1500 to 30 000 rev/min provided heating-up is not excessive and also by lowering the temperature to $-40\,°C$ (Fig. 9). In many high-temperature tests there is no apparent tendency for the fatigue limit to be developed. The fatigue limit of low-carbon steel reaches a maximum value in the temperature range 200–300 °C, the temperature at which peak fatigue strength occurs being dependent upon the frequency at which the alternating stress is applied. The effect is due to dynamic strain ageing.

(8) Fatigue cracks normally start at the surface of a specimen, consequently surface treatments are important. The endurance limit is decreased by a *weak skin* such as in decarburised steel or aluminium clad duralumin. Removal of a thin surface layer (e.g. 0·05 mm) after exposure to alternating stresses is beneficial. An abraded surface is superior to one electrolytically polished. A beneficial hard skin can be produced by nitriding and carburising of steel.

(9) The endurance limit is increased by *compressive* stresses in the surface skin, produced by cold working, e.g. shot peening or skin rolling fillets, and also by heat treatment at the appropriate stage in manufacture. For example a gear heat treated *after* machining will have compressive surface stresses, but machining a quenched blank can sometimes expose a zone of tension stress which added to the operating stress can initiate fatigue failure.

A large increase in fatigue strength of specimens containing discontinuous longitudinal welds can be obtained by inducing residual compressive stresses at the site of the most dangerous notch (i.e. ends of weld) by locally heated or pressed spots. Preloading of a notched specimen sufficient to cause plastic flow at the root of the notch also introduces compressive stresses there and greatly increases endurance. The removal of residual stresses can produce a large increase in fatigue strength of specimens subjected to *alternating* loading but has little effect under *pulsating* tension stresses.

Fig. 14 Fatigue fracture of automobile axle shaft ($\times$ 2)

(10) Non-metallic inclusions just under the surface may act as starting points for cracks and undoubtedly influence the statistical variability of steels. Easily deformed inclusions such as MnS or lead have little effect but hard inclusions such as $Al_2O_3$, $SiO_2$ are detrimental. Inclusions in basic electric steels seem to be worse than those on open hearth steels.

(11) Evaluation of the fatigue strengths of whole structures is an important aspect. Attempts are being made to relate these to the fatigue strength of the basic material of construction by the introduction of an 'equivalent stress-concentration factor' for the whole structure.

*Fundamental aspects of fatigue.* Progress in the understanding of the fundamental mechanisms involved in fatigue failure has been made by detailed structural examination, more recently by transmission electron microscopy. Differences exist between the structure of fatigue-hardened metals and the same material hardened by a single application of the same stress. In many fatigued metals, a high density of small dislocation loops is observed, accounting for some extra temperature-sensitive hardening compared with unidirectionally stressed metal. Coarse slip bands (striations) observed on the surface of some fatigued metals have been identified with interior 'mats' of dislocations. In face-centred cubic metals, cross-slip is thought to be important in the development of fatigue cracks; f.c.c. alloys of low stacking fault energy (p. 112) in which cross-slip is difficult are more resistant to fatigue. Alloys which rely on an unstable precipitate formed during age hardening for their static strength often have relatively poor fatigue strengths in relation to their static strengths. Included in this group of alloys are many of the high strength aluminium alloys (e.g. Al–Zn–Mg–Cu alloys). In small localized regions of these alloys subjected to fatigue, resolution of the precipitates or over-ageing due to enhanced diffusion rates causes local softening and damage spreads catastrophically. Alloys which undergo strain ageing (e.g. low-carbon steel) are among those with the best fatigue strengths in relation to static strengths.

On the microscopic scale fatigue cracks generally start at the surface by shear fracture on a slip plane (Stage I fracture (Fig. 54($d$)) from an intrusion (Fig. 62). It slows down and at a depth of $10^{-3}$ to $10^{-4}$ cm becomes almost 'dormant', sometimes for three-quarters of the total life and with low stresses may become a 'non-propagating fatigue crack'. Usually this is followed at a later stage by the crack turning out of the slip plane into a direction generally perpendicular to the applied stress (Stage 2 fracture). Detailed examination of Stage 2 fracture surfaces at high magnification in the electron microscope reveals striations marking the progress of the crack, each striation being associated with a single cycle of the applied stress, Fig. 5($d$). The crack spreads at an increasing rate until the remaining cross-section becomes small enough to break completely in a single tensile stroke.

*Corrosion fatigue*

This occurs when a metal is subjected to repeatedly applied stress under conditions favourable to mild corrosion, and the endurance limit is lowered although the loss of metal by corrosion may be negligible. Water vapour in the atmosphere reduces the fatigue strength of aluminium alloys appreciably and impervious coatings are helpful. Ordinary commercial steels appear to possess no corrosion fatigue limit and, given sufficient time, failure will occur at very low stresses. The main effect of the corrosion is to produce sharp pits from which numerous cracks form, frequently meeting to form a serrated fracture.

The reduction in fatigue strength is illustrated in Table 2:

TABLE 2

| Steel | Tensile Stress N/mm² | Endurance Limit ± N/mm² at 5 × 10⁷ cycles | | |
|---|---|---|---|---|
| | | Air | Fresh Water | Salt Water |
| 0·16% C steel | 453 | 247 | 138 | 62 |
| 3·7% Ni Steel | 625 | 340 | 156 | 110 |
| Stainless iron (12·9% Cr). | 618 | 378 | 263 | 200 |

See also fretting corrosion, p. 151.

*Damping capacity*

This is the capability of a material to damp out vibratory stresses by conversion of the energy into heat. An index of its value is frequently obtained from a bar set into torsional vibration, the decreasing amplitudes being recorded. There are large differences in the damping capacity of riveted and welded articles.

The most important cases are those in which owing to speed of rotation, resonance of some vibration occurs which builds up a dangerous movement. It is often possible to modify the natural frequency of the various parts by changing the design, dimensions and masses.

Although felt, rubber, cork and plastics are commonly used for damping the 70/30 manganese/copper alloy offers interesting possibilities as shown in Fig. 15. Sonoston (Pat) has a composition of 40 Cu, 4 Al, 3 Fe, 1·5 Ni, balance Mn and about 20 times the damping capacity of mild steel. Tensile strength is 580 N/mm², El 30%, Izod 27 J. It has been used for marine propellors but it is susceptible to stress corrosion which can be avoided by cathodic protection.

## Transverse test

This is commonly used for cast iron. A round bar, its size varying with the thickness of the casting it represents, is supported at the ends and loaded at the centre until it breaks. The maximum load, deflection and sometimes the Modulus of Rupture are reported.

$$\text{Modulus of Rupture} = \frac{\text{Ultimate bending moment}}{\text{Modulus of section}} = \frac{WL}{0.3928 d^3}$$

where W is the load, and L the distance between the supports and $d$ the diameter of the bar.

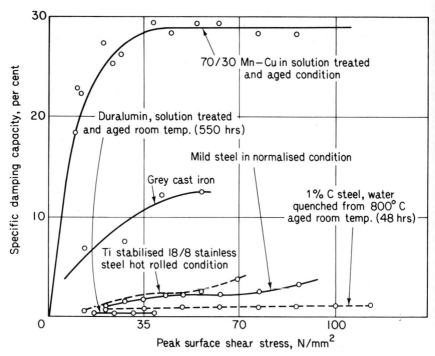

Fig. 15 Torsional damping capacity of some materials at 10 Hz and room temperature

Bending tests are also carried out on other materials, such as cast steel and welded joints, in order to indicate the degree of ductility. The angle of bend or the elongation of the outer skin is measured.

### Cupping tests

Simple mechanical tests have proved insufficient to allow prediction of performance in deep drawing and pressing operations due to the importance of such factors as the *shape* of the tools and *friction*. Many tests have been devised which attempt to simulate the conditions in metal forming operations more closely. Examples of these are Erichsen and Olsen tests in which a 20 mm diameter ball or plunger is pressed into a sheet held between smooth circular clamps, until the dome produced just begins to split. The depth of the cup is the index, but the appearance of the dome is also important. A roughened appearance indicates a large grain size and if the cracks occur in one direction it indicates variation in ductility in different directions.

Such biaxial stretching tests simulate deformation in stretch forming operations (e.g. motor car panels) rather than true deep drawing operations (bullet-case). For the latter, cup drawing tests, e.g. Swift tests are applied. These involve finding the maximum diameter of the blank which can be successfully drawn into a cylindrical cup.

## Non-destructive tests

The development of highly stressed components such as welded drums in high-pressure steam and nuclear energy plant and aluminium castings and forgings in aircraft has focused attention on means of detecting flaws in the finished article without destroying any part of it.

*The magnetic particle method* consists of magnetising the steel article and pouring over it kerosene containing fine iron oxide. If a crack lies across the path of the magnetic flux each side of the crack becomes a magnetic pole which collects iron oxide. Surface flaws such as grinding cracks are revealed by the accumulations of oxide.

*Penetrant test*, which is not limited to magnetic materials, consists in dipping the article in various liquids some of which contain a fluorescent substance to fill the cracks. The surplus liquid is then removed from the components which, after drying and dusting with fine absorbent powder, are examined under normal or ultra-violet light. Cracks are shown up by strains on the absorbent powder. Radioactive isotopes in conjunction with a photographic film or a geiger counter can be used in an analogous manner.

*Ultrasonic test*

Pulses of high-frequency sound waves are applied to the part under test by a piezo-electric crystal. In the intervals between pulses a crystal detects echoes reflected either from the far edge of the test part or from any flaws in the path of the beam. The signals received are shown on a cathode-ray tube, which also has a time base connected to it, so that the position of the signal on the screen gives an indication of the distance between the crystal generator and the surface from which the echo originates. For application to large components (e.g. plate for aircraft wings) automatic scanning methods are used, involving immersion in water tanks.

*Other methods* include the use of eddy currents induced in the sample, remote TV scanning and $\beta$ rays for gauge control.

## X-rays

*Radiography*

This method consists in passing a mixed beam of X-rays through the object and studying the shadow formed either on a fluorescent screen for visual examination of thin sections, or on a photographic film placed at the back of the object. Blowholes, cracks and such defects absorb the rays less than does sound metal, and their position is marked on the film by dark areas, but as light areas on a positive print.

Defects as small as 2% of the thickness of the object can be detected, and

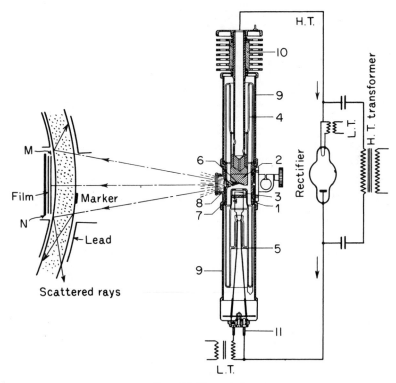

FIG. 16 X-ray tube

1. Chromium–iron Cylinder
2. Anticathode
3. Filament
4 and 5. Glass insulating Cylinders
6. Target
7. Lead Jacket
8. Aperture Cap
9. Ray-proof Bakelite Cylinders
10. Aluminium Radiator
11. Plug Connection

this sensitivity is checked by placing on the side of the object a stepped piece of metal, with steps 0·013, 0·026, 0·05, 0·10, 0·13 mm with holes 0·064 mm in diameter—called a penetrameter. Opaque lead numbers suitably placed serve to identify the position of the faults.

The modern apparatus for the production of X-rays consists of an X-ray bulb of the hot cathode type operated by a high-tension transformer with sometimes a valve rectifier, as shown diagrammatically in Fig 16. The

# GAMMA RAYS

*quantity* of X-rays emitted depends on the electron current, measured by a milliameter. The *quality* of the radiation depends on the voltage applied to the tube. If the voltage is increased, the radiation is said to be harder and its penetration power is also increased. Penetration power is given by the thickness of material required to halve the amount of radiation, for example for steel.

| | | | |
|---|---|---|---|
| Tube voltage, kV | 150 | 250 | 1000 |
| Half-value thickness, mm | 4 | 6 | 15 |
| Maximum thickness of steel mm | 38 | 89 | 300 |

The *contrast* between two areas of different thickness or density on the radiographic image is greater as the energy of the radiation is reduced to a minimum compatible with reasonable exposure time.

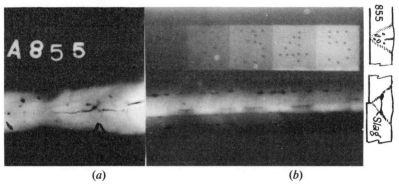

(a)  (b)

FIG. 17 Portions of two radiographs of welds. (*a*) Crack and porosity; (*b*) Penetrameter, lack of penetration and slag lines

Two typical radiographs of welds are shown in Fig. 17; (*a*) illustrates porosity and a crack with identifying numbers; (*b*) penetrameter and slag inclusions at sides and lack of fusion at the centre.

Since conventional X-ray tubes do not operate at voltages greater than about 400 kV, high-energy X-radiation necessary for the inspection of heavy steel castings is supplied by machines in which the electrons are accelerated by more complex methods. Two such machines are the betatron and the linear accelerator.

## Gamma rays

In operation it is only necessary for the specimen to be placed at a definite distance from the isotope with a photographic film behind the object. The photographic emulsion absorbs only a small proportion of the rays and intensifying screens are used on both sides of a double-coated film.

Elaborate screening to prevent scattering effect is unnecessary. The gamma rays are produced by the decomposition of radioactive substances, such as radium or radioactive isotopes from the atomic pile. The strength of the source decreases with time, e.g. iridium 192 is roughly equivalent to a 500 000 volt X-ray set as regards penetration power and has a useful life of 40 days (i.e. time to fall to half its initial radiation value). Cobalt 60 has a half life of 5·3 years and is useful where a greater penetration is required, e.g. 53 mm or more of steel.

The method is very suitable for outdoor work and in confined spaces, since it requires neither electric power nor water supply for operation.

As compared with X-rays, gamma rays often involve greater exposure times and yield poorer contrast and definition. The latter is due to the 'fixed quality' of the source so that the penetration power of the radiation cannot be matched to the absorbing power of the article as is possible with X-rays (i.e. lowest voltage); and also it is difficult to approach a *point* source.

**Selection of material**

The choice of a material for a particular article depends on its overall cost and on one or more outstanding properties combined with others at a lower level, e.g. a high tensile structural steel should preferably be as weldable as mild steel with increased corrosion resistance to allow a thinner section to be used. The main properties to be considered are:

Strength at room and elevated temperature
Toughness and notch sensitivity
Embrittlement in service
Brittleness at cryogenic temperatures
Rigidity in a column and elastic resilience in a spring
Size and mass effect
Fatigue strength
Surface hardness
Strength/weight ratio
Strength/Specific gravity ratio
Cost/kg; Cost/cm$^2$
Density
High or low melting point
Corrosion and oxidation resistance
Stress corrosion
Corrosion fatigue
Compatibility e.g. can and fuel in nuclear pile or bearing and shaft
Electrical and magnetic properties, neutron cross-section
Fabrication properties, wastage, die costs, dimensional accuracy
Complexity, section and thickness, weldability

TABLE 3  STRENGTHS AND COSTS OF SOME MATERIALS

| Material | TS N/mm² | Density g/cm² | £ Cost/ 1000 kg | Cost/cm² pence | Strength/ Cost/cm² | Specific Strength | Specific Young's mod E/sp gr kN/mm² |
|---|---|---|---|---|---|---|---|
| Low alloy steel | 975 | 7·8 | 90 | 0·173 | 5610 | 125 | 26·5 |
| Mild steel | 380 | 7·8 | 50 | 0·196 | 3800 | 48 | 26·5 |
| Reinforced concrete | 14 | 2·4 | 8 | 0·046 | 305 | 6 | — |
| Timber | 7 | 0·5 | 33 | 0·031 | 226 | 14 | 26 |
| Grey iron | 187 | 7·3 | 90 | 0·17 | 110 | 25 | 12 |
| Al castings | 210 | 2·7 | 290 | 0·194 | 1080 | 78 | 25·8 |
| Mg castings | 187 | 1·7 | 400 | 0·17 | 1100 | 110 | 25 |
| Brass strip | 462 | 8·6 | 350 | 0·79 | 580 | 53 | 10 |
| Zinc die casting | 28 | 6·6 | 225 | 0·36 | 78 | 4·2 | 12 |
| Copper strip | 28 | 8·9 | 456 | 1·0 | 28 | 3·1 | 14 |
| PVC | 62 | 1·4 | 200 | 0·062 | 1000 | 44 | 2 |
| Polythene | 14 | 0·9 | 200 | 0·045 | 311 | 15 | 0·02 |
| Fibre glass | 765 | 1·87 | 2000 | 0·96 | 795 | 410 | 22 |
| Nylon 66 | 77 | 1·15 | 800 | 0·20 | 380 | 67 | 4 |

Cost is not limited to the intrinsic cost of the raw material but must include fabricating costs (dies, wastage, etc.) and operating costs such as protection from corrosion. Such fabrication costs may be so high that an expensive material is feasible if it cheapens fabrication costs. Small gears can be made by machining from phosphor bronze, sintering bronze powders and injection moulding plastics depending on the precise gear form, size, operating conditions and quantity required. A few machined gears are economically possible, but several thousand would be necessary for the other processes. Mechanical strength is the most common requirement and cost per kg can be misleading. Cost per unit volume puts Al, Mg and Ti in a better position. The strength divided by the cost of unit volume gives a strength/cost ratio which is revealing as shown in Table 3, and steel, concrete and timber head the list.

Wherever inertial or gravitational forces play an important part, such as in aeroplanes, turbines, etc., the strength/weight ratio is more important than strength alone, and the reduction in operating costs may enable an engineer to use expensive lightweight structural materials, e.g. Ti, fibre composites, etc. Table 3 shows specific strength of materials, i.e. tensile strength divided by specific gravity. Elastic stiffness is often of importance and although a higher tensile material may enable thinner section to be used, elastic buckling of a column or the wrinkling of an aircraft wing may prevent commercial use. Aluminium and magnesium, despite low Young's moduli, have advantages over heavier materials because thicker walls can be used.

The maximum working stress is limited by Young's modulus $\left(\sim \dfrac{E}{100}\right)$ which does not change very much even if strengths are increased, and consequently elastic deformation may become excessive for the engineering part, e.g. an aeroplane wing could not be allowed to deflect elastically many inches. The limiting factor is therefore specific Young's modulus, i.e. E divided by specific gravity (Table 3). Steel, aluminium, magnesium, titanium and wood have similar values and are closely competitive for aircraft. Low Young's modulus is an advantage in collisions where greater energy is absorbed.

For static applications, design is often based on the yield point or proof stress, but for dynamic conditions involving fatigue, the ultimate tensile strength is used.

# 2 The Macro-examination of Metals

By the examination of fractured pieces and of small prepared sections of metals the metallurgist can obtain vital information regarding their properties and the treatment to which they have been subjected.

*Fractures*

When a failure occurs in service the first test should be to examine the fracture. For instance, slag particles and blow-holes developed during the casting process are usually immediately evident, while incorrect pouring temperature or faulty heat-treatment are often revealed by a coarseness of grain. Fractures produced normally in works processes or in testing the metal are often ready means of detecting the use of improper material. In the manufacture of tools, for example, when the rolled bars are sized by shearing, internal flaws open up under the applied stress and are clearly visible to the naked eye; much expense can be saved by rejecting such defective material at this early stage. The characteristics of a fatigue fracture given in Chap. 1 (p. 24) should be specially noted.

Full information regarding structure cannot be obtained without the examination of prepared sections, the method adopted for the visual examination of which can be divided into two groups, as follows:

(1) *macro*-examination either with the naked eye or under a very low magnification ($\times$ 5);
(2) *micro*-examination at high magnification ($\times$ 20–2000). (See Chap. 3.)

**Macrography**

Macro-examination requires no elaborate apparatus and determines the character of quite large areas of metals. This method will indicate:

(1) Non-uniform composition due to the segregation of alloying elements such as antimony in bearing metals, lead in bronze and phosphorus in steel.
(2) Non-metallic inclusions such as slag, sulphides and oxides.
(3) Size of crystals and the mechanism of growth.
(4) Methods of manufacture, e.g. casting, forging, welding and brazing.

(5) Physical defects due to manufacture such as pipes, blowholes, seams, laps and rokes. Uniformity of heat treatment.
(6) Mechanical strain.

*Sulphur print*

Sulphides, when attacked with dilute acid, evolve hydrogen sulphide gas which stains bromide paper and therefore can be readily detected in ordinary steels and cast irons. While sulphur is not always as harmful as is sometimes supposed, a sulphur print is a ready guide to the distribution of segregated impurities in general. In order to obtain a true interpretation the surface examined must be chosen with care, and quite frequently it is necessary to prepare more than one surface which may include sections taken at different angles.

The process of sulphur printing is carried out as follows: Grind flat the surface chosen and smooth it by machine or by means of emery paper to number 0 grade. Clean the surface from dirt and grease. Soak pieces of bromide or gaslight photographic paper (matt surface) in a 3% solution of sulphuric acid in water for two minutes; longer time causes the gelatine to swell too much and become slippery. Take the paper out of the acid and remove the excess solution with blotting-paper so that none remains except that which has been absorbed by the gelatine. Place the sensitised surface in contact with one edge of the prepared surface of the specimen and stroke it into position so as to prevent air bubbles remaining. Medium-sized specimens can be held in a vice and a squeegee used to press the paper into intimate contact with the specimen. The time of contact should be 1 to 2 minutes, but it varies with the steel; the corner of the paper can be lifted to examine the intensity of the brown stain. Remove the paper, swill in water for 3 minutes and immerse in 20% hypo solution for 5 minutes. Swill in water for 20 minutes and dry.

No dark room is required for this process. Sulphur prints cannot be obtained from high alloy steels or non-ferrous metals. Fig. 27 shows the sulphur print of a steel ingot.

*Macro etching*

When the prepared surfaces of metals are subjected to the action of suitable etching reagents, crystalline structure, different constituents or parts of the metal in differing degrees of strain are attacked selectively and reveal themselves as a pattern. While naturally many reagents have been suggested, the ones here described will meet general requirements.

The usual procedure is to wet the surface with a swab, or to immerse the specimen in the solution which is contained in a porcelain dish, until the structure is evident. The specimen is removed, swilled in water, then in methylated spirit, and dried. The dish can be dispensed with for large specimens by surrounding the surface to be etched with a ridge of plasticine. The etching time varies from a few seconds to several minutes. A film is frequently formed on the surface of the specimen, which is often helpful in revealing structure; but sometimes the reverse is the case, and the film should be rubbed off.

Success in etching depends largely on experience, and a reader should not be discouraged by a few initial failures. With a little practice good results can be obtained.

## Macro etching reagents

*To reveal segregation and crystal structure in steel*

| Iodine | 10 g |
| Potassium Iodide | 20 g |
| Water | 100 cm$^3$ |

Dissolve the potassium iodide in a small quantity of water, add iodine, and when dissolved add remaining water. Immerse specimen in solution.

*To reveal variations in crystal structure in steel*

| Ammonium persulphate | 10 g |
| Water | 90 cm$^3$ |

The above solution must be made up fresh.

| Nitric Acid (conc) | 10 cm$^3$ |
| Water | 90 cm$^3$ |

*Deep etch for steel*

| Hydrochloric acid | 140 cm$^3$ |
| Sulphuric acid | 3 cm$^3$ |
| Water | 50 cm$^3$ |

Immerse specimen in this solution at 90 °C for about $\frac{1}{4}$ to $\frac{1}{2}$ hour. Polish is not important, in fact defects are revealed in sheared surfaces. To prevent rapid rusting, dip in dilute ammonia solution, swill and dry.

The impurities are attacked severely especially slag in wrought iron, and care is necessary in interpreting the results.

*To reveal deformation lines in steel* (Fry)

This test indicates regions which have been exposed to strains beyond the elastic limit. It is limited to mild steels, which contain a certain amount of nitrogen.

| Cupric chloride | 90 g |
| Hydrochloric acid | 120 cm³ |
| Water | 100 cm³ |

Before etching it is essential for the specimen to be heated to 200–250 °C for ½ hour. The time for etching is 2 to 24 hours.

*To reveal dendritic structure in steel* (Humphrey)

| Copper ammonium chloride | 9 g |
| Water | 91 cm³ |

The time required varies from ½ to 4 hours, and the solution should be agitated. The copper deposited on the steel should be removed by rubbing with cotton wool. The addition of hydrochloric acid to the solution increases the etching contrast, but the operation must always be started in a neutral solution to prevent the copper adhering to the steel. Lightly rub the dried specimen with fine emery paper.

*Stainless and austenitic steels*

| Hydrochloric acid | 15 cm³ |
| Nitric acid | 5 cm³ |
| Water | 100 cm³ |

*Aluminium and alloys*

| Hydrofluoric acid | 10 cm³ |
| Nitric acid | 1 cm³ |
| Water | 200 cm³ |

The black film can be removed by dipping in dilute nitric acid.

*Brass, bronze and copper*

| Nitric acid | 45 cm³ |
| Water | 50 cm³ |
| Potassium bichromate | 0·2 g |

*Zinc*—50% hydrochloric acid in water.
*Magnesium*—10% tartaric acid in water.

If a *permanent record* of the macro structure is required the etched surface can be photographed; alternatively, an ink print can be taken if the specimen is etched by Humphrey's solution. To obtain an ink print the specimen is held firmly in a vice, and a *thin* film of printer's ink applied by means of a rubber roller. The printing should be done on a glossy paper or cellophane, pressure being applied by a squeegee or by a letter press.

The various stages in the manufacture of a cold-headed bolt, together with a macrograph of a sectioned finished bolt, are shown in Fig. 18.

# EXAMINATION OF FAILURES

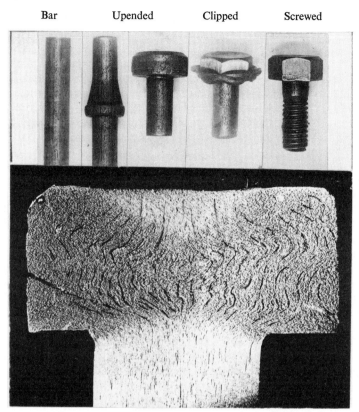

FIG. 18 A cold-headed bolt: stages in manufacture
Macrograph ($\times$ 4) shows flow lines

Lines which reveal the direction of flow of the metal are evident and also a crack. The regions of maximum distortion have etched dark, leaving the shank and a cone-shaped mass at the top light. If the bolt had been hot forged or annealed the flow lines, which are due to impurities, would still be seen, but the dark and light regions would not be present.

## Examination of failures

A considerable amount of experience is necessary when investigating the cause of failures and it is advisable to examine first the working conditions, past history of the article and the stresses it has sustained before undertaking metallurgical examination; otherwise, misleading conclusions may sometimes be obtained. The classification of defects and failures is extremely difficult, but the following points for consideration may prove helpful.

(1) *Composition* may be unsuitable for the design in hardness, resistance to wear, shock resistance, corrosion, oxidation and permanency of dimensions. Faulty composition may arise due to incorrect mixture, loss during melting, or contamination, especially by sulphur from furnace gases.

(2) *Abuse in service* may be the cause of failure owing to overloading, looseness (connecting-rod bolts), misalignment of shafts and rough handling (initial overstressing of screw-threads by the use of excessive leverage on spanners).

(3) *Fabrication* may be the cause of failure due to rough machine marks, local cold work due to punching of holes, overheating during excessive grinding.

(4) *Fatigue failures* (p. 24) are extremely common and are clearly indicated by the characteristic fracture. Abrupt changes in section, rather than defective material, are usually the cause of the failure. Contrast with *brittle fracture* with chevron (p. 15).

(5) *Coarse crystalline fractures* may be due to over-heating during heat-treatment, high finishing temperature (p. 127), high casting temperature or critical cold work followed by annealing (p. 125).

(6) *Intercrystalline fractures* may be due to brittle impurities at grain boundaries, such as FeS in steel, Bi in copper; weld decay (p. 256); season cracking (p. 120); caustic embrittlement (p. 121); white-metal penetration in brass, steel and aluminium alloys (p. 121); impurities in zinc-base die-castings; burning, and gas penetration of copper (p. 303).

(7) *Defective structure* may cause failure due to blowholes, segregation, local concentrations of impurities, precipitation of slag particles in steel sheets, decarburisation, re-carburisation from oil flame, strings of beta constituent in brass sheets, incorrect or unsuitable heat treatment, brittle or low melting-point constituent at grain boundaries ($Fe_3C$, FeS).

(8) *Cracks* may start from slag particles, brittle micro-constituents, such as cementite networks in case-hardened steel (p. 205); quench cracks (p. 189); local differences in expansion due to uneven heating of large masses; sudden release of casting stresses; too rapid heating of high-speed steel; concentration of hydrogen.

(9) *Ageing* due to changes in properties which occur with time (p. 170) may be the cause of failure.

(10) *Surface roughness* may be due to stretcher strains (p. 119), large grains (p. 125), incorrect die design.

(11) *Corrosion* may be cause of failure due to parasitic currents near electrical equipment, dezincification of brass, air bubble impingement and pitting (p. 144).

# 3 Micro-examination

Macro-examination of metals should normally be supplemented by a study of the *micro*-structure which is concentrated upon much smaller areas and brings out information which can never be revealed by low magnification. Macro-structure depends for its development essentially upon the presence of comparatively coarse segregates of impurities or alloying elements formed during casting and which persist in varying degrees even after heat-treatment involving the formation of extremely fine constituents.

## Selection and preparation of specimens

When investigating the properties of a metal it is essential that the specimens used be representative of the whole mass. Three ways of taking samples from a sheet are illustrated in Fig. 19, and if the material is perfectly uniform each will yield the same result. This is suggested at X by sections through grains assumed to be spheres. When the sheet is rolled 50%

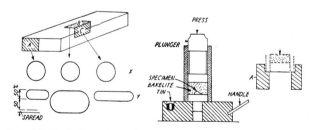

Fig. 19 Variation of structure with position of sample (left)
Press for mounting small specimens in Bakelite (right)

the structures revealed at planes A, B, C are shown at Y, and each is different. This may lead to a wrong interpretation of the grain size and degree of cold work. Where only one specimen is taken it should represent position C in order to reveal any directional effect and any variation in structure through the material.

In cutting or grinding a specimen avoid disturbing the structure of the metal by excessive pressure or heating. A deformed layer may exceed 0·005 mm in depth and will affect subsequent etching.

*Mounting small specimens*

Very small specimens are difficult to polish, but this can be greatly facilitated by mounting them in thermoplastic resin, polystyrene or methylmethacrylate. For doing this excellent machines are on the market, but if one cannot be obtained a simple apparatus can be readily constructed as illustrated in Fig. 19. The apparatus is heated to about 230°C, the temperature being indicated by the melting of the tin contained in the drilled hole. The specimens are placed on the anvil, resin powder added to the cylinder and compressed into a solid mass by a suitable press, such as can be adapted from a hydraulic motor-car jack. The mounting can be ejected from the cylinder by replacing the anvil by the stand A and applying pressure.

In cases where neither high-pressure nor heating are desirable a cold setting thermoplastic resin can be cast round the specimen.

*The polishing operation* can be divided into two stages:

(1) *Fine grinding of the levelled specimen* is carried out on a series of silicon carbide papers of increasing fineness, e.g. Nos. 280, 400, 600. The operation is as follows: place the No. 280 paper on a glass plate; hold the specimen so that the scratches from the abrasive will be formed in one direction only, at right angles to the file-marks. Rub down until the latter and the deformed layer are removed, swill the specimen, transfer to the No. 400 paper and, after turning the specimen through a right angle, rub down until all scratches from previous paper are removed. Repeat on remaining paper using very light pressure. In polishing soft metals, such as aluminium, paraffin lubrication should be used. Quicker grinding, without the risk of embedding abrasive in the specimen or affecting it by frictional heat is possible with waterproof silicon carbide papers, on an inclined glass plate down which water is continuously flooding.

(2) *Polishing* The fine scratches should now be removed by fine flaky powders on broad-cloth, chamois leather or selvyt stretched on a flat disc (Fig. 20). The specimen is held against the rotating disc, which is kept

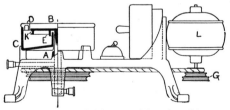

FIG. 20 Polishing machine (*Beck.*)

A—Spindle  
B—Disc  
C—Sump  
D—Guard ring  
E—Lip  
G—Pulley  
K—Groove, for fixing cloth  
L—Motor

moist with a mixture of powder and water. Suitable powders are: diamond dust, γ-alumina, green chromic oxide and magnesia. These abrasives can be prepared by roasting ammonium alum (alumina), ammonium bichromate (chromic oxide) and magnesium carbonate (magnesia). The residue is crushed and used as a suspension in distilled water. Diamond dust ($6-\frac{1}{2}$ μm) is used as a paste and is particularly useful for polishing hard specimens and for minimising the production of relief effects. The final polishing of copper, brass and soft metals is often carried out by means of metal polish (Sylvo) on selvyt or on chamois leather.

*Electrolytic polishing* is the preparation of smooth, distortion-free surfaces by anodic corrosion in a suitable solution. The roughly polished specimen is connected to the positive terminal (anode) of a 15–100 volt d.c. supply, the negative terminal being connected to a sheet of aluminium or stainless steel to form a cathode. By correct adjustment of the composition of the solution, temperature, voltage, current density and time, the protruding portions of a specimen can be dissolved, leaving a plane surface without deformation effects. Specific procedures are required for each material and reference should be made to American Society for Testing Material Standard, E3–4.4T.

After polishing, the specimen should be examined under the microscope in order to detect any cracks or coloured constituents. The most common of these are:

(1) Constituents harder than the background or matrix which stand out in relief, e.g. phosphide in cast iron.
(2) Blowholes and spongy places, which should be confirmed by focusing on the side or bottom of the cavity.
(3) Non-metallic inclusions: (*a*) Manganese sulphide (dove-grey masses), round in castings, elongated in forgings (Fig. 104); (b) iron sulphide (brownish yellow), in films round grain boundaries (Fig. 107); (*c*) slag (Fig. 106), Mn–silicate has greenish tinge, iron-silicate is black and greasy $Al_2O_3$ occurs as tiny hard, rounded grains; (*d*) copper oxide (blue); (*e*) graphite in cast iron (grey).

The matrix of a specimen prepared as described above will exhibit no structure, appearing very bright, after the fashion of a mirror.

## Etching

To make its structure apparent under the microscope it is necessary to impart unlike appearances to the constituents. This is generally accomplished by selectively corroding or etching the polished surface.

The main difficulties in getting a satisfactory etch are:

*Uneven attack*, most frequently due to grease on the surface, which may be removed by soap or caustic soda solution.

*Stains*, frequently caused by oxidation due to too long a delay before drying, or by oozing of reagent from cracks.

The polished specimen is immersed in the etching solution for a given time, removed by means of nickel tongs, swilled in running water, rinsed in methylated spirits and dried in hot air or on a soft towel. The specimen is then placed face downward on a glass plate, surrounded by an accurately parallel brass ring, and mounted on a glass slip with plasticine.

Etching is occasionally facilitated by making the specimen the anode, and using a platinum or carbon cathode with a voltage of 2–6.

## Etching reagents for micro-examination

*Steel and cast iron*

(*a*) Concentrated nitric acid   2 cm$^3$
    Absolute methyl alcohol 98 cm$^3$
(*b*) Saturated solution of picric acid in alcohol.

Time required in each case is 10–30 seconds. Solution (*a*) is used preferably for grain boundary etching and solution (*b*) for pearlite. A 10% solution of hydrochloric acid in alcohol is useful for etching hardened steels.

*Etching of cementite*

Picric acid    2 g
Caustic soda 24 g
Water        73 cm$^3$

(For temper brittle steels, see p. 221).

Time required is 10 minutes' boiling, and at conclusion the solution should be displaced by running water without exposing the surface of the specimen. Cementite, iron tungstide and $Fe_2W_2C$ are darkened.

*Etching carbides in chromium steels*
    (Murakami)

Potassium ferricyanide 10 g
Potassium hydroxide    10 g
Water                  100 cm$^3$

A 10% aqueous solution of sodium cyanide darkens carbides without attacking austenite or grain boundaries. Use electrolytically, cathode of similar material 25 mm from specimen with 6 volts. Time, 5–10 min.

## ETCHING REAGENTS FOR MICRO-EXAMINATION

*18/8 Austenitic steels, and inconel*

Hydrochloric acid 10 cm$^3$
Nitric acid 3 cm$^3$
Water 100 cm$^3$
Pickling restrainer (flour) trace
Use solution at 80°C for 15–60 seconds.

*Copper and alloys*

(a) 25% aqueous solution of ammonium hydroxide with a few drops of hydrogen peroxide.

Suitable for high magnifications, but must be freshly made.

(b) Saturated solution of chromic acid in water.
(c) 10% aqueous solution of ammonium persulphate (freshly made).

Solutions a, b and c are very suitable for brasses and alpha bronze.

(d) Ferric chloride 10 g
Hydrochloric acid 30 cm$^3$
Water or 50:50 water-alcohol mixture 200 cm$^3$

Suitable for bronzes, aluminium bronze, high zinc brasses and copper alloys containing nickel.

*Aluminium and alloys*

Nitric acid 5 cm$^3$
Hydrofluoric acid 2 cm$^3$
Water 100 cm$^3$

Solution should be freshly made. Time of etching is 15–60 seconds.

*Magnesium and alloys*

Ethylene glycol 75 cm$^3$
Distilled $H_2O$ 24 cm$^3$
Nitric acid (conc) 1 cm$^3$

*Nickel monel and cobalt*

Nitric acid (colourless) 50 cm$^3$
Glacial acetic 50 cm$^3$

*Zinc, white metals and magnesium*

2% nitric acid in alcohol.

*Lead*

Remove scratches with 5% ammonium molybdate solution, repolish on selvyt and immerse in

Glycerol      40 cm³
Acetic acid   10 cm³
Conc nitric acid 10 cm³

*Titanium and zirconium*

1–5% aqueous solution of hydrofluoric acid with sometimes a little nitric or phosphoric acid to retard staining. For alpha alloys use 2% HF in saturated oxalic acid with a crystal of ferric nitrate.

*Identification of non-metallic inclusions in steel.* Non-metallic substances, such as slag particles, metal oxides and sulphides, are found in steel and a broad classification of such inclusions is revealed by the following tests.*

(1) Immerse the clean polished specimen for 10 minutes in 10% chromic acid. Inclusions attacked are mainly manganese sulphide (dove grey) although, due to mutual solubility, iron sulphide and oxides may also be present.
(2) Next immerse the sample in 1% oxalic acid for 1½ minutes. The yellowish-brown particles are mainly iron sulphide.
(3) Immerse the specimen in 2% hydrofluoric acid for about 3 minutes. The silicate base inclusions are attacked while oxides are untouched.

## Appearance of micro-structure

The polished surface of a uniform specimen appears bright without any detail. The reason for this is indicated in Fig. 21(*a*), which shows that the light rays from the vertical illuminator strike the surface of the specimen normally, with the result that they retrace their path, finally passing into the eye. The slight chemical attack or etching of the polished surface first reveals the grain boundaries. Further attack produces shades of varying degrees in the grains. This is due to the fact that the etching reagent has produced, not a general solution of the surface, but a series of small but well-defined facets upon each grain. These facets have the same orientation in each grain. In adjacent grains, however, the inclination to the surface changes. Thus, one grain may reflect the light up the microscope tube, while the adjacent grains reflect most of the light in other directions and so appear darker. The truth of these statements can be verified by changing the angle of the incident light or by rotating the specimen, when the bright

*For further information, see C. BENEDICKS and H. LÖFQUIST. *Non-metallic Inclusions in Iron and Steel* (Chapman & Hall, 1931).

grains becomes dark. Fig. 21(b) illustrates this effect, and Fig. 21(c) indicates the cause of the apparent difference in size of duplex constituents. These constituents are common in alloys, frequently consisting of bands of

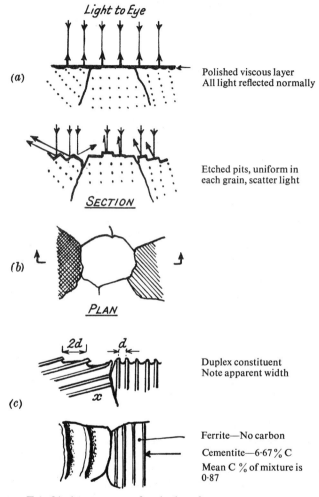

FIG. 21 Appearance of etched surface

material having widely different degrees of either hardness or rate of corrosion. Hence, during polishing or etching, either the soft or the soluble constituent is partially removed, consequently the micro-structure appears banded. It will be noticed, however, from Fig. 21(c) that grain $x$ appears to possess laminations of width $2d$ owing to the angle at which they cut the surface, whereas the true distance is still the same as in the adjacent grain. This is important in the case of pearlite in steel and cast iron.

*Relation between micro-structure and mechanical properties*

This entails the consideration of four factors:

(1) *The number of micro-constituents present.* With one constituent, consisting of a pure metal or a solution of two metals in the solid state, the properties are affected by:

(*a*) the *size* of the crystals. Small grain size is associated not so much with increase of tensile strength as with increase in resistance to shock; cleavage strength is related to $\frac{1}{\sqrt{\text{Dia}}}$.

(*b*) the *shape* of the grains. Rounded grains indicate soft annealed metal; whereas elongated crystals are associated with hard cold rolled or drawn metal.

With two micro-constituents we have also to consider:

(2) *The intrinsic properties of the two phases.*
(3) *The relative volumes of each.*
(4) *The distribution of the phrases*—for example, globules or plates in the interior of the grain; films at grain boundaries (Figs. 103 and 107); coherency and stress fields.

For example, the structures of steels and white cast iron consist of mixtures of almost pure iron called ferrite and a compound of iron and carbon called cementite. The cementite is very hard but brittle, while the ferrite is soft and relatively weak. The cementite may be embedded in the ferrite in the form of globules which cause a stiffening of the ferrite. This reinforcement of the soft ferrite, however, is far more effective when the cementite occurs as thin plates alternating with layers of iron, the continuity of the iron being interrupted. Maximum tensile strength is obtained when the structure consists entirely of such an intimately stratified mixture, at 0·9% carbon (Fig. 100). Increase in volume of the hard cementite beyond this point results in the presence of membranes of excess cementite at grain boundaries or of massive plates in the grains (Fig. 117). While such a volume produces increased hardness, the tensile strength falls due to the initial failure of the brittle cementite (Fig. 100). White cast iron (Fig. 185) exhibits brittleness to a great degree owing to the large volume of free cementite.

Special consideration must be given to brittle or low melting-point constituents which occur as membranes round the grains, since the cohesion of the aggregate may be impaired, for example, bismuth (p. 302), and iron sulphide (p. 169).

# The metallurgical microscope

Metallic specimens, in contrast to transparent specimens used in microscopic work on plant and animal life, have to be examined by reflected light. Consequently, the substage condenser and mirror underneath the table are not required. A diagrammatic arrangement of a metallurgical

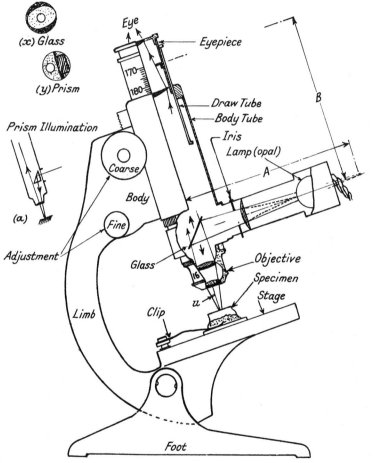

FIG. 22 Details of a metallurgical microscope

microscope is shown in Fig. 22. The instrument consists of a three-limbed foot of sufficient size and shape to support the body. The limb, pivoted in order to obtain suitable inclinations of the body, carries at the upper portion a coarse adjustment, consisting of a rack and pinion device, and a fine adjustment to facilitate final focusing of the object. These adjustments operate on the body tube which contains a draw-tube by means of which the distance between the eyepiece and the objective can be varied. The draw-tube carrying the eyepiece (23·3 mm diameter) is engraved in millimetres which indicate the distance from the back of the objective to the end of the tube.

The body tube carries a suitable vertical illuminator, on to the end of which the objective screws. In the more expensive microscopes a revolving nose-piece carrying three objectives enables a quick change of objective.

The two most important optical parts of the microscope are the objective for resolving the structure of the metal, and the ocular or eyepiece for enlarging the image formed by the objective.

The *eyepiece* is so named because it is nearest the eye and for bench work the Huygenian form is most common. This is illustrated in Fig. 19. It is made in a variety of powers, such as $\times$ 5, $\times$ 8, $\times$ 10, $\times$ 12·5, marked on the top, these designations meaning that the lenses magnify the image formed by the objective by these amounts. Other types known as compensating and projection eyepieces are employed in photo-micrography.

*Objective*

The combination of lenses nearest the specimen is called the 'objective', their quality determines the final result. High-class lenses are corrected for chromatic and spherical aberrations and produce an image free from colour (apochromatic) when used with a correct tube length. The magnification depends upon the focal length, such that the shorter the focal length the higher the magnification. The focal length of an objective is usually marked on the side and, in rarer cases, the numerical aperture (NA). The focal length, however, does not indicate the distance of the objective from the specimen, as Table 4 shows.

TABLE 4

| Focal Length mm | Working Distance, mm | Numerical Aperture (NA) | Approx. No. of Lines per mm Resolvable NA × 100 000 | Useful Magnification Limit NA × 1000 |
| --- | --- | --- | --- | --- |
| 16 | 7 | 0·25 | 1000 | × 250 |
| 8 | 2 | 0·50 | 2000 | × 500 |
| 4 | 0·3 | 0·75 | 3000 | × 750 |
| 2 | 0·15 | 1·20 | 4700 | × 1200 |

The following properties of an objective are important:

# THE METALLURGICAL MICROSCOPE

## Magnifying power

This is always expressed in linear diameters and offers no difficulty when the image is projected on to a screen, since the enlarged image of an accurately divided millimetre scale can be measured. In visual examination we find more difficulty, mainly due to the fact that the apparent image seen by the eye cannot be measured. This image is assumed to exist at 254 mm from the eye, which for normal sight is the least distance for distinct vision. If an object 1 mm diameter is placed at 254 mm from the eye it looks a certain size and a magnification of × 15 means that the microscope makes it appear as if it were an object 15 mm diameter at 254 mm distance.

A rough value of the magnification of a bench microscope is given by:

$$\text{Magnification} = \frac{D}{F} \times \text{power of eyepiece}$$

where D = distance from back of objective to eyepiece
F = focal length of objective.

## Resolving power of objective

Resolving power is the property by which an objective shows distinctly separated two small adjacent bands in the structure of an object. This is usually expressed as the number of lines per mm that can be separated (vide Table 4). It depends on the numerical aperture* and the wavelength of the light employed, e.g. a larger number of lines are resolved with blue than with yellow light. Resolution is particularly important when dealing with certain microconstituents of metals consisting of fine laminations, for with poor resolution these appear as one uniform area; whereas an objective with high numerical aperture (NA) reveals the duplex nature of the structure. Naturally the magnification must be sufficient to render the resolved lines visible (that is *useful* magnification) but a higher magnification obtained by extending the tube length, or by a high-power eyepiece, does not increase the resolution, and is really *empty* or useless magnification. These points are illustrated in Figs. 98 and 99, which shows two views of pearlite in steel. Fig. 98 was obtained with a 16 mm lens, and Fig. 99 with a 2 mm oil immersion lens which has been sufficient to resolve the structure.

*Numerical aperture (NA) = $n \sin \mu$, where $n$ is the refractive index of the medium between the specimen and the objective (air = 1, cedar oil = 1·5); and $\mu$ is half the angle of the cone of rays entering the objective.

The smallest distance between two lines for them to be seen separately is $\delta = \frac{0 \cdot 5\lambda}{NA}$ where $\lambda$ is the wavelength of light.

The use of oil immersion lenses, therefore, increases the resolving power. To obtain the highest resolution from an objective it must be used with a suitable tube length and this sometimes differs from the values stated by the maker.

## Vertical illumination

The simplest form of illuminator consists of a thin glass slip held in an adjustable mount situated just behind the objective (see Fig. 22). The incident light strikes the plate at 45° and is partially reflected down onto the specimen. The rays are returned by reflection, passing through the objective and glass plate to form the final image.

In certain cases a right-angled prism is substituted for the glass plate, being arranged to cover only half the area of the tube (Fig. 22(*a*)). The prism transmits much more light than the glass plate, but the resolution is not so great. The correct appearance of the illuminator with the eyepiece removed is shown in views $x$ and $y$.

From Fig. 22 it will be noticed that an iris diaphragm is fitted to reduce glare, its opening should be the smallest at which it will work without encroaching on the field of view (i.e. just fills the back lens of the objective). For best results, also, critical illumination should be employed such that distances A and B are equal.

The *photography* of the specimen entails a few modifications in the arrangement of the equipment. A high-power illuminant such as an arc light is necessary together with a camera placed behind an eyepiece of the projection type. In most cases the microscope is arranged in a horizontal position and for metallography the 'inverted' type with the stage on top is favoured.

The procedure consists in focusing the image on a ground glass screen which is then replaced by a photographic plate contained in a holder. Exposure is made while using a definite colour screen (green, yellow), with subsequent development and printing.

## Phase contrast microscope

In the phase contrast microscope the specimen is illuminated by a hollow cone of light and the objective incorporates a glass phase plate having an annulus which advances by quarter of a wavelength the directly reflected light passing through it relative to the scattered light transmitted by the rest of the plate, so that high areas in the specimen appear light. In this way very small differences in level of the order of $50 - 250 \times 10^{-10}$ metres can be detected and this may also be helpful in interpreting electron micrographs.

## Electron microscope

The optical microscope reaches its limit of performance at about $\times 2000$ dia. owing to the detail size being comparable with the wavelength of light. The use of ultra-violet light, with its shorter wavelength, extends the magnification limit somewhat, but the use of electrons with a wavelength some

# THE METALLURGICAL MICROSCOPE

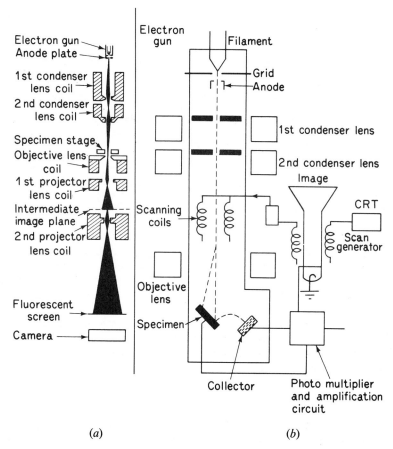

FIG. 23 (a) Diagram of electron microscope
(b) Diagram of scanning electron microscope

40 000 times smaller still vastly extends the useful range to a present commercial limit of × 75 000. The 'optical' systems in the electron and optical microscope are analogous except that in the former the focusing is done with energised coils, Fig. 23(a). The main limitation from the metallurgical aspect is the fact that the electron beam must pass through the object and this necessitates the use of a thin plastic or carbon replica of the etched surface. The anodic film on aluminium can be separated from the parent metal and be used to show fine detail.

Inclusions can be fixed in a replica by dissolving the matrix after the replica has been fixed on the surface. With such an extraction replica electron diffraction photographs can be taken of particular particles to enable their identity to be deduced, in addition to the normal electron micrograph. Recent development involves the use of thin foils, formed by anodic dissolution of the sample, for direct observation of internal structure and imperfections and is complementary to replica technique which is limited to surface topography.

An important extension of electron microscopy is the scanning microscope which is basically the electron-optic equivalent of the reflecting-light microscope. The most common use of the SEM is concerned with the examination of surfaces, and magnifications of $\times$ 80 000 are possible, maintaining a remarkable depth of focus (e.g. Fig. 84).

A finely focused beam of electrons is moved by the scan coils in a square-raster over the specimen surface. Secondary electrons generated in the surface layers of the specimen are collected in a scintillator/detector system and the detected signal is displayed on the observation cathode-ray tube which is monitored in step with the scan-signal, Fig. 23(*b*).

# 4 Solidification of Metals

Solid metals in bulk may be produced by a variety of ways, but the melting and freezing of a liquid is by far the most common method of preparing metals used in engineering, either as castings with the desired final shape or as ingots which are subjected to subsequent shaping processes.

*Crystallisation of a pure metal*

Most metals used industrially are alloys of two or more pure metals, but the process of crystallisation can be well illustrated by the example of a pure metal. When a pure molten metal solidifies, it does so like other pure elements at a fixed temperature, known as its freezing-point. If the temperature of the metal is observed at intervals and plotted against time, a *cooling curve* is obtained, as shown by *a* in Fig. 24. The horizontal portion

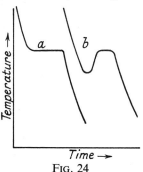

Fig. 24

of the curve is due to the evolution of *latent heat* at the freezing-point. This type of cooling curve is obtained only if solid foreign particles, described as heterogeneous nuclei, are present in the liquid metal and the rate of cooling is relatively slow. When there is no suitable solid matter present, a liquid experiences difficulty in starting to crystallise and may cool 0·1–100 °C below its real freezing-point, i.e. *nucleation undercooling*. Nuclei or 'seed' crystals then form, followed by their *growth* which is the second stage of freezing and the temperature may rise to the true freezing-point. The type of cooling curve in this case is shown by *b* in Fig. 24.

When heat is abstracted rapidly, however, solidification (i.e. growth) occurs at a temperature intermediate between that of nucleation undercooling and the equilibrium freezing point. This gives rise to *solidification undercooling* which is significant in die castings and in spheroidal cast iron. It leads to fine structures due to a decrease in diffusion rates. Most commercially 'pure' metals contain impurities which may have significant secondary effects on solidification. In practice it is possible to add to the liquid metal elements or compounds which will promote heterogeneous nucleation. These additions, which often have a crystallinity similar to that of the solidifying metal, are described as inoculants or grain refining agents, as they lead to an increased number of crystals per unit volume of solidified metal, e.g. TiC in aluminium alloys.

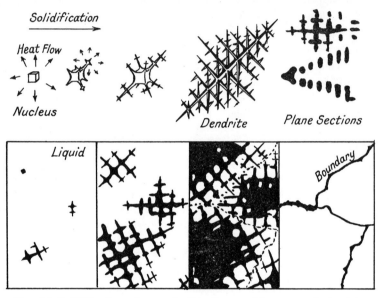

FIG. 25 Solidification of a metal. For f.c.c. and b.c.c. metals dendrite arms extend in the cube face [100] directions

During solidification, heat evolution raises the temperature at the solid–liquid interface, but protruding solid fingers may extend into a cooler, undercooled region and their growth is accelerated. Radial arms are formed which send out secondary arms at right angles to them. This is repeated until a fir-tree type of crystal is produced, known as a *dendrite* (Fig. 25). Spikes grow in definite crystal directions, e.g. in lead direction is along cube edge $<1007$ and preferred orientation is along heat flow direction. This type of growth is very common in metals and alloys but growth can also proceed by a curved or a plane surface propagation into the liquid, which is particularly frequent with intermetallic compounds (Fig. 254).

Figs. 39 and 228 are particularly interesting in showing the tendency to grow in certain directions and yet form angular masses.

The dendrites grow outwards until contact is made with neighbouring growths; this contact surface becomes the *boundary* of the crystal or grain. The dendrite arms then become thickened until finally a solid crystal remains with no indication of the dendritic growth, except where shrinkage occurs, with the formation of interdendritic porosity. In practice dissolved or insoluble impurities in the metal are frequently expelled by the growing crystal to the grain boundaries and are also trapped in between the arms of the dendrites. The distribution of these impurities, therefore, betrays the dendritic growth (Fig. 113).

It should be noted that

(1) Since each grain starts from a nucleus the final crystal size is dependent on the number of effective heterogeneous nuclei;
(2) Orientation of the atomic pattern or crystal lattice is constant in a given crystal but varies from grain to grain.

## Structures of ingots and castings

One of the early stages in the manufacture of most metals is the pouring of molten metal from the furnace or a ladle into a hollow mould, usually made in sand or in suitable metal, the latter being known as a chill mould or die. On solidifying, the metal object is called a *casting* if the shape is such that no further shaping process is necessary; simple blocks of cast metal are called *ingots*, which differ from a casting in that they have to be rolled, forged or extruded in order to produce semi-finished products.

In both ingots and castings a small uniform grain size is generally desirable, as weakness is often associated with coarse structures. The shape and size of the grains depends mainly on

(1) Number and distribution of seed crystals or nuclei;
(2) The direction and rate of crystal growth.

These factors are influenced in practice by the casting temperature, mass of metal, thermal conductivity of mould material and the composition of the alloy, so that in general the purer the metal the larger the grains formed.

*Uniform slow cooling* produces a few nuclei which are fairly evenly distributed, and growth is roughly the same along various crystal axes. The result is a coarse structure of equi-axed grains, about the same size but differing in orientation. This condition is found frequently in sand castings in which there is very slow heat dissipation (Fig. 26).

With a more rapid cooling, a large number of heterogeneous nuclei are effective and consequently small equi-axed crystals result. This condition arises when a metal is cast, from just above its freezing temperature, into a chill mould ore by adding inoculants.

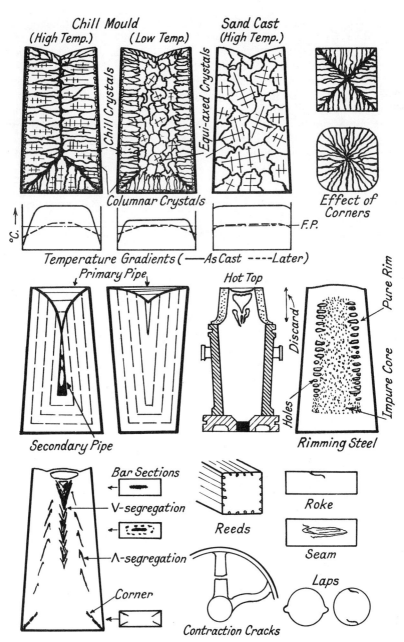

FIG. 26 Structure, contraction effects and segregation in ingots

## Ingots

The structure of ingots is more varied and will be examined in more detail. When molten metal, at a temperature well above its freezing-point, is cast into a metal mould, the portion in contact with the cold surface will be almost instantly cooled below its freezing-point, while the metal at the centre will be little affected. A shower of seed crystals is therefore formed only in the outer skin of the cast metal. Growth takes place in all directions but contact is soon made with adjacent crystals, with the result that very small equi-axed crystals are formed, known as *chill crystals*. If the casting temperature is very high these chill crystals may re-melt. It is more often, however, that some of the inner layer of chill crystals, which have a favourable growth axis, continue to grow in the opposite direction to the abstraction of heat because the chance of crystal growth is greater than the tendency to form fresh nuclei. Side growth is prevented by collision with neighbouring crystals, hence long thin crystals are formed, called *columnar crystals*, which, if the initial temperature of the metal is high, may reach the centre, as in Fig. 26. If the mould has sharp corners, *planes of weakness* are formed at the junction of the crystal growths. Such ingots may break in rolling, but the difficulty is often avoided by rounding off the corners of the mould (Fig. 26).

On the other hand, with a lower casting temperature, the interior portions of the ingot may be cooled sufficiently to crystallise from a large number of centres before the columnar grains reach the centre. Such an ingot will consist of an outer shell of chill and columnar crystals with a core of equi-axed grains (Fig. 26), and generally possesses good rolling and mechanical properties.

## Test bars

A test bar is a small casting of specific design for various alloys and is used mainly to obtain information regarding the mechanical properties of the alloy for specification or quality control purposes. Engineers sometimes fail to appreciate that a test bar is a means of controlling production quality of castings at a consistent level and that test-bar results do not necessarily correspond to the properties which would be obtained from test specimens prepared from various parts of the casting. These properties frequently depend on the section of the casting (e.g. large propeller).

## Shrinkage effects

If molten metal is poured into a mould and allowed to solidify it is found that it shrinks and in the end the solid metal does not completely fill the mould. Such shrinkage of the metal in a mould is made up of three parts:

(1) *Contraction in the liquid state* between the casting and freezing temperatures.
(2) *Shrinkage owing to solidification*. This varies from 0 to 7% (by volume) for metals and alloys except bismuth, which expands. Graphite in cast irons also tends to counteract the shrinkage and produces good surface impressions. Both (1) and (2) may be obviated by 'feeding' additional metal.
(3) *Contraction in the solid state*, which is allowed for by making the mould larger than the desired casting, using a contraction allowance by means of a pattern-maker's rule. In uneven sections stresses may be set up in the casting, and may be sufficiently high to cause either warping or fracture. Curved (S-shape) spokes in cast-iron pulleys are used to allow them to accommodate these stresses (Fig. 26).

Typical examples of contraction are:

TABLE 5   PERCENTAGE OF VOLUME OF SOLID AT $20\,°C$

|  | Mild Steel (low carbon) | Copper | Aluminium |
|---|---|---|---|
| Liquid contraction per $100\,°C$ | 0·8 | 1·9 | 1·3 |
| Freezing   ,, | 7·0 | 4·1 | 6·5 |
| Solid         ,, | 7·35 | 6·4 | 5·5 |

Solidification shrinkage is important since it often produces defects in ingots and castings. Let us consider the solidification of metal in a slightly tapered ingot mould, placed narrow end uppermost, as used in many steel works.

*Piping*

A layer of solid metal forms round the ingot walls, contraction occurs, resulting in a fall in the level of the liquid. Successive layers of solid form, each accompanied by a fall in the liquid level, the fall increasing as the volume of liquid decreases. In this manner a central cavity of the shape shown in Fig. 26 is formed, known as the *primary pipe*.

With the ingot mould used narrow-end-up a conical volume of metal still remains liquid after the top portion of the ingot is solid. Solidification of this metal will give rise to further cavities, known as *secondary piping*. The latter can be prevented by using the mould 'wide-end-up' since the shrinkage is counteracted by the feeding down of molten metal from the top, the region last to solidify. The primary pipe can also be reduced by using a brick-lined top or other types of feeder designs on the mould, in which the metal remains molten for a prolonged period, and consequently 'feeds' the ingot (Fig. 26).

In practice crusts form over the primary pipe owing to air cooling.

During the rolling of the steel the portion of the billet containing the primary pipe is discarded, but the secondary piping still remains, and, although the cavities are closed up by the rolling, their surfaces may not be welded together and are liable to open up again when the steel is quenched, thus constituting a dangerous flaw in good-quality steels. Such steels are, therefore, cast in moulds, wide-end-up fitted with a properly designed feeder head.

A method of ingot casting used widely for ingots in non-ferrous alloys and more recently for steel is that of semi or continuous casting. In this case the base of the shallow mould is closed only at the beginning of the pour by a plug which is then withdrawn. The outgoing ingot is rapidly cooled by quenching below the mould. Owing to the unidirectional method of freezing in this case the piping problem is readily controlled by the speed of ingot withdrawal.

Central shrinkage cavities are obviated also in centrifugal castings owing to the unidirectional crystallisation.

*Shaped castings* are also subject to piping (or sinking) in the thicker portions, but generally most of the solidification shrinkage in shaped castings tends to be enclosed in the casting as either coarse, localised or fine distributed porosity. Most shaped castings have, therefore, feeders or risers, or reservoirs of molten metal to compensate for liquid and solidification shrinkage. Efforts are also made to delay cooling in the feeder by either insulation or generation of heat by exothermic walls. Alternatively chills can be used to accelerate the cooling at thick junctions which are liable to be unsound.

*Interdendritic sponginess* (see below) often exists in the centre of ingots and castings and is the frequent cause of leaky castings. It is most prevalent in alloys which freeze over a wide range of temperature, since the feeding of the spaces enclosed by the long interlacing dendrites is difficult.

Moderate amounts of dissolved gas may be helpful in redistributing local concentrations of shrinkage cavities and thus improving pressure tightness in some castings (e.g. valve bodies).

## Gas holes

Gas holes are due to the trapping of gases in the casting. The gases may arise in different ways such as:

(1) *Gases dissolved* during melting are evolved after solidification has started. Hydrogen is a common source of gas unsoundness, e.g. the even distribution of fine pinholes in aluminium alloys. The gas may enter the melt from hydrated corrosion products, oil, etc., on the surface of the raw materials, from furnace atmosphere, from damp flux, crucibles and sand moulds. The amount of gas which may dissolve in an alloy is influenced by

the presence of other elements. For example, additions of tin, zinc and aluminium decrease the solubility of hydrogen in copper; nickel increases it and silver has little effect. Dry nitrogen does not give rise to gas holes and the bubbling of nitrogen through most metals tends to get rid of the other gases, by reducing the partial pressure of the gas above the melt. Pre-solidification, vacuum melting and casting are also effective in removing dissolved hydrogen.

(2) *Gases generated* by the mould dressing or by the mould walls (i.e. reaction with constituents in the mould material) and alloy reaction with moisture in sand moulds. Small blowholes can form just under the surface and in the case of steel become oxidised during soaking prior to rolling. They are subsequently elongated into *reeds* as shown in Fig. 26. When stressed at right angles to their length the reeds open up into splits which extend into the material, causing marked deterioration.

(3) *Gases produced by chemical reaction* during solidification within the melt. In the case of copper dissolved hydrogen reacts with dissolved copper oxide to form steam (p. 301), while in steel carbon monoxide is formed by the reduction of oxides by carbon. In 'killed' steel the oxygen is reduced to negligible quantities by deoxidisers, such as aluminium, ferro-silicon and ferro-manganese. The soluble ferrous oxide originally present is thereby replaced by insoluble oxides which largely enter the slag before carbon is added.

The bulk of the low carbon steel for sheets, nails, etc., however, is only partially deoxidised, e.g. 500 gm Al per tonne. When this partial deoxidation is properly controlled a thick rim of exceptionally pure metal solidifies. The residual liquid is enriched sufficiently in carbon and oxide for the reaction $FeO + C \rightarrow Fe + CO$ to occur. Deep-seated blowholes are formed by the carbon monoxide at a uniform depth below the surface and piping is counteracted. The central core is rich in impurities uniformly distributed and it corrodes more rapidly than the pure rim. This type of steel is called 'effervescing' or '*rimming*' steel; a typical macro structure is shown in Figs. 26 and 27(*b*).

The surface of the blowholes formed in the ingot by carbon monoxide (a non-oxidising gas) is not oxidised and appears silvery white. Such blowholes weld up more or less completely during subsequent hot rolling. Owing to the absence of piping, practically no material is discarded. Rimming steel can be readily identified by the characteristic segregated core. The high reputation of this material for deep drawing and pressing is due to the ductile properties of the skin which, in most operations, is the most highly stressed region. Semi-killed or *balanced steel* is made by adding more aluminium to the ingot than in the case of rimming steel so that the gas holes just balance the shrinkage. *Stabilised* steels are sometimes made by completely deoxidising the molten core with aluminium after the rim has formed ($Al = 0.03\%$).

(a)         (b)

FIG. 27 Sulphur prints of killed steel (a) and of rimming steel ingot (b)

## Chemical heterogeneity—segregation

It is found that the composition of castings and particularly of metal ingots is not uniform throughout and that the concentration of impurities and sometimes of alloying components (see also coring, p. 86) is greater in certain parts of the ingot than in others. This subject is particularly important in steel production owing to the large ingots used (up to 200 tonnes) and it has been studied by the Heterogeneity Committee of the Iron and Steel Institute, who have issued reports from 1926 onwards.

Molten steel contains soluble impurities—sulphur, phosphorus—and soluble alloying elements together with insoluble impurities or slag particles in suspension. As will be seen in Chap. 5, the first crystals to separate contain less impurity than is indicated by the average composition of the alloy, and those elements which lower the freezing-point, such as sulphur, phosphorus, carbon, manganese and silicon, collect in the last portions to solidify. This separation of various substances into different places in the ingot is called *segregation*, and for *killed steel* and also other metals it can be classed as follows:

*Interdendritic or micro-segregation*, caused by the concentration of impurities or alloying components between the arms of the dendrites (Fig. 65). On annealing, elements, such as carbon, will diffuse and become uniformly distributed, but phosphorus and non-metallic inclusions will be only slightly affected.

*Normal or macro-segregation* in the case of steel ingots is the concentration of sulphur, phosphorus and carbon in the centre and upper portions of the ingot. It is largely associated with the pipe and is mainly removed when this is discarded. The direction of flow of the metal during feeding is shown by the V-form of this segregate (Fig. 26).

*Inverse or Λ segregation* is the concentration of fusible elements near the outer zones of an ingot, and in extreme cases of inverse segregation exudation occurs on the surface of bronzes, particularly those containing lead (tin sweat). In 'killed steel' this type of segregation occurs about midway between the centre and the outside, in the form of an 'inverted V' (Fig. 26).

*Ingot corner segregation* is the entrapping of impurities at the junction of two sets of crystals growing from the sides and bottom of the mould.

*Gravity segregation* occurs when the primary crystals differ in density from the liquid. A good example is the rising of the cuboids in tin-antimony alloys (Fig. 254).

*Seams—rokes—shells—laps in steel ingots*

Although the ingots should be cast at as low a temperature as possible in order to get good working properties, it is just as important that the metal should be sufficiently fluid to prevent semi-solid skins forming on the surface as the metal level rises in the mould. Periodically the liquid metal bursts through the oxidised skin and covers it. When rolled a *seam* is formed, as shown in Fig. 26. Smaller defects starting from the surface, but due to similar causes, are called *rokes* (Fig. 26).

A defect, somewhat similar to a roke, is caused by poor roll design or by rolling at too low a temperature. The metal spreads to an extent greater than the design pass and form fins on opposite sides of the bar, which in subsequent passes are lapped over to give the *lap* illustrated in Fig. 26.

Splashes of oxidised metal on the surface of the mould, particularly near the bottom, appear as scales or shells on the ingot after it is stripped.

Surface defects such as mentioned above tend to cause cracks during subsequent working and heat-treatment and have a harmful effect on the fatigue and other properties of the material. They can be revealed by macro-examination or various non-destructive tests e.g. penetrants or magnetic.

*Non-ferrous ingots* are usually subject to dross and dirt inclusions at the surface and are frequently surface scalped. Internally they are often prone to macro- or micro-segregation of alloying elements.

*Sand moulds*

Castings are made both in permanent moulds (p. 374) and in sand moulds which necessitate the use of patterns, except in the case of loam moulds of simple curved shapes, which only require templates in their construction.

The essential properties of a moulding sand are resistance to high temperatures, satisfactory bond strength and permeability to gases. The red sands, consisting of silica grains bonded together with natural clay, are often used for non-ferrous and iron castings. For cast iron, coal dust is added to the sand or coated on the mould surface (i.e. blacking) in order to improve the surface finish of the casting, hence the black colour of the sand. For the same purpose moulds for casting aluminium are frequently dressed with french chalk. Sand used in the steel foundry must be capable of resisting very high temperatures and a refractory sand or a 'compo' consisting of crushed silica bricks, old crucibles, china clay, etc., is often used alternatively, zircon sand is used instead of silica sand.

Silica sand also can be bonded with 3–4% bentonite to make a *'synthetic'* mould sand; with 10% cement; and with 3–5% water glass which is hardened by passing carbon dioxide through the mould ($CO_2$ *process*).

'Green sand' moulds are made from sand which has been made sufficiently damp to hold together, while 'dry sand' moulds are dried in stoves before casting. Cores, which have to be strong enough to be handled and yet easily removed from hollow parts of a casting, are frequently made from silica sand bonded with linseed oil and dried. Many other types of organic binders can also be used for cores:

*Shell* moulding is a process of making mould shells or biscuits which are then glued or clamped together to make a mould. The shell is obtained by holding a sand containing 4–5% of synthetic resin on a hot metal pattern, which 'hardens' the sand up to a required thickness and then the remainder of the sand is 'dumped' by turning over the assembly. At present this process is used for making light castings, ferrous and non-ferrous, when saving in machining or improved surface properties can be obtained.

*The lost wax process* of precision casting to within $\pm 0.05$ mm per mm has recently become more important for gas turbine blades and other intricate parts which would otherwise entail much expensive machining. It consists in making a wax pattern in a master mould. A suitable mineral such as molochite, made into the consistency of cream with ethyl silicate binder, is poured around the wax pattern, and vibrated. When the mould has set the wax is melted out and the jointless mould baked and the metal poured normally into a hot mould.

*Design*

In designing castings, engineers should avoid abrupt changes of section and massive junctions; recessed parts or undercuts; rigidity of thin sections liable to tearing; small openings which restrict venting. Core stability and fettling factors should also be considered. See also pp. 185, 350.

# 5 Equilibrium Diagrams and their Interpretation

The development of the atomic bomb has convinced even the layman that the structure of the atom is of more than theoretical interest. The subject is abstruse, but to understand some of the characteristic properties of crystals of metals and alloys it is essential to consider the form and arrangement of the atoms, of which there are 101 types.

**Atoms**

Any atom consists of a *heavy positively* charged core or nucleus and a number of *light negative* electric particles or *electrons* which revolve round the core, analogous to the planets revolving round the sun. The atom as a whole is electrically neutral. Hydrogen has the simplest atom containing one planetary electron, while uranium, a complex stable one, has 92 planetary electrons.

These planetary electrons revolve in groups of orbits called shells spaced from each other so that the energy of the electron increases in unit steps from an inner to an outer shell group. There is a limit to the number of electrons which may be contained in any one shell, e.g. 2 in first shell, 8 in second, etc., as illustrated in Fig. 28. This gives rise to the arrangement of the elements in a periodic table in which groups of elements have similar proportions. (See Appendix 1.)

It is found that the chemical properties of the elements are closely related to the number of electrons in the outer orbit group. These are called *valency electrons*. A completely filled outer shell renders the element chemically stable such as in the inert gases helium and argon. Where the outer shells are incomplete, there is a tendency for the elements to combine chemically to form stable outer shells by pooling electrons in the incomplete shells, e.g. sodium chloride, NaCl in Fig. 28.

The elements can be divided into two classes:

(1) Electro-negative elements that *gain* a few negative electrons to form stable rings.
(2) Electro-positive elements that *easily part with* one or more electrons and thus change from neutral atoms to positively charged ions.

The majority of the elements are electro-positive: they are the metallic elements (see Table 9).

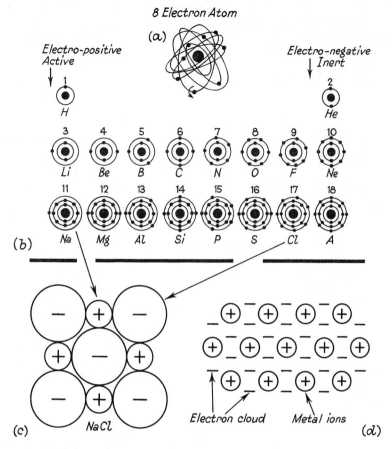

FIG. 28. (*a*) Schematic representation of orbits in an 8-electron atom; (*b*) the planetary electrons around cores of first 18 elements, arranged in periodic order of the number of electrons in outer shell (valency electrons); (*c*) association of sodium and chlorine to form common salt; (*d*) linking of atoms in metallic crystal

*Association of atoms*

Atoms can associate in the following ways:

(1) e.g. inorganic compounds such as common salt, NaCl, in which the chlorine receives one electron from the sodium and the resulting positive sodium ions and negative chlorine ions are held together by electro-static attraction and the relative numbers of atoms are fixed.

$$\text{Na}\cdot + \cdot\overset{..}{\underset{..}{\text{Cl}}}\!: \rightarrow \text{Na}:\overset{..}{\underset{..}{\text{Cl}}}\!: \quad \text{(see p. 89)}$$

This is called *ionic* bonding, and such crystals are non-conductors of

electricity because the electrons are firmly bound to individual ions.
(2) Strong covalent bonding can occur between elements which are adjacent to one another in the periodic table rather than on opposite sides for ionic bonding; or when a molecule or a crystal of a single element is formed, e.g. $Cl_2$ may be written:

$$:\!\ddot{C}\!l. + .\ddot{C}\!l\!: \rightarrow :\!\ddot{C}\!l\!:\!\ddot{C}\!l\!:$$

Further, with a polyvalent atom the several bonds usually have a characteristic arrangement in space. For example, the four covalent bonds from a carbon atom in a diamond are found to be directed towards the corners of a regular tetrahedron, Fig. 29(a). The complete pairing of electrons gives rise to extreme hardness and lack of electrical conductivity. Carbon can also exist as graphite in which the carbon atoms are arranged in sheets of regular hexagons (Fig. 29(b)) and these layers are held together by weak binding forces so that they can slide over each other easily, i.e. a greasy lubricating characteristic.
(3) Organic compounds in which definite numbers of hydrogen and electro-negative carbon, oxygen, chlorine, fluoride and nitrogen atoms associate in family parties or long chains—see polymers.
(4) Metals in which the few loosely held electrons are shared by several atoms. This is equivalent to a number of positive spheres (ions) surrounded by a gas or cloud of negative electrons free to wander. This electron cloud gives metals their characteristic high electrical conductivity.

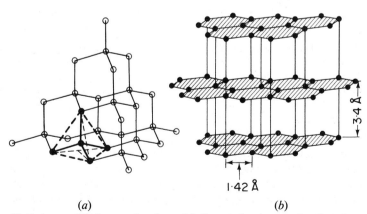

(a)          (b)

FIG. 29 Polymorphic forms of carbon: (a) diamond structure showing that each atom has four near neighbours located at the corner of a tetrahedron (b) graphite structure

There are no bonds between atom and atom in a metal, in fact, the positive charged spheres (ions) repel each other. The atoms are free to alter their relative positions without breaking any bonds and the marked plasticity of metal (p. 104) is due to this property. On the other hand, the whole mass is made coherent by the common negative cloud which permeates it, and causes the atoms to pack together very closely.

An arbitrary choice of atomic size is the closest distance of approach of the atoms, which can be represented by spheres just touching one another (Fig. 30). An atom is about one thousandth of a mm in diameter. Alternatively, the atoms can be represented as mere points arranged periodically in

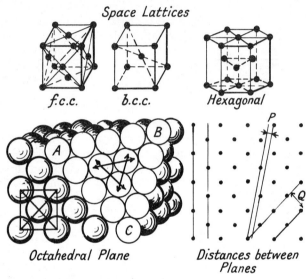

FIG. 30 Atomic arrangements (see also pages 104 and 113 for additional explanation)

f.c.c. is face centred cube; b.c.c. is body centred cube

three dimensions, known as a *space lattice* (Fig. 30). Most solid metals form atomic patterns in which the atoms are arranged at the corners and centres of cubes of a constant size or upon a hexagon as shown in Fig. 30. These patterns possess high symmetry and the metals are ductile and malleable. Manganese, bismuth and antimony have arrangements of lower symmetry and are somewhat brittle.

In an alloy the atoms (or molecules) can exist in different arrangements. In the gaseous form the atoms fly about without restriction, in a haphazard manner (Fig. 24(*a*)). In the liquid form the atoms are still without regular geometrical arrangement, but are closer together and restricted to a definite closeness of packing. When a homogeneous liquid solution of two metals is formed the different atoms—denoted by ○ and ● in Fig. 31—are intimately mixed together without preference as to position. Visible or mechanical separation is impossible and we have one liquid with properties differing from those of the individual components.

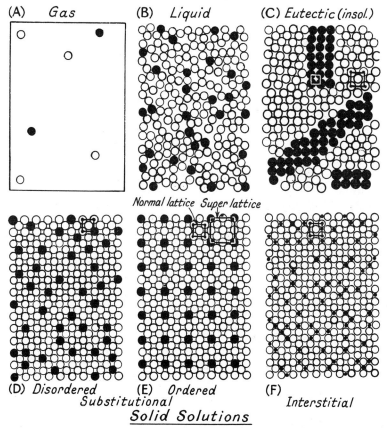

FIG. 31 Diagrammatic representation of atomic arrangements in gaseous, liquid and solid alloy

When a molten metal or alloy solidifies the atoms take up a *definite geometrical arrangement and this is the fundamental difference between the crystalline form and the liquid or vapour state*. When the metals are insoluble in the solid stage the different atoms form two kinds of crystal, each with its own atomic arrangement. Frequently, the micro-structure consists of fine strata of the two kinds of crystal forming a constituent known as a eutectic (Fig. 31(C)).

## Solid solutions

In certain cases the solidification of an alloy results in the formation of *one* kind of crystal in which both metals are present, but they cannot be detected by the microscope, although the properties of the crystals are profoundly altered. In such a case we have a solid metal in which the interatomic state which existed in the liquid solution has been preserved after solidification, and it is known as a *solid solution*.

In a solid solution the atoms occur in a definite geometrical pattern, which is usually a slightly distorted form of that of one of the constituent metals.

The various patterns have been investigated by X-ray methods and in the more general case of *substitutional solid solution* atoms of the parent metal are replaced, indiscriminately, by the atoms of the second metal in the proportion required by the chemical composition of the alloy (Fig. 31(D)).

In certain alloys—gold–copper, β-brass, β-bronze—the atoms have this random distribution at high temperatures. At low temperatures, however, such alloys which have simple ratios of the atoms show a tendency for the different atoms to take up certain preferential positions as illustrated in Fig. 31(E). The atoms of the second metal form a pattern superimposed on the primary pattern, and the structure is called a *super-lattice*. Except in a few alloys, the alteration in mechanical properties is unimportant.

Two metals can be completely soluble in each other only if they have the same lattice patterns, nearly equal atom diameters, and equal numbers of valency electrons. Where the difference in atomic diameters of the two metals exceeds 14–15% the 'size factor' is unfavourable and the range of solution is restricted. The reason is that the greater the difference in size, the greater is the distortion of the atomic pattern and the more likely it is that the atoms can adopt some other arrangement which is more stable.

The more electro-negative the solute element and the more electropositive the solvent, or vice versa, the greater is the tendency to restrict the solid solution ranges and to form intermetallic compounds—known as the *chemical-affinity effect*.

A metal of lower valency tends to dissolve a metal of higher valency more readily than vice versa—known as *relative valency effect*. Silicon with a valency of 4 dissolves little copper with a valency of 1; but copper dissolves an appreciable amount of silicon.

For a more detailed consideration of the various factors reference should be made to Hume-Rothery's publication (p. 429).

Of less common occurrence is the *interstitial solid solution*, in which the atoms of the second element fit into the spaces between those of the primary metal. One atom is much smaller than the other; elements which form this type of solid solution are hydrogen, boron, nitrogen and also carbon in γ-iron which forms the basis of steels.

# X-ray spectra

In addition to finding faults, described in Chap. 1, a second method of using X-rays is one which reveals details about the metal itself, such as grain size, orientation of the crystals, internal stresses and recrystallisation. This method is essentially a laboratory process and suffers in some ways from the defect that it is essentially concerned with the surface.

## Back reflection method

A *monochromatic* X-ray beam defined by pinholes is directed onto the specimen (Fig. 32). The X-rays are reflected from the various planes of atoms which are at a suitable angle to the axis of the beam. With coarse crystalline material only one or two grains are in the narrow beam of X-rays, consequently only a few spots are found on the film (Fig. 32). As the grains become smaller a larger number of suitably oriented rows of atoms reflect the X-rays and the number of spots on the film increase. These spots fall on a concentric ring round the beam, and with very fine grains the spots overlap to form a continuous ring. On close examination it will be found that the ring consists of a doublet due to the monochromatic K radiation having two components, $K_{\alpha_1}$ and $K_{\alpha_2}$. Cold work causes the spots to arc and widen radially, thus producing continuous rings which, due to fuzziness, mask the doublet. With much cold work the exposure time is increased and a considerable scattering of the beam produces a heavy background on the photograph. These characteristics are illustrated in Fig. 32.

The details may be summarised:

| State of Metal | Back Reflection Photograph |
|---|---|
| Coarse grains > 0·04 mm | Few clear spots |
| Fine grains 0·004 mm | Large number of spots in ring |
| Very fine grains 0·0004 mm | Continuous rings, doublet resolved |
| Cold work | Continuous rings, no doublets |

## Powder method

This is used in investigations on alloy phases. A sample of the alloy, in powder form on a hair, is placed at the centre of a flat cylindrical camera, around the rim of which a photographic film is placed (Fig. 33). The specimen is usually rotated. The narrow pencil of monochromatic X-rays is diffracted from the powder and recorded by the photographic film as a series of lines of varying curvature. The phases are recognised and their structures analysed by means of the characteristic pattern of 'lines'. From the spacing of the lines on the pattern and the known dimensions of the

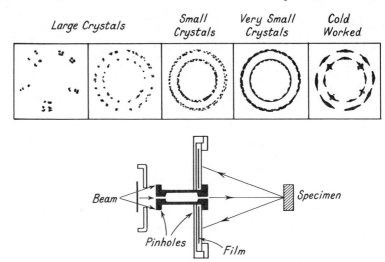

FIG. 32 Back reflection X-ray photographs and apparatus

camera, the apex angles of cones are calculated and from these are obtained the angles of reflection $\theta$ of the Bragg equation which relates the *spacing of the atomic planes* ($d$), the wavelength of the X-rays ($\lambda$) and the angle of reflection ($\theta$) ($n$ is an integer)

$$n\lambda = 2d \sin \theta$$

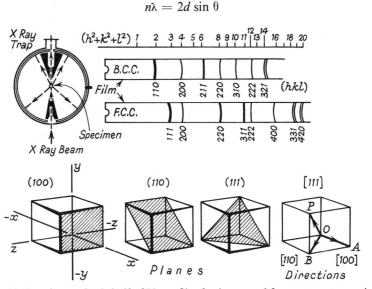

FIG. 33 Powder method; half of X-ray film for b.c.c. and f.c.c. structures typical planes and directions in a cubic lattice

Hence spacing ($d$) of the various atomic planes and also the unit size of cell or parameter may be obtained. Each crystal type has an arrangement of planes of atoms which gives a characteristic spacing to the lines on the spectrogram and with a little practice it is possible to recognise the characteristic patterns. Typical patterns for face centred and body centred cubic lattices are given in Fig. 33, while Fig. 34 shows planes for h.c.p. metals.

Many laboratories nowadays use counters for the recording of X-ray diffraction data instead of film. In the X-ray diffractometer a flat specimen is rotated about the diffractometer axis and a counter (usually a proportional counter) rotated at twice the angular velocity. The intensity of the diffracted beam is obtained directly from a chart record.

A modification of the X-ray diffractometer allowing specimens to be rotated about other axes is used for the determination of quantitative pole figures.

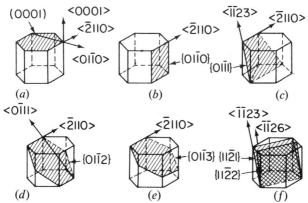

FIG. 34 Important planes and directions in a h.c.p. metal

*Crystallographic plane notation*

Planes are usually represented by Miller indices, e.g. (111) which are the smallest integers proportional to the reciprocals of the intercepts, in terms of the parameters of the plane on the three crystal axes ($x, y, z$). Take the faces of a cube. Since each of these intersects only one axis the intercepts are $(1, \infty, \infty), (\infty, 1, \infty), (\infty, \infty, 1)$. The reciprocals of these numbers are $(1, 0, 0), (0, 1, 0)$ and $(0, 0, 1)$ usually written (100), (010) and (001). The three opposite faces would have negative signs thus ($\bar{1}$00) (see Fig. 33). The general example is ($hkl$) and {$hkl$} to denote family of planes.

*Directions of planes*

Direction is represented by an index in the general form [$pqr$], i.e. square brackets to differentiate an axis from a plane. They are derived in a way that is different from that used for planes, as reciprocals are not involved. A direction OP is defined by the vector translations parallel to the axes of the cell to get from O to P. For example, in Fig. 33 direction OP is given by co-ordinates OA, AB, BP, i.e. [111]. In cubic crystals the direction [$pqr$] is parallel to the normal to the plane ($pqr$) but this is not generally true in other crystal systems.

*Representation of crystal orientations*

More and more frequently, the literature contains diagrams representing the orientation of single crystals or the limits of variation of orientation of crystals in a polycrystalline aggregate. Fig. 35 shows how such a diagram is derived for a *single* crystal. Imagine the unit crystal of Fig. 33 to be at the centre of a sphere. Draw normals from the planes such as (110) to cut the sphere surface at P, which is the pole of the plane. Draw a line from P to the south pole S, to cut the equatorial plane at $P_1$, which is the *stereographic projection* of the pole P. $P_2$ is similarly the projection of a (111) plane and 0 the projection of the (100) cube face. If this procedure is repeated for all possible planes of these types, the equatorial plane is covered with a symmetrical array of spots (Fig. 35(*b*)). For the cubic system, this array consists of 24 repeated triangles, and it is therefore possible to represent a unique axis such as tension axis or compression axis by

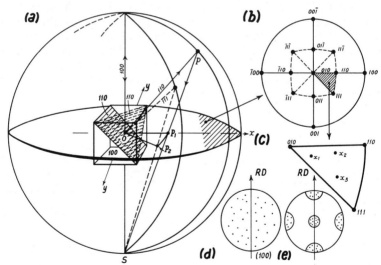

FIG. 35 (*a*) Stereographic projection; (*b*) equatorial plane; (*c*) unit stereographic triangle; (*d*) random orientation; (*e*) preferred orientation shown by (100) pole figures for sheet material
RD = rolling direction

its position within a *unit stereographic triangle* (Fig. 35(c)), e.g. $x_1$, $x_2$, $x_3$ are orientations of three samples.

Processing of metal often produces changes in crystal orientation (Fig. 54(D) and p. 118) and these changes are readily depicted on a *pole figure*. In this case, the single crystal at the centre of the sphere is replaced by wire or sheet with, say, the rolling direction parallel to the $y$ axis. Because the specimen is a polycrystalline aggregate, only one plane is projected, but from many crystals. If these crystals have a random orientation, the poles plotted on the equatorial plane will be scattered at random, Fig. 35(d). If, however, the metal has developed preferred orientation, this will give rise to regions of high pole density on the stereographic projection. A particular example of this is shown in Fig. 35(e), which represents the distribution of (100) poles in heavily rolled and annealed copper sheet. The crystal cube faces are parallel to the plane of rolling—this is the well-known 'cube-texture'.

The type of preferred orientation developed during fabrication depends on many factors including the method of working, the crystal structure of the metal and the presence of alloying additions in solution. Crystals tend to reach their final orientations when slip planes and directions are symmetrically oriented with respect to the direction of working. Preferred orientation developed during working is usually referred to as a *deformation texture*, while preferred orientation present when a heavily-worked metal is recrystallised is called a *recrystallisation texture*. Although the best way of describing preferred orientation is by means of a pole figure, an alternative method is to describe the texture in terms of *'ideal orientations'*, i.e. crystallographic planes and directions parallel to the directions of applied stresses. When this method of description is used, it must be understood that usually many crystals will deviate from these ideal positions so that there is a spread of orientations about them.

A list of ideal orientations which approximately describe the textures is given in Table 6.

TABLE 6

|  | Wire & tension // axis | Compression // axis | Rolling // plane ( ) // rolling direction | Recrystallization // plane ( ) // rolling direction |
|---|---|---|---|---|
| Al, Cu, Ni | [111] + [100] | [110] + [100] | (110) [1$\bar{1}$2] to (112) [11$\bar{1}$] | (100) [001] + (123) [41$\bar{2}$] |
| 70/30 brass | [100] + [111] | [110] + [111] | (110) [1$\bar{1}$2] | (113) [$\bar{2}$1$\bar{1}$] |
| Fe ( ) | [110] | [111] + [100] | (001) [110] to (111) [110] | (111) [110] + (111) [11$\bar{2}$] |
| Mg | [10$\bar{1}$0] |  [0001] | (0001) [10$\bar{1}$0] | (0001) [11$\bar{2}$0] |
| Ti, Zr | [10$\bar{1}$0] |  | (0001) tilted, [10$\bar{1}$0] |  |

## Equilibrium or constitutional diagrams

The solidification of an alloy generally occurs as a continuous process over a range of falling temperature and even after becoming solid, constitutional changes of far reaching importance may continue to take place. In a series of alloys such as the cupro-nickels or the carbon steels, it is possible by plotting temperature against composition to represent graphically the changes* which take place during and subsequent to solidification. Such graphs, suitably annotated, are known as equilibrium diagrams, and a knowledge of them is useful in the control of casting operations and heat-treatment processes. An objection to these diagrams, which is sometimes raised, is that they deal with alloys in a stable or equilibrium condition, only realised by either extremely slow cooling or very prolonged annealing, a condition rarely obtained in practice. *Nevertheless they do show the direction in which changes are likely to occur.*

While often appearing complicated, the sections covering industrial alloys can be interpreted on the basis of three simple types—eutectic, solid solution and peritectic, each characterised by a definite arrangement of lines. They do not indicate the structural arrangement of the phases, i.e. laminae, globules, films; nor do they indicate the velocity of reactions.

## Eutectic

For the present purpose the lead-antimony system may be considered a simple eutectic, in which the metals are completely soluble in the liquid and entirely insoluble in the solid state. The equilibrium diagram is shown in Fig. 36. Alloys within the temperature-composition limits above AEC are liquid and the boundary lines AE and EC, obtained from cooling curves, are known as the *liquidus*. A characteristic of this diagram is that the liquidus consists of two branches falling in temperature from the freezing-points of the two metals to a minimum intersection point, known as the eutectic point (E).

All the alloys become completely solid at the same temperature, corresponding to the temperature at E, and the boundary ADEBC is known as the *solidus*. The areas ADE and CBE represent alloys in a pasty condition —solid dendrites and liquid.

Consider the *solidification of an alloy* of composition $x$ (40% lead). At temperature $t_1$, seed crystals of antimony are formed. The melt is consequently enriched in lead and is represented by a point, such as $y_1$ (55% lead). Its temperature must therefore fall to $t_2$ if the dendrites are to continue to grow by further deposition of antimony and consequent enrichment of the liquid in lead. This process is repeated continuously until the

---

*The changes are indicated by cooling and heating curves, electrical conductivity measurements and by the microscopic examination of quenched specimens.

composition and temperature are given by point E. Further separation of antimony proceeds along an extension of CE (shown dotted), reaching a point $E_1$ which is supercooled sufficiently below curve AE* to cause solidi-

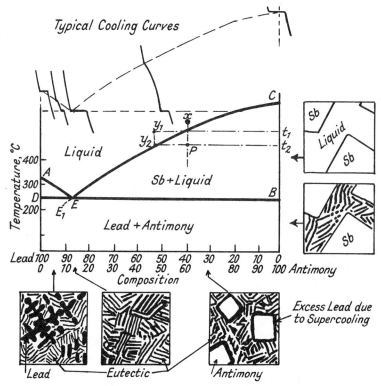

FIG. 36 Eutectic diagram structures and cooling curves

fication of lead with heat burst and cessation of growth of antimony. Lead crystals can continue to grow for a very short time only since its separation along line AE soon enriches the alloy in antimony to a point at which antimony must again separate. This process of alternate solidification of minute quantities of lead and antimony maintains the alloy at an almost constant temperature, until it is completely solid, then normal cooling is resumed as indicated by the cooling curves in Fig. 36.

Similarly, any alloy between E and A (13 to 0% antimony) commences to solidify on reaching line AE by forming lead dendrites and the residual melt is gradually enriched in antimony until point E is reached, when it

*This supercooling causes the primary dendrites to be larger than they should be and to balance matters an excess of second metal next occurs.

freezes as an eutectic mixture. Fig. 37 shows a similar structure in a copper–silver alloy. The alloy of eutectic composition E solidifies entirely as eutectic at a constant temperature. The eutectic in all the other alloys of lead and antimony has the same composition, E.

As a rule, one constituent forms a continuous matrix in which the other constituent is dispersed. The dispersed constituent may consist of spheres, rods, plates and angular shapes but all tend to be fine in the centre of the grain, thus giving rise to a 'colony' effect, seen in Fig. 38, which shows a copper–silver eutectic of the globular type. Fig. 35 illustrates 'angular' dendrites of $CuAl_2$ embedded in $CuAl_2$–Al eutectic. Other eutectic structures are shown in Figs. 183 and 228.

A *eutectoid*, as its name implies, is closely related to a eutectic and the eutectoid diagram, so important in the case of steels and aluminium bronze, can be interpreted in a manner similar to the eutectic diagram just discussed. Just as the liquid may be considered to break up into two kinds of crystal on solidification, so the eutectoid is formed from the breaking up of a *solid solution* into two different, but intimately mixed, constituents.

## Interpretation of equilibrium diagrams

At this stage it is convenient to summarise a few rules which help in interpreting equilibrium diagrams for two elements.

(1) The diagram consists of solubility lines which divide it into areas, known as *phase fields*, representing either single or double constituents, such as ADE in Fig. 36. Single phase fields are always separated from one another by a two-phase region, and three phases can only exist at a single temperature (e.g. E).
(2) If a vertical line, drawn to represent the composition of an alloy, crosses a line in the diagram, it denotes that some change occurs in the alloy.
(3) At any point in a two-phase field the

 (*a*) *Composition* of the constituents, in equilibrium with each other, is given by the intersection of the boundaries by the horizontal line drawn through the point. For example, at point P in Fig. 36 the intersections $t_2$ and $y_2$ represent solid antimony and liquid containing $y_2$ per cent of lead;
 (*b*) *Relative weights* are given by the relative lengths of the lines, $Pt_2$ and $Py_2$.

$$\frac{\text{weight of liquid}}{\text{weight of solid antimony}} = \frac{Pt_2}{Py_2}.$$

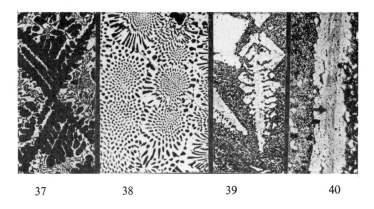

FIG. 37 Copper dendrites in copper–silver eutectic; 35% copper (× 80)
FIG. 38 Copper–silver eutectic. Note colonies (× 80)
FIG. 39 $CuAl_2$ dendrites in eutectic; 45% copper (× 100)
FIG. 40 30% copper–tin alloy showing effect of peritectic reaction (× 80)

**Solid solution diagram**

A typical diagram for two metals completely soluble in both the liquid and solid states is shown in Fig. 41 for copper and nickel. Again there is a range of solidification, but in this case not pure metals, but solid solution always solidifies and in certain cases the temperature of solidification may be higher or lower than that of the pure metals.

Consider the solidification of an alloy of any composition, say $x$ (50% copper). When it reaches a temperature $t$, freezing starts as shown by the liquidus curve, the solid formed, however, is not of the same composition as $x$, but of composition $y$ (35% copper), on the solidus at temperature $t$, which of course shows the temperature at which a solid alloy of composition $y$ would commence to melt on heating, forming a liquid of composition $x$.

Since the first dendrites are richer in nickel the remaining melt will have a composition such as $x_1$, richer in copper than the original composition $x$. At temperature $t_1$ the composition of the solid in equilibrium with the liquid is $y_1$ and fresh deposition of solid will adjust this composition. At the same time original dendrites should absorb copper from the liquid to change their composition from $y$ to $y_1$. Finally, the liquid $x_2$ vanishes at temperature $t_2$ and the solid has the uniform composition of $x$.

This state of affairs rarely occurs in commercial alloys owing to the rapid cooling which prevents the attainment of equilibrium.

*The effect of rapid cooling* is to prevent the first dendrite $y$ from absorbing the proper amount of copper as the temperature falls to $t_1$. Hence the average composition of the solid is not $y_1$, but one between $y$ and $y_1$. As a result the liquid contains more copper than that indicated by $x_1$. *Each layer of solid which is deposited on the dendrite skeleton differs in composition from the preceding layer* and when temperature $t_2$ is reached there still remains some liquid richer in copper than $x_2$ which solidifies at some temperature $t_3$. The alloy consists of a number of *cored* crystals whose average composition is $x$, but in each the primary skeleton is rich in nickel and the interdendrite spaces are rich in copper. Many commercial alloys have this type of structure in the cast form and a typical photograph is shown in Fig. 65. Suitable annealing eliminates the cored structure by allowing diffusion of the atoms to occur (Fig. 67).

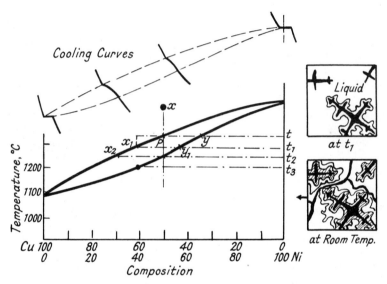

FIG. 41 Equilibrium diagram for two metals completely soluble in the liquid and solid states

*Zone refining* is a process used for purifying metals, particularly silicon and germanium for transistors. It uses the fact revealed in Fig. 41, namely that the first crystals deposited are much purer in one component than the liquid at the same temperature. A narrow zone in a bar of the impure metal is melted at one end and caused to move along the bar so that metal is melting into the zone on one side and freezing out at the other. The first solid frozen out is purer than the average composition while the liquid becomes enriched in solute which is deposited at the other end of the bar. Repeated cycling in one direction can reduce the impurity to below 1 part in $10^8$.

## Combination types

Where, among other factors, the two kinds of atoms are very different in size, each metal may be capable of dissolving only a limited percentage of the other. Then the equilibrium diagram formed may be classed as intermediate between the two types just discussed, and consists of two parts, one resembling a simple eutectic and the other which may be regarded as a part of a complete solid solution curve.

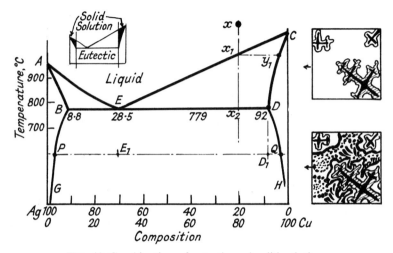

FIG. 42 Combination of eutectic and solid solutions

*Eutectic and solid solutions*

A typical diagram is shown in Fig. 42 of the copper–silver system. The small inset drawing shows the components. The liquidus AEC is precisely similar to that shown in the simple eutectic (Fig. 36). The solidus ABEDC, however, has two sections, AB and CD, resembling parts of the solidus in Fig. 41. In fact, alloys containing less than 8·8% or more than 92% copper will solidify forming cored solid solutions in a manner exactly as has been already described in the case of a simple solid solution series.*

The point B denotes the maximum amount of copper which can dissolve in silver when the alloy is just solid, while the point D gives the corresponding information regarding silver soluble in copper. All alloys between B and D solidify as described for the simple eutectic, except that, *instead of pure metals separating from the melt we can only have solid solutions B and D, which will be cored unless the cooling is exceptionally slow.*

*Rapid cooling will tend to prevent equilibrium and a minute amount of eutectic may be present.

## Changes in the solid state

The solubility of the metals in each other after solidification will undergo changes upon further cooling, as shown by the curves BG, DH.

Consider the changes occurring during the solidification of an alloy of composition $x$ (80% copper).

(1) Solidification will commence at a temperature represented by point $x_1$ on liquidus, and solid of composition $y_1$ will separate out.

(2) The temperature will fall to that of the eutectic E. Ignoring the effect of coring, the alloy will now consist of a solid solution whose composition is denoted by D and liquid of composition E (28·5% silver), the relative proportions of liquid to solid being in proportion to the lengths of $x_2D$ to $x_2E$.

(3) The liquid will solidify completely at this temperature into a eutectic composed of two solid solutions, B and D, in the proportions of ED to EB.

(4) Upon further cooling the solid solutions B and D in the eutectic will become respectively poorer in copper and silver. Their compositions at any temperature, say 600 °C, are P and Q and the relative proportions are given by the lengths of $E_1Q$ and $E_1P$.

(5) Similarly, the primary dendrites of composition D will split up into two solid solutions of Q and P in the proportion of $PD_1$ to $QD_1$. These changes in the solid state take considerable time to complete themselves and may be partially if not completely prevented by quenching. Herein lies a fact of which advantage is repeatedly taken in heat-treatment processes (p. 97).

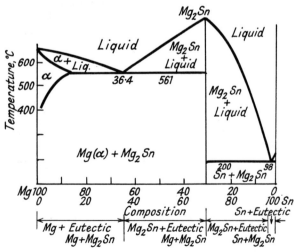

FIG. 43 Equilibrium diagram showing presence of intermetallic compound

## Intermediate phases

In many binary alloy systems mutual solubility of the metals is limited and intermediate phases (called β, γ, δ, etc.) are formed, which may be classed as two types.

(1) *Intermetallic compounds of fixed composition* which obey the usual valency laws, like ordinary chemical compounds (e.g. NaCl). A typical example is $Mg_2Sn$ containing 29·08% magnesium. The constitutional diagram is shown in Fig. 43. The compound has a definite melting-point which is lowered by addition of excess magnesium, or of tin. The diagram consists in fact, of two eutectic diagrams placed together, each resembling Fig. 36 and not therefore calling for further observation.

(2) *Intermetallic compounds of variable composition* which do not obey the valency law; known as *electron compounds*. Many of these fall into three classes according to the ratio of valency electrons (see p. 71) to the number of atoms:

  (*a*) Ratio $\frac{3}{2}$—beta (β), e.g. CuZn, $Cu_3Al$.
  (*b*) Ratio $\frac{21}{13}$—gamma (γ), e.g. $Cu_5Zn_8$, $Cu_9Al_4$.
  (*c*) Ratio $\frac{7}{4}$—epsilon (ε), e.g. $CuZn_3$, $Cu_3Sn$.

## Peritectic diagram

In many systems of alloys a reaction takes place, at a definite temperature, between the definite proportion of the solid already deposited and the residual melt (BP:BA in Fig. 44) to form another solid solution or compound of a composition intermediate between that of the first solid and the liquid. This is known as a peritectic reaction and it gives rise to a diagram such as is shown in Fig. 44.

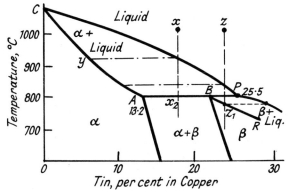

FIG. 44 Equilibrium diagram showing peritectic reaction (Cu–Sn)

If the effect of coring is ignored any alloy up to 13·2% tin (A) will solidify as described for a solid solution, as will also alloys with more than 25·5% tin (P).

Consider alloy $x$; primary crystals $y$ form and as the liquid cools the solid ($\alpha$) changes to A, while the liquid changes its composition along the liquidus CP to P. The liquid which remains ($Ax_2$ to $Px_2$) reacts with the solid crystals A to form a new intermediate phase (B). Since $x_2$ lies between A and B, all the liquid P will be used up before the crystals A, and the alloy will then consist of solid $x$ and $\beta$ crystals of composition A and B respectively. An alloy of composition Z again commences to solidify by depositing $\alpha$ solid solution which changes in composition to A at 798 °C. Theoretically, these $\alpha$ crystals are then completely converted to $\beta$ crystals of composition B, since Z lies between B and P, and the excess liquid P solidifies by depositing $\beta$, which changes in composition along solidus BR to $Z_1$.

In practice these peritectic reactions rarely go to completion owing to the fact that primary $\alpha$ crystals become coated with a skin of $\beta$, which prevents the diffusion of the tin rich material to the central regions. Even in alloys of composition Z, a core of $\alpha$ remains. A typical example of this effect is shown in Fig. 40 for an alloy of 30% copper in tin. The $Cu_3Sn$ ($\varepsilon$) is coated with $Cu_3Sn_3$ ($\eta$) and the background is eutectic of $\eta$ + Sn.

## Ternary equilibrium diagrams

The necessity for a more complete knowledge of ternary systems is now generally recognised and an increasing number of ternary diagrams occur in the technical literature. A few details are given below, although it is beyond the scope of this book to deal adequately with such diagrams.

The complete equilibrium diagram of a three-component system necessitates a three-dimensional figure, i.e. a solid model, Fig. 45($a$). On paper, therefore, it is convenient to use sections through such a solid model, the most convenient being an equilateral-triangle plot, representing boundaries at one temperature only. The three pure metals are represented by the corners of the triangle and binary alloys of the metals are represented by points along the sides of the triangle. The location of a point within the triangle fixes the composition of a ternary alloy; thus alloy O (Fig. 45($b$) and ($c$)) contains amounts of A, B and C in proportion to the perpendicular distances, OI, OH and OG respectively.

In the binary diagram, the liquidus and solidus are indicated by lines but in a ternary system *surfaces* are required. A binary eutectic in a ternary system is represented by a *line* formed by the intersection of the liquidus surfaces and this slopes down to the ternary eutectic *point*. A binary eutectic can, therefore, form over a range of temperature and only the ternary eutectic forms at a constant temperature.

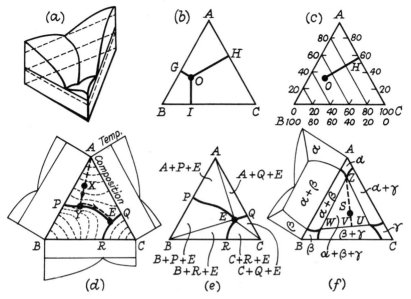

FIG. 45 Ternary equilibrium diagrams

Fig. 45(*d*) shows projected on to the triangular base the liquidus boundaries in a ternary system in which the metals A, B, C form simple binary diagrams shown on the sides of the triangle. The view shown in Fig. 45(*d*) is not a true isothermal section since the temperatures of the liquidus boundaries are continuously decreasing to that of the ternary eutectic. The temperatures of these boundary lines and of the whole liquidus surface are conveniently represented by temperature contours shown dotted as height is represented on a map.

Consider the solidification of an alloy represented by point X. In area APEQ the alloys are rich in metal A.

Consequently primary crystals of A solidify, leaving a residual liquid enriched in B and C but without altering the proportion of B to C, i.e. along a line AX extended to meet the binary eutectic trough PE at Y.

Along this trough the binary eutectic P is formed and the liquid enriched in C until E is reached.

Solid alloys whose compositions lie on lines AE, BE, CE contain only a primary metal and ternary eutectic. A section of the ternary model at a temperature lower than that of the eutectic E may be divided into the phase areas shown in Fig. 45(*e*), although essentially the whole diagram is a three-phase area consisting of A + B + C.

Fig. 45(*f*) shows the phase fields in *solid* alloys in which the component metals have a limited mutual solubility in the solid state. The ternary eutectic consists of three solid solutions $\alpha$, $\beta$, $\gamma$, instead of pure metals as in previous example.

*Rules* for interpreting ternary diagram bear a similarity to those for binary systems discussed on p. 84.

(1) Three-phase fields are always triangular; with boundaries in common with two-phase fields and corners that touch one-phase fields.
(2) The relative proportions of co-existing phases in a two-phase field are given by the lever relation given on p. 84, section 3(*b*).
(3) The relative proportions of the phases in a three-phase field, say S, is given by drawing a line through S and any corner, say Z, to intersect the base of triangle at V. Then equations are

$$\frac{\text{Amount of } \alpha}{\text{Amount of } (\beta \times \gamma)} = \frac{\text{length SV}}{\text{length SZ}}$$

and

$$\frac{\text{Amount of } \beta}{\text{Amount of } \gamma} = \frac{\text{length VU}}{\text{length VW}}$$

*Constitution and properties*

The properties of an alloy are determined by its crystalline structure and in general similar structures are alike in general characteristics. Although there are exceptions, certain generalisations can be made.

*Eutectics.* The mechanical and electrical properties are a linear function of composition (by volume) except at the eutectic composition when somewhat higher values are obtained for mechanical properties. The mechanical properties depend on (1) fineness of the state of dispersion, (2) individual properties of the phases, (3) properties of the continuous phase. Eutectics are of common occurrence in castings.

*Primary solid solutions* are usually malleable, ductile and also stronger than the pure metals. Many commercial forged alloys consist of such solutions. Electrical conductivity and the temperature coefficient decreases sharply with alloying addition.

*Intermediate phases*

*Compounds.* These are usually hard, brittle, having no useful properties in themselves, but are useful in certain commercial alloys when mixed with a large proportion of soft material, for example, $Fe_3C$ in steels, $CuAl_2$ in aluminium alloys. Electrical conductivity is usually high relative to neighbouring alloys.

*Beta phases*, in copper alloys, are reasonably strong and, at certain temperatures, can be readily hot rolled. They are somewhat more brittle at room temperature.

*Gamma phases*, in copper alloys, are usually brittle in hot and cold states.

# 6 Hardening of Metals

So far crystals have been treated as perfect arrays of atoms with all lattice sites occupied. There are good reasons, however, for considering that a number of imperfections exist in a metal crystal in addition to those caused by impurity atoms.

In a real crystal the atoms are not fixed rigidly but vibrate about their mean position, the amplitude increasing with temperature. Under some conditions (see nuclear bombardment, p. 361) some atoms can be displaced to an interstitial position (p. 76). They are then known as an *interstitial defects*. Another defect is an atom site which is unoccupied by an atom, i.e. *vacancy*. These two imperfections and also substitution atoms (p. 76) are sometimes called *point defects*. Such defects can interact with each other, for example an interstitial atom may fill a vacancy or vacancies can condense in one zone to form a pore. Other imperfections are *stacking faults*, *edge* and *screw dislocations* (line defects) and *small angle boundaries*, which are groups of dislocations between mosaics of smaller crystals differing slightly in orientation within the main crystal boundary. The stacking fault is a discrepancy in the packing sequence of the layers of atoms although all the lattice sites are occupied, e.g. instead of a regular sequence A B C A B C . . . there is a change A B C B A C . . .

### Methods of strengthening and hardening material

At this stage it is convenient to group together the various metallurgical techniques for increasing strength and toughness of metals. Metals are weaker than they should be theoretically because of dislocations and the ease with which these move under applied shear stresses. The primary principle for strengthening is therefore to retard the movement of dislocations. Complete barriers can be dangerous because pile up of dislocations can lead to a catastrophic crack.

Crystals such as sapphire, quartz, diamond and titanium carbide are held together by a dense network of strong covalent bonds, i.e. requiring 3 or 4 valency electrons, are loosely packed and are light but elastically stiff. Dislocations are immobile except at high temperatures.

Whiskers of metals are single crystals free from dislocations and therefore possess almost ideal theoretical strengths. They are expensive.

Hexagonal materials such as graphite have a restricted number of slip planes and a suitable texture can reduce slip. (For use of whiskers and graphite fibre see p. 365.) The main forms of hardening are:

(1) *Work hardening* increases the number of dislocations which mutually entangle one another.
(2) *Solid solution hardening* which distorts the lattice offers resistance to dislocation movement which is greater with interstitial elements which cause asymmetric lattice distortion, e.g. C in steel.
(3) *Order hardening*, i.e. definite periodicity of solute atoms through lattice. Important in gold alloys (p. 394).
(4) *Dispersion hardening* (including ageing and secondary hardening in steels, p. 188).
(5) *Transformation hardening.* High carbon martensite being tetragonal has fewer available slip planes than a b.c.c. lattice.
Acicular ferrite or lath martensite has a high density of dislocations which interfere with one another (p. 107).
(6) *Fine grains.* Grain boundaries are effective barriers to dislocations but before pile up reaches a catastrophic value slip is nucleated in adjacent grain. Strength and toughness are increased (p. 167).
(7) *Thermomechanical treatment* (p. 225). Dislocation arrays influence subsequent transformed product.
(8) *Irradiation hardening* by producing vacancies or point defects.
(9) *Stacking faults* increase irregularities of lattice to produce obstruction and also provide sites for precipitation of solute atoms which lock dislocations.
(10) *Fibre reinforcement.*

*Diffusion in the solid state*, which is the movement of atoms from one lattice site to another, is an extremely important physical process and it enters into many metallurgical phenomena such as:

(*a*) homogenising treatment;
(*b*) annealing and recrystallisation (p. 122);
(*c*) phase changes, e.g. $\gamma - \alpha$ iron (p. 162) malleablising of cast iron (p. 283);
(*d*) precipitation of phases in age-hardening; tempering of alloy steels (p. 97);
(*e*) formation of bonds between metals in powder metallurgy (p. 135), pressure welding (p. 394);
(*f*) oxidation of metals and surface treatments.

Naturally, in view of its vital importance in metallurgy attempts have been made to explain diffusion in terms of movement of atoms through the lattice. Experiments on the mutual interdiffusion of brass and copper showed that zinc diffused out of brass much faster than copper diffused into it, the junction moves and holes are often formed (Fig. 247) (known as Kirkendall effect). It is now thought that diffusion of substitutional atoms involves lattice defects of which the most important are the vacant holes

# METHODS OF STRENGTHENING

not occupied by atoms. Diffusion of vacancies in one direction will be accompanied by diffusion of atoms in the opposite one and this is supported by some self-diffusion work using radio-active isotopes. Fig. 46 illustrates this. An increase in vacancies in a sample should therefore be associated with increased diffusion rate. The number of vacancies in a metal increases with temperature and it is also found that the *rate of diffusion increases* with temperature. Drastic quenching of an alloy can

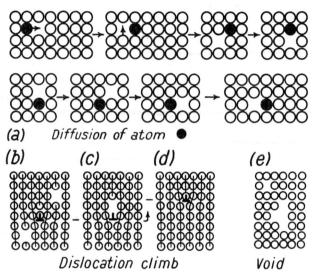

FIG. 46 (*a*) Diffusion of substitutional atom by utilising the movement of a vacancy; (*b*) edge dislocation with two vacant sites; (*c*) vacancies diffuse into preferred positions; (*d*) dislocation climbs by collapse of vacancies; (*e*) vacancies coagulate to form pores

retain a greater number of vacancies in a sample and subsequent changes, such as precipitation hardening, are accelerated. Irradiation and plastic deformation also produce non-equilibrium concentrations of vacancies. Diffusion rates are also higher through a grain boundary region than through the bulk of the metal; are also higher in cold worked metal where there are more imperfections in the lattice. Therefore cold working is an advantage before a homogenising anneal.

In phenomena where diffusion of atoms is a controlling factor the rate of the reaction can be expressed in equation:

$$y = Ae^{\frac{-Q}{RT}} \text{ or } \log y \propto \frac{1}{T}$$

where    $y$ = fraction transformed, e.g. Austenite.
         A, R = constants, T = absolute temperature.

A plot of log $y$ against $\frac{1}{T}$ is a straight line and the slope gives a value of Q, the energy or heat of activation, which by comparison with values for known reactions may indicate the controlling factor in a given transformation. This kind of treatment is found in the literature dealing with all the phenomena mentioned above.

## Age-hardening

The first discovery of this hardening was due to A. Wilm (1906), who observed that when commercial aluminium, alloyed with copper, was quenched from a relatively high temperature, it increased in hardness on standing subsequently at room temperature. We now know that a similar phenomenon occurs in numerous alloys, ferrous and non-ferrous, and it is known as age-hardening.

Fig. 47 shows the falling solubility curve AB which is characteristic of all

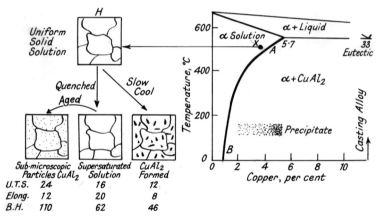

FIG. 47 Solubility of copper in aluminium

alloys which age-harden, and the aluminium–copper–alloys can be used as a typical example.

When an alloy containing 4% of copper is annealed for a sufficient time at temperature X (500 °C) all the compound $CuAl_2$ is dissolved to form a homogeneous solid solution (structure H). When the alloy is slowly cooled from 500 °C, the compound is precipitated as relatively coarse particles, visible under the microscope, until at room temperature only 0.5% of copper remains in solution.

If on the other hand the alloy is quenched from 500 °C a supersaturated solid solution is retained at room temperature, which is slightly harder but more ductile than the slowly cooled sample. This condition is unstable and the re-precipitation of the excess constituent from the supersaturated solution can in a large measure be controlled at will by heat-treatment to give an increase in strength and hardness. The maximum hardness is attained, however, before any precipitate is visible under the microscope, although some atomic change must have occurred. In quenched aluminium alloys hardening takes place to some extent spontaneously at room temperature, although heating to temperatures up to 200 °C accelerates the rate of hardening. Other aluminium alloys require precipitation hardening at about 165 °C. To prevent the onset of normal age-hardening storage in a refrigerator at temperatures down to −20 °C is used, especially where riveting or pressing operations are involved on solution treated alloys.

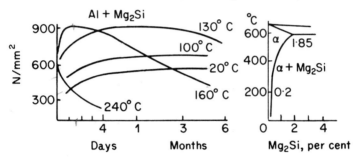

FIG. 48 Effect of ageing temperature and time on the tensile strength

The general effect of the *temperature* of ageing is shown in Fig. 48 which is characteristic for all alloys except for the actual numerical values. As the temperature of ageing is raised the time required to reach maximum hardness is decreased, but at 130 °C a longer period of tempering results in a drop in the curve due to coalescence of the particles. At 240 °C the rate of softening is very great. Cold working of the quenched alloy and drastic quenching accelerates the speed of the subsequent ageing (i.e. more vacancies and quicker diffusion).

The actual cause of age-hardening is the obstruction to slip set up by a critical dispersion of fine particles sometimes with coherent stress fields around them.

Similar precipitation phenomena occur in copper steels, copper alloys (p. 325) in tempering of alloy steels, and sometimes the properties are affected in a deleterious manner in some materials such as quench ageing and nitrogen embrittlement of steels (p. 171).

## Theory of age-hardening

The actual mechanism of age-hardening often differs in various alloys. The aluminium–copper system has been studied in great detail and the results will be considered briefly because some of the conclusions are of general importance.

The first step in the precipitation from the supersaturated solid solution is the segregation of copper atoms into 'clusters' or platelets a few ångströms in thickness and a hundred in diameter, but still part of the parent lattice. These platelets are often called Guinier–Preston zones, or GP1 zones. In the next step copper atoms diffuse to the GP1 zone to form a larger one called GP2 zone or $\theta''$, 8 Å thick and 150 Å in diameter.

The $\theta''$ precipitate has a tetragonal structure which fits perfectly with the aluminium unit cell in two directions but not in the third. For the precipitate and the matrix to maintain fit or coherency the aluminium planes have to be distorted and thus give rise to coherency strains, Fig. 49(b). Such

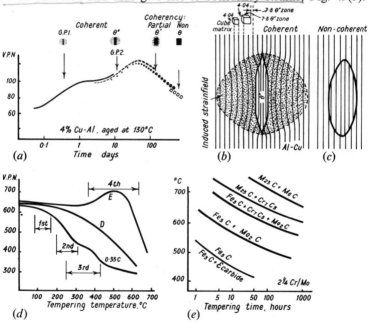

FIG. 49 (a) Ageing curve for 4% copper–aluminium alloy showing types of precipitates formed; (b) diagram showing distortion of the (100) planes near a $\theta''$ zone. Dotted line shows induced stress field; (c) illustrates non-coherent precipitate—no fit of planes; (d) tempering curves for 0·35% C steel and die steels D and E in Table 33. Shows stages of tempering and secondary hardening; (e) shows effect of time and temperature of tempering on carbides formed

strains show up as dark regions around the precipitate in the thin foil electron micrograph (Fig. 52). These strain fields, which have a much larger effective size than the precipitate, oppose the movement of dislocations, i.e. harden the alloy.

At a still later stage a new transition phase θ', still partially coherent with the matrix, is precipitated and produces maximum hardness. Over-ageing converts this transition precipitate into the equilibrium precipitate θ or $CuAl_2$ *without* coherency (Fig. 49(c)), although the precipitate is oriented in a definite direction with respect to the solid solution in which it forms (e.g. Widmanstätten form). The sequence of intermediate decomposition products in supersaturated alloys is GP1 zone → θ″ → θ' → θ ($CuAl_2$). Lower alloy content and high ageing temperature favour fewer intermediate phases, e.g. θ' → θ only.

The form of the hardness curve associated with these intermediate forms of precipitate, shown in Fig. 49(a), is characterised by two stages with a flat. An alloy hardened at a low temperature to the flat (i.e. GP1) and then heated for a short time at a higher temperature softens by solution of GP1 and can be re-hardened by low temperature ageing. This is known as *reversion*. This resolution of an initial precipitate and the renucleation of a second is explained by the addition of extra metastable solubility curves to Fig. 47 as shown in Fig. 50. The resolution of ε-carbides in tempering steels is another example.

*General case*

As in the Al–Cu example, nearly all age-hardening systems show an ageing sequence:

zones → intermediate precipitate → equilibrium precipitate

The zones are coherent with the matrix and as they form the alloy becomes harder. Discs cause least strain, needles intermediate and spheres most strains. The intermediate precipitate may be coherent or partially coherent and further hardening or softening may result. The formation of the equilibrium incoherent precipitate always leads to softening.

The type of dispersion depends on the degree of *supersaturation* of the matrix. With low supersaturation nucleation of the precipitate is difficult. It is easy with high supersaturation and a uniform size and spacing, depending on the ageing temperature, is formed throughout the matrix. *Prior plastic deformation* can markedly change the time cycles for heat treatment. The most dramatic effects are produced in weakly supersaturated solutions in which few nuclei form, but strain induced dislocations provide increased numbers of nucleation sites, although such changes may not always be to the engineer's benefit. For example, high temperature failures have occurred in Austenitic 347 steels (18/9 Cr Ni Nb) in

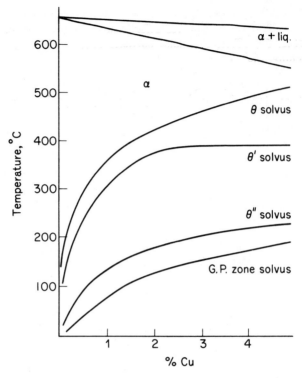

FIG. 50 Solubility curves for stable θ and metastable phases

which creep ductility has been reduced by introduction of welding strains. In highly supersaturated solid solutions, ageing time to reach peak hardness may be shortened $\frac{1}{4}$—$\frac{1}{10}$th of that normal in the absence of prior plastic strain.

In addition to the above general case of *continuous precipitation* some systems experience local or *discontinuous precipitation* around the grain boundaries surrounded by a layer of depleted matrix.

In some systems, too, fine precipitates are formed with no coherency strain fields but which still offer resistance to dislocation movement.

### Dispersion hardening

Ageing or precipitation of particles from supersaturated solution is one way of producing dispersion hardening of a matrix by impeding the movement of dislocations. Under stress dislocations can either cut through closely spaced small particles, which form on slip planes or loop between large particles, leaving dislocation rings or debris around particles (Fig. 51).

# DISPERSION HARDENING

With large particles shape is important, and this can range from spheres, discs, cubes to needles. Spheres offer least resistance to slip. Coherent strains (Fig. 52) effectively make particles larger than they seem. For strengthening, the particles should be small—ideally about $10^{-7}$ cm dia,

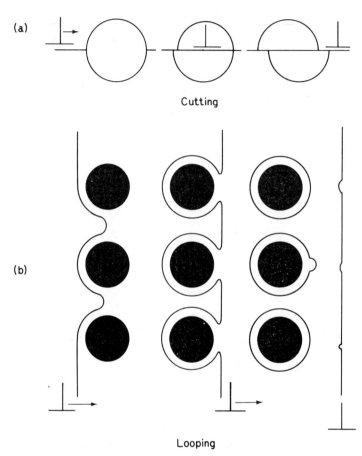

FIG. 51 (*a*) Shearing of a particle by a dislocation
(*b*) Dislocation looping around a particle

spaced $10^{-6}$ cm apart—and have a high cleavage strength. The disadvantage of precipitated particles is that as temperature rises, they can coarsen, producing greater spacing, react with the matrix and eventually dissolve. Dispersion hardening can also be produced by introducing *insoluble* particles, such as oxides, nitrides and carbides, and properties are then retained to temperatures well above the normal softening temperature because dislocations are impeded, and softening by recrystallisation or grain growth is prevented by the pinning effect of the particles. The creep properties are high.

It seems important that the particles should be wetted by the matrix, otherwise on deformation, separation of the matrix at the interface will occur causing premature ductile fracture (p. 11), resulting in poor ductility.

There are various methods of producing a dispersion-strengthened material. The simplest case is undoubtedly that of the pure metal which has a relatively stable oxide, such as aluminium, magnesium and lead. The

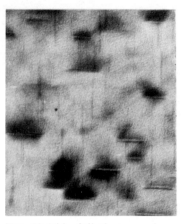

Fig. 52 Thin foil electron micrograph showing coherency strain fields around $\theta''$ precipitates in Al–4% Cu aged 1 day at 130 °C (*J. Nutting*) ($\times$ 400 000)

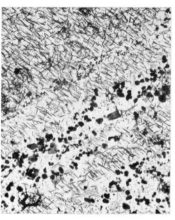

Fig. 53 Steel, 0·2 C, 3·8 Mo, 0·22 Ta quenched and tempered 100 hours at 600 °C. ($\times$ 12 000). Carbon replica areas of fine $Mo_2C$ with massive $M_6C$ (*R. W. K. Honeycombe*)

metal powder is partly oxidised and then consolidated by pressing, sintering or, more usually, hot extrusion, e.g. SAP (Sintered Aluminium Product). This principle has been applied to lead strengthened with $1\frac{1}{2}$ vol% PbO and its creep strength at 20 °C is trebled.

Another relatively simple technique is to mix and consolidate a finely powdered metal with an even more finely powdered oxide or other refractory phase, or to mix two oxides and reduce one to metal, e.g. NiO + ThO to form TD nickel which has a stress rupture at 800 °C, about ten times that of ordinary nickel, although its strength at room temperature is modest. Internal oxidation can be used to form oxide particles of reactive alloys, e.g. Si and Al in copper, producing 1–2% vol of $SiO_2$ or $Al_2O_3$ in this foils.

## Widmanstätten structure

When a solid solution breaks down into two phases in age hardening systems and in many other cases, e.g. aluminium bronze (p. 319), steel (p. 165), the crystal structure of the matrix influences the orientation of the particles of the new phase. There is a tendency for the boundaries of the two phases to have similar atomic spacing and pattern (i.e. matching or coherency) and this gives rise to geometric arrangement of precipitate known as Widmanstätten structure. For example $\alpha$-brass forms with (111) plane parallel to the (110) plane of $\beta$-brass and [110] direction of $\alpha$ parallel to [111] direction of $\beta$.

An important feature of the precipitation hardening reactions in ferrous and non-ferrous alloys described is that the optimum engineering properties are obtained before the precipitate is visible under the optical microscope. The electron microscope, however, has revealed particles as small as 20–100 Å, which have a major influence on properties. Fig. 53 illustrates the fineness of some of the carbides present in alloy steel.

# 7 Deformation and Annealing of Metals

The majority of the processes used for shaping metals involve hot or cold working and annealing. It is therefore convenient to discuss some general aspects of deformation and annealing.

Deformation is frequently divided into elastic strain which disappears when the stress producing it is removed and plastic strain which persists after the stress ceases to act. Elastic strain produces no noticeable effect on a polished surface even under high magnifications.

When considering plastic deformation a distinction can be made between amorphous and crystalline materials.

A liquid increases in viscosity as the temperature falls, and in the case of glass the viscosity becomes so great that the material appears as a solid at room temperature although its atoms have a random or amorphous pattern. Such amorphous materials have mechanical properties which vary widely with temperature and rate of loading. Under fast loading they have no ductility while under small stresses, applied over a long period, they exhibit infinite ductility and flow like liquids, e.g. pitch.

**Plastic flow by block-slipping**

As shown in Chap. 5, a metal crystal is built up of atoms arranged in a regular pattern, the unit of which for most metals is cubic. Atoms may also occur at the centre of the cube, known as body centred cubic (b.c.c.), or at the centre of the face of the tube, known as face centred cubic (f.c.c.). These are illustrated in Fig. 30, which also shows the close packed hexagonal arrangement found in zinc and magnesium.

The atoms can be regarded as being arranged on certain crystallographic planes, along which a gliding or slipping movement occurs during plastic deformation. This phenomenon of 'slip' is illustrated in Fig. 54(a), which shows, diagrammatically, a section through two crystals before and after straining. Slip occurs on only a few planes, which are in a favourable position relative to the shear stress (Fig. 55(a)); with the result that movement of the metal occurs in 'blocks' which form a series of steps on a polished surface. When viewed under the microscope the steps appear as black lines, known as *slip-bands*. In practice, of course, slip occurs in three dimensions. Fig. 56 shows the types of slip-bands found in iron. When the specimen is re-polished the steps are removed and no evidence of slip-bands is seen, even after etching. After a finite amount of slip on one set of planes movement ceases and recommences on another set which have rotated into a favourable position. Severe strain causes slip to occur on a large number of planes and produces regions which dissolve more rapidly than the original metal, etching therefore reveals *strain-bands* which change direction at grain boundaries, as shown in Fig. 72. The original crystals are distorted and broken up into sub-grains, or cells containing few dislocations bounded by regions of high dislocation density although in some

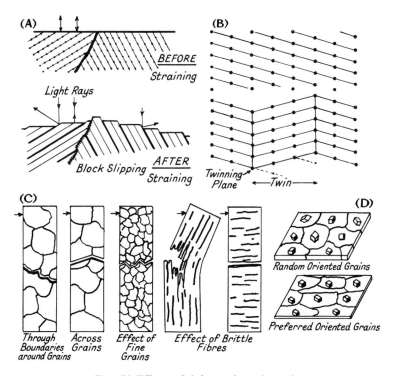

FIG. 54 Effects of deformation of metals

metals a complete tangle of dislocations form and gliding is rendered more difficult; in other words, the material is work-hardened. Fracture occurs when the metal has work-hardened so much that the stress required to produce further slip is greater than the cohesive strength.

With *creep*, slip is spread over more planes with less movement on each, being rarely visible and supplemented at elevated temperatures by a concentration of flow in the boundaries (p. 113). With *fatigue* forward and backward slip can produce extrusions and intrusions in the slipped region and these lead to cracks, Fig. 62(*d*).

## Explanation of mechanism of deformation

Finality has not been reached in regard to theories of deformation, but the following is suggested as one hypothesis. The *minimum* movement involved during slip is the distance between neighbouring atoms, and the

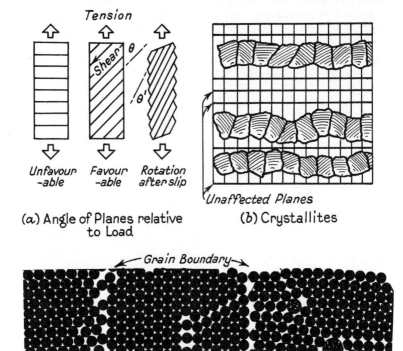

FIG. 55 (*a*) Block slips; (*b*) crystallite formation; (*c*) the change in regularity of the atom pattern at grain boundaries and around strange atoms

White line shows changes in direction at boundary and at distorted lattice due to presence of large atom

slip of one plane over another must be an integral number of these quanta.

The slip, however, does not take place over the whole plane simultaneously but is supposed to start at one point by the atoms moving one place onwards and this movement then propagates itself along the plane; thus less energy is required. The point of initiation is at a lattice imperfection— called dislocation (Fig. 55(c)).

This concept of an imperfection is necessary because it has been shown that ordinary metals are weaker than they should be theoretically by a factor of 100–1000 times. Most crystals in their as-grown state contain dislocations as a consequence of the processes of crystal growth ($10^4 - 10^6$ per cm$^2$); for example, atoms can attach themselves readily to steps on a crystal surface provided by a dislocation. Although a theoretical concept

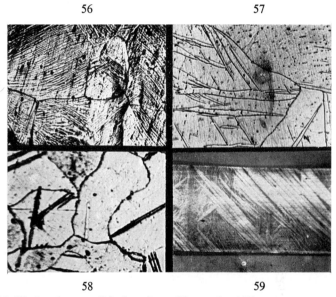

FIG. 56 Slip bands on polished surface of iron ($\times$ 100)
FIG. 57 Strain twins in cast zinc, squeezed in vice, etched ($\times$ 80)
FIG. 58 Neumann bands in pure iron, caused by sudden shock, etched nitric acid ($\times$ 80)
FIG. 59 Stretcher strains in steel strip ($\times \frac{3}{4}$)

at first, the thin foil technique has enabled dislocations and their movement to be observed in the electron microscope. Figs. 60 and 61 show typical dislocation networks, which are three dimensional as shown by the line drawing.

Two types of dislocation are represented diagrammatically in Fig. 62. The *edge dislocation* is formed when the top half above a slip plane contains one more vertical sheet of atoms than does the bottom half. This results in a line of axis *a* of maximum lattice disturbance. The atoms are squeezed together in the top half and spread out below—represented as ⊥.

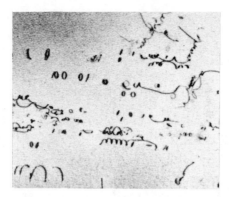

Fig. 60 Thin foil electron micrograph, dislocation loops in Al–4% Cu alloy quenched 525°C in brine ( × 10 000) (*R. E. Smallman*)

From the elastic forces which result it is apparent that dislocations of similar sign (⊥ ⊥) in a given slip plane tend to repel each other, whereas those of opposite sign may emerge together, i.e. annihilation ⊥ ⊤. In Fig. 62(*a*) slip has occurred to right of axis *a* and the *degree* (i.e. one atomic plane spacing) and *direction* of the translation in the displaced zone is represented by *b*, known as Burgers vector. The direction of the vector is in direction of slip and perpendicular to axis.

In the screw *dislocation* the directions of the vector, of slip and of the axis are the same. In Fig. 62(*b*) slip has occurred on the right-hand side of

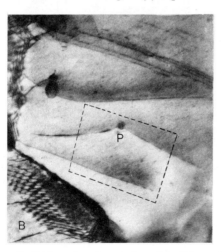
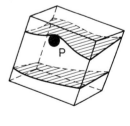

Fig. 61 Dislocations in a sub-grain of iron, Sub-grain boundaries are revealed by networks and dislocations B. P is particle pinning a dislocation ( × 80 000)

axis *a*. The shear force F tends to move the dislocation axis *a* along the slip plane to the left, and each row of atoms moves in direction of the force, the successive positions being shown. The curved arrow round the dislocation axis indicates that a complete revolution about the axis does not return to the point of origin, that is the atomic planes describe a *helical* surface, i.e. a screw. Dislocation loops consist of a combination of edge and screw types.

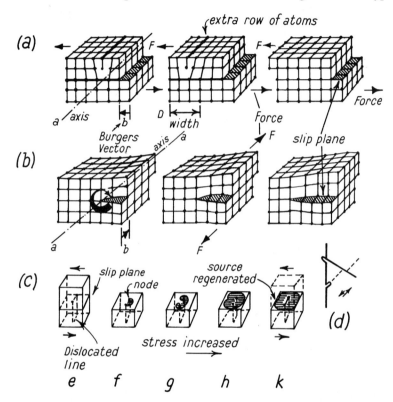

FIG. 62 (*a*) Edge dislocation. Increased stress from left to right; (*b*) screw dislocation; (*c*) Frank-Read source. Grown in dislocation loops formed by missing patch of atoms has a part in a slip plane. Stress causes loop to bow outward but it is anchored at the ends; (*d*) extrusions and intrusions caused by to and fro slip in fatigue

Slip occurs intensely on a small number of crystal planes, in which some hundreds of dislocations move, hence some effective creators of dislocations must exist. These are known as *Frank-Read sources* which are illustrated in Fig. 62(*c*). A dislocation loop has a mobile portion with fixed ends (nodes) lying in the slip plane, and this is able to generate dislocations when appropriate stress is applied by bowing ($f$) and moving progressively across the plane ($g$), joining up behind the source ($h$) to form a closed loop which can continue to expand under the action of the applied stress. Meanwhile, the source then regenerates ($k$). During cold work Frank-Read sources come into operation on each active slip plane and the total number

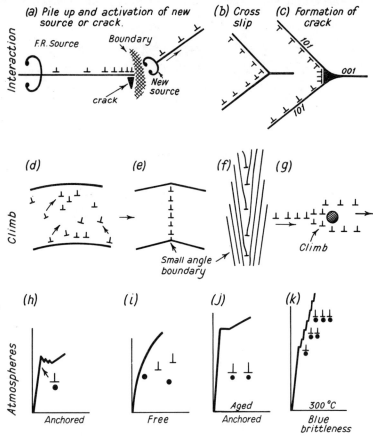

FIG. 63 ($a$), ($b$), ($c$) Interaction of dislocations; ($d$), ($e$), ($f$), ($g$) climb of dislocations to form small angle boundary and to bypass obstacle; ($h$), ($i$), ($j$), ($k$) formation of Cottrell atmospheres to give yield point and blue brittleness in steel

of dislocations is increased (up to $10^{12}$ per cm$^2$). Their free movement, however, is *impeded* by interaction with each other in cross slip, in which a dislocation changes from one slip plane to another intersecting slip plane and with other barriers such as grain boundaries and precipitates. This produces *strain hardening*. These effects are illustrated by (*a*) in Fig. 63 which shows a *pile-up* of dislocations at a boundary which generates a large stress in the adjacent crystal. This activates another source and causes slip in the second grain or it may initiate a crack. (*b*) shows interaction of dislocations on two interacting slip-planes and under some conditions this may lead to the initiation of a crack (*c*).

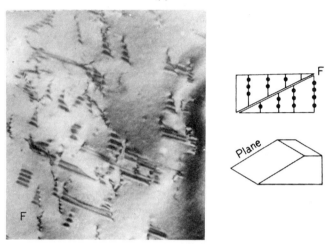

FIG. 64 Stacking faults in 18/8 austenitic stainless steel. Line diagram shows displacement of reflecting planes at a stacking fault F. Interference between waves from above/below fault gives a fringe pattern  ($\times$ 80 000)

*Dislocation geometry—stacking faults*

A dislocation with a particular Burgers vector is sometimes able to lower its energy by splitting into two partial dislocations with different Burgers vectors. In f.c.c. metals these partial dislocations are known as Shockley partial dislocations, and the region between them is a faulted ribbon. Fig. 64 shows the appearance of a stacking fault.

In a f.c.c. lattice the packing sequence of atoms is ABC ABC. The partially slipped plane has moved from the B to the C position, thus:

```
B B B B B B B B      C C C C|B B B B            B|C C C C C|B B
A A A A A A A A      B B B B|A A A A            A|B B B B B|A A
C C C C C C C C      A A A A|C C C C            C|A A A A A|C C
B B B B B B B B      C C C C|B B B B     ⟶     B|C C C C C|B B
A A A A A A A A      A A A A A A A A            A A A A A A A A
C C C C C C C C      C C C C C C C C            C C C C C C C C
B B B B B B B B      B B B B B B B B            B B B B B B B B
A A A A A A A A      A A A A A A A A            A A A A A A A A
       I                    II                         III
```

The line separating the A and C planes is a stacking fault. The stacking fault is corrected by the passage of the second partial dislocation (III). The line of the original dislocation has spread into a ribbon. This stacking fault ribbon has area and hence energy per unit area—the stacking fault energy which varies with the equilibrium separation distance between the partials. In metals of low stacking fault energy $\gamma$ (e.g. silver, $\gamma \simeq 20$ ergs/cm$^2$) the ribbon width will be greater than in metals with higher stacking fault energies (e.g. copper, 80 ergs/cm$^2$, and aluminium $\gamma \simeq 200$ ergs/cm$^2$). Consequently, it is more difficult for cross slip to occur in metals of low stacking fault energy, and these metals strain-harden more rapidly, they show a different temperature dependence of flow stress, they form different deformation textures and they form twins more easily on annealing compared with metals with narrow stacking faults.

The stacking fault energy of solid solution f.c.c. alloys is usually lower than that of the parent metal, e.g. 70/30 brass has a stacking fault energy of $\sim$ 10–15 ergs/cm$^2$. In brass there is a tendency for recrystallisation to be 'in situ' and textures developed in early stages of working are not always removed by intermediate annealing. Final rolling should be limited to 40–50% and penultimate annealing temperature should be similar to the final annealing temperature.

Dissociation of dislocations does not normally recur in b.c.c. metals.

*Dislocation climb, polygonisation, recovery*

When cold worked metal is heated the dislocation loops re-arrange themselves, the internal stresses, hardness and electrical resistivity are progressively reduced. This is known as *recovery* and is brought about by the movement of dislocations due to vacancy diffusion, so that the dislocations effectively *climb* out of their slip planes and tend to collect into stable arrays of lower energy, i.e. form walls of dislocations, one above the other, Figs. 63(*e*) and 46. This constitutes a boundary (*f*) at which the atomic planes meet at a low angle, and which form sub-structures or veining within the normal high-angle boundaries. This is known as *polygonisation*.

*Creep*

Dislocation climb is important in high temperature creep. Steady state creep can be regarded as a balance between strain hardening and recovery going on together. In this way the dislocations continuously generated by plastic flow are removed from the interior of the sub-grains to the sub-grain boundaries. Thus the dislocation density and hence the resistance to deformation within the grain is maintained substantially constant, giving steady state creep. Since rate of dislocation climb depends on diffusion of vacancies which occurs more rapidly in the irregular structure of normal boundaries, climb and hence creep is greater in fine grained material which contains more grain boundary area. Fig. 63(g) also indicates how climb gives dislocations more freedom to get round a precipitate and to move on other planes.

At recrystallisation temperatures large-angle boundaries are formed which move through the whole crystal structure, sweeping up the excess dislocations and leaving behind less distorted material.

*Yield point and strain ageing*

One of the impressive achievements of the dislocation theory is the explanation of yield point phenomena in mild steel. Foreign atoms interact with dislocations in such a way as to reduce their energy. To do this large atoms can be sited in the dilated regions of the lattice and small atoms in the compressed region. Once such atmospheres of solute atoms are formed the movements of dislocations are hindered, i.e. higher stress must be applied. Movement of the atmospheres depends on diffusion rates and will be slow at room temperature. In this hypothesis the upper yield corresponds to the applied stress at which the anchored dislocations can be separated from their atmospheres of solute atoms under the conditions of the test ($h$). When free the dislocations can move at a lower stress, which is related to the lower yield point. Once yielding has occurred, immediate reloading of the specimen produces no yield point ($i$) because low diffusion has not allowed the atmospheres to reform and dislocations are unanchored. Ageing, however, allows the atmospheres to reform on even more dislocations, hence the yield point returns at a higher stress, i.e. strain ageing ($j$). As little as 0·003 % nitrogen or carbon is sufficient for these effects. Blue brittleness at 300 °C can be explained in a similar way. At this temperature there is repeated rapid formation of atmospheres and breaking away of dislocations leading to a succession of yield points on the stress-strain curve ($k$). (See also p. 119).

*Planes and direction of slip*

The spacing of the atoms in the lattice is distorted to some extent during plastic deformation, giving rise to the effect known as *lattice distortion.*\*

Fig. 30 shows planes of atoms represented by lines and it is evident that some (P) are close together but sparsely populated with atoms, while the densely populated planes (Q) are farthest apart. The latter planes are important, since slip usually occurs on them, owing to the binding forces being weaker where the atoms are farthest apart.

\*Owing to the formation of numerous crystallites, X-ray back-reflection photographs show an arcing of the spots, which eventually overlap to form continuous rings. Lattice distortion causes a broadening of these rings.

Fig. 30 shows the atoms, represented by spheres, in a face-centred cubic pattern and seven atoms have been removed from one corner of the pile, thus revealing a plane of atoms packed closely together. This process could be carried out from each of the corners of the pile, to form *four octahedral planes*; so called since they form the faces of a regular octahedron (Fig. 105). In each plane there are three lines of closest packing, AB, BC, CA, that is twelve possible directions on the four planes. In a metal such as aluminium or copper, having a f.c.c. lattice, *slip occurs on these octahedral planes and in the direction of the line of the close packed row of atoms which is nearest to the direction of maximum shear stress.*

In general, metals having a body-centred cubic lattice possess at least two types of planes of easiest slip, although these are not *closely* packed. The absence of close-packed atom planes makes plastic deformation of metals with b.c.c. lattices take place less readily than with f.c.c. metals. Iron at room temperature is exceptional. It has three types of plane, giving a total of 42, practically equivalent, slip planes. Slip bands in iron are often forked, irregular or wavy and apparently are unrelated to any principal plane.

In the case of zinc, cadmium and magnesium, which crystallise in a hexagonal lattice, only *one* plane of easy slip, namely, the base of the hexagon, is available at room temperature, but other planes operate as well. These metals are therefore less ductile than those having a cubic lattice.

*Effect of the number of crystals*

Industrial alloys consist of an aggregate of crystals, with different orientations relative to one another. Only a few grains may be in a position favourable for slip to occur and even then deformation can only proceed a little way before it is blocked by neighbouring grains having a different orientation of planes. Once movement in a crystal occurs, however, the stress distribution is altered and other crystals slip, until eventually all are deformed.

The nature of the atomic arrangement at the grain boundaries also hinders slip. The boundary region consists of a number of atoms, each acted upon by forces arising from both lattices, so that they occupy intermediate positions and form a *transitional region* between one lattice and the other. The geometric regularity of the atoms is therefore disturbed at the grain boundary and gliding across this region is rendered difficult (Fig. 55($c$)). Hence the finer the grain size (i.e. the greater the number) the harder, tougher and more uniform is a material, since the direction of active slip changes frequently. This effect is pronounced in a notched-bar test as shown in Fig. 54(C). The optimum grain size varies with particular processes and service conditions.

*The effect of impurities* is also illustrated. It will be noticed that failure is usually across the grain (transcrystalline), but when metals are stressed slowly at elevated temperatures they frequently fail at the grain boundaries (intercrystalline). Pure metals and homogeneous solid solutions can usually be deformed a greater amount than alloys containing two phases of different hardness, since minute cracks tend to develop at the boundary planes between the hard and the soft constituent which flows readily. Cracks tend to propagate along brittle constituents such as slag fibres and when these occur parallel to the line of stress a 'short' facture results; but when the laminations are at right angles to the stress axis the crack is deflected from its normal course and a fibrous fracture is formed (Fig. 54(c)).

*Brittleness*

We can now see why a metal is strong and how its strength can be increased. It is strong if each little slip process is confined to a short run in the metal and not allowed to spread over large crystals. A body which cannot slip is brittle and it fractures by one plane of atoms losing its cohesion on an adjacent plane (i.e. cohesive failure).

To summarise, there are five ways in which brittleness can be caused, by:

(1) Extreme cold working which leaves no planes unaffected by slip.
(2) Presence of obstacles to movement of dislocations at intervals along the plane, e.g. inclusions, segregations of stranger atoms (p. 98) and precipitated compounds and their associated stress fields.
(3) Films of brittle, weak material surrounding the crystals, Iron sulphide (p. 169), Bismuth in copper (p. 302).
(4) Application of equal tension stresses in three directions at right angles at the same time.
(5) Alteration in ratio of plastic yield strength to cohesive strength (p. 14).

**Twinning**

In certain metals, notably zinc and tin (tin-cry), deformation can also occur by a process known as twinning, which may be regarded as a special case of slip movement. Instead of whole *blocks* of atoms moving various distances along the slipping planes, each plane of atoms concerned moves a definite distance, so that the total movement at any point relative to the twinning plane is proportional to the distance from this plane. In block slipping the two parts of the crystal retain the same general orientation of atoms, but twinning produces two parts symmetrical about the twinning plane, each a mirror image of the other. Fig. (54B) illustrates these effects and Fig. 57 shows *strain twins* in cast zinc produced by squeezing the sample in a vice. Parts of three grain boundaries are shown and the small particles of impurity betray the different orientation of atoms in the three grains. A special form of twinning is found in $\alpha$-iron when the deformation is caused by a *sudden blow* at room temperature or slow deformation at low temperatures. The twins, known as *Neumann bands*, are illustrated in Fig. 58.

In certain face-centred cubic metals (copper alloys, austenitic steels) twins are formed when cold-worked material is annealed. These are known as *annealing twins* and are characterised by flat twinning surfaces as shown in Fig. 71. Their presence thus indicates that the sample has been deformed previously, since an annealed casting shows no such twins (Fig. 67).

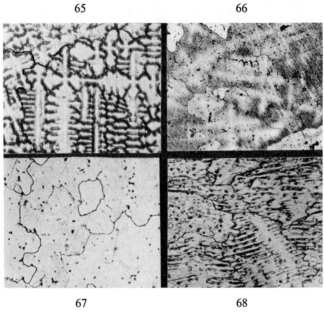

4% tin bronze, etched—ammonia + hydrogen peroxide  (× 80)
  Fig. 65 As cast: cored
  Fig. 66 Casting partially annealed: traces of coring
  Fig. 67 Casting fully annealed: boundaries and porosity
  Fig. 68 Casting lightly rolled: distortion of dendrites

Annealing twins are not deformation twins formed by shear but form as part of grain growth where a twinned configuration can have lower energy relative to a formed boundary. The number of twins in a grain increases in proportion to the number of new grains encountered with a decrease in stacking fault energy. Thus brass is profusely formed whereas aluminium is not.

Other aspects of cold working on the microstructure are illustrated in Figs. 68, 69 and 72. Light deformation produces a bending of the dendrites

TWINNING 117

in a casting (Fig. 68) or the twin bands in an annealed metal containing them (Fig. 72). The grains are slightly elongated and strain markings become evident. Heavy working produces crushed crystals with needle-like form (Fig. 69), which tend to give a fibrous effect.

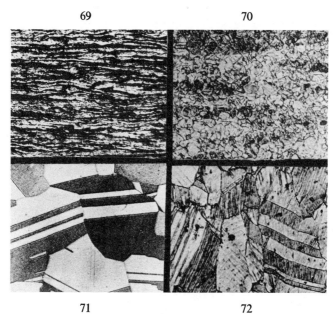

69      70

71      72

4% tin bronze, etched (× 80)

FIG. 69 Ingot heavily rolled: grains elongated into fibres
FIG. 70 As Fig. 69, but annealed at 600°C, 5 minutes. Recrystallisation with traces of coring
FIG. 71 Worked and fully annealed at 800°C, 1 hour. Twinned equi-axed grains
FIG. 72 As Fig. 71, but lightly cold rolled: bent twins: strain bands

*Fibre*

The production of fibre and directional properties in a worked material may be due to

(*a*) *Second phases*, e.g. impurities, such as non-metallic inclusions and segregation of phosphorus in steel or β particles in brass containing less than 64% of copper. When such material is rolled the impurities are elongated and frequently remain so even after annealing. The material then exhibits properties which differ in different directions. In wrought iron and many steel forgings these directional properties are used to advantage, since the flow of the metal is so regulated that the working stresses are exerted at right angles to the grain flow (see vacuum melting, p. 159).

(b) *Preferential orientation of the crystals*, which is brought about by the process of slip causing the rotation of the planes of slip in all the crystals into more favourable directions with respect to the direction of maximum shear stress. Only an approximate orientation of the planes is obtained; its character depends upon the cold-working process and becomes important when the material is heavily deformed.

After annealing a metal with a marked texture due to heavy deformation, the recrystallised structure may also show a preferred orientation which may be the same as the deformation texture or a different one (see Table 6). This influences the tensile properties in different directions and produces 'ears' on cups pressed from the metal and sometimes causes local thinning of the pressing. In brass, low strength and high ductility is frequently found at 45° to rolling direction. Typical variations, which vary with previous treatment, are given in Table 7. (See also stacking faults p. 93.)

TABLE 7  DIRECTIONAL PROPERTIES OF 70/30 BRASS STRIP

| Treatment | Tensile N/mm$^2$ | | | Elongation % on 50 mm | | | Reduction of Area, % | | |
|---|---|---|---|---|---|---|---|---|---|
| Angle to Rolling Direction | 0° | 45° | 90° | 0° | 45° | 90° | 0° | 45° | 90° |
| Rolled 90% | 740 | 770 | 850 | 3 | 3 | 2 | 47 | 35 | 22 |
| Rolled 90% and Ann 725°C | 294 | 278 | 294 | 61 | 64 | 60 | 68 | 72 | 65 |

In the case of 3% silicon–iron and certain iron–nickel alloys preferred orientation can produce a desirable increase in magnetic qualities in one direction. A close connection has been established between the deep drawing performance of sheet metal and the plastic anisotropy of the sheet, which is a consequence of the development of preferred orientation or texture. An index of drawability is the limiting drawing ratio (LDR), the ratio of the diameter of the largest circular blank that can be drawn to the punch diameter. The required anisotropy is that the strength in the through-thickness direction of the sheet should be high compared with strengths measured in uniaxial tests in the plane of the sheet. Since it is difficult to measure the through-thickness strength directly, it is often assessed by strain-ratio measurements made on tensile test-pieces cut in different directions relative to the rolling direction. The strain-ratio, $\bar{r}$, is defined as the ratio of width strain to thickness strain, which would be equal to one in isotropic material. An average value of the strain ratio in different directions, $\bar{r}$, is linearly related to the LDR. Fig. 73 shows the depth of cups related to $\bar{r}$.

In cubic metals, high $\bar{r}$ values are obtained when a large proportion of crystals have {111} planes parallel to the sheet surface and a small proportion have {100} planes in this orientation. This situation can arise during the normal processing of low carbon steel strip, more especially in Al-killed steel rather than in rimming steel. The better drawing performance of the former is therefore due to the formation of a particularly favourable texture, arising, it is thought, from the interaction of Al N particles with grain boundaries during the early stages of recrystallisation, giving a pancake structure.

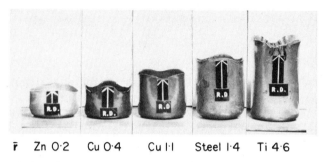

r̄    Zn 0·2    Cu 0·4    Cu 1·1    Steel 1·4    Ti 4·6

Fig. 73 Relationship between deep drawability, shown by the height of the largest cup that can be drawn, and plastic anisotropy of the sheet, characterised by average strain-ratio, r̄

### Stretcher strains or Lüder's lines

Most annealed mild steel shows the unique property of yielding in certain locally stressed regions, giving rise to differences in level of the surface when stretched less than 4% elongation. This effect is frequently encountered in drawing or stamping operations, and the surface markings in relief are called stretcher strains (Fig. 59). The remedy consists in slightly deforming the sheet by a pinch pass or roller levelling so that the pronounced yield point is obviated (Fig. 2). This effect, however, wears off with time and pressing is usually carried out within 24 hours of the treatment. Special 'non-ageing' steel is immune from such markings.

In certain steels, similar markings can be developed by Fry's reagent (Chap. 2). The dark markings outline regions of deformation. This is due to preferential precipitation of some compound after treatment at 200 °C, which causes more rapid etching.

*Evidence of cold work is indicated therefore by slip bands and stretcher strains on the polished surface; strain bands, Lüder's lines, bent annealing twins, distorted dendrites, elongated grains and strain twins on etched specimens.*

## Changes in properties due to cold working

Plastic deformation raises the proof stress, tensile and fatigue strengths, electrical resistance, and increases the hardness and rate of chemical solution, but decreases the values of elongation, reduction of area and creep properties.

In steel the modulus of elasticity is unaffected by cold working, but it may be increased by up to 20% for other metals. With heavy cold working the proportional limit is raised almost to coincide with the ultimate tensile strength. In such a condition the material is useful for springs, and phosphor bronze and low carbon steel springs are typical examples.

## Internal stresses

After plastic deformation a material may be left in a state of internal stress, the intensity of which varies from place to place. Such stresses can be estimated roughly by machining layers from the surface of a bar and noting the subsequent changes in length and diameter of the bar, these being used in a suitable formula. The surface layers are frequently in a state of tension and the removal of this material permits the compressed core to expand and the piece to lengthen. A ring may be split and the opening or closing of the slot noted.

Internal *compression* stresses produced in portions of the surface subjected to service tension stresses can be beneficial, e.g. increasing the strength of guns and pressure vessels by plastic expansion. (See also peening, p. 28.)

Objects in a state of internal tension stress are liable to crack, often after an interval, when subjected to:

(*a*) annealing, i.e. fire-cracking;
(*b*) mild corrosive media—stress-corrosion cracking;
(*c*) liquid metals.

*Season cracking* of brass pressings, spinnings, condenser tubes and cartridge-cases is liable to occur when they are used in a work-hardened condition; 60/40 brass bolts can also fail when in the tightened condition. Failure takes place when the internally stressed article is subjected to mild corrosive media, such as industrial atmosphere, ammonia, mercurous nitrate or sea-water. A characteristic of the failure is that the cracks usually pass round the grains (intercrystalline) and the tensile stresses cause the fractured surfaces to separate to a great extent. However, there are many cases now found where the crack is transgranular (e.g. Mg alloys and austenitic stainless steel) in which a slow repairable surface film is ruptured by slip, tunnel corrosion starts and leads to ductile fracture. Susceptible alloys often have low stacking fault energies.

A test for susceptible material consists in immersing it in an aqueous solution containing 1% mercurous nitrate and 1% of nitric acid. Highly stressed brass will crack within a few minutes. An example of 70/30 tube is shown in Fig. 74.

The defect may be largely overcome by annealing at a low temperature (200–300 °C) insufficient to bring about appreciable softening, but which evens up and partially removes the internal stresses.

The intercrystalline cracking of steel boiler plates, especially round punched rivet holes, is often due to the combined effect of internal stress and caustic alkali in the water. This is known as *caustic embrittlement* and

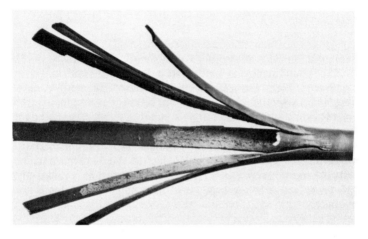

FIG. 74 70/30 condenser tube tested for season cracking

can be avoided by annealing at 650 °C. Other examples of stress-corrosion are stainless steels (p. 258), beta-brass (p. 310) and aluminium–magnesium alloys (p. 414).

Molten solder and tin *penetrate* the crystal boundaries of β-brass and nickel–chromium steel when a tensile stress exists in the sample. Soft soldering has caused the failure of nickel–chromium axle-tubes used in aeroplanes.

## Annealing of cold-worked material

Cold working hardens a metal and while advantage is taken of this increased strength in certain applications, in others it is frequently necessary to soften the material in order to allow further forming processes (e.g. deep drawing) to be carried out. This softening is produced by heating the metal to a temperature at which the disturbed material reverts to a more stable condition. The process is known as *annealing*.

*Stress relief*

Even if the metal is heated below this softening temperature a certain relief of internal stress takes place together with a partial restoration of some physical properties, without change in optical micrograph. This is known as *recovery* (p. 112). The material is safer for use and less liable to distortion than in the original condition. The stresses in castings and welds, due to contraction during solidification and cooling, are removed by annealing at temperatures high enough for the material to be somewhat plastic, e.g. 650 °C for steel.

*Grain refinement*

Large grains in metals can be converted to small crystals by two methods. One depends on (the allotropic) transformations occurring in the solid state. This phenomenon is limited to a few metals and alloys, notably steel and aluminium bronze, whereas the following method of refining coarse-grained structures is applicable to all metals. It consists in annealing heavily cold-worked material at a suitable temperature so as to form small equi-axed crystals which are usually independent of the old shattered grains. This phenomenon is called *recrystallisation* and it is used extensively in non-ferrous metals and a great deal in the heat-treatment of steel sheets (see process anneal, p. 175). If the temperature is raised these new crystals grow in size at the expense of their neighbours; this is known as *grain growth*.

*Factors affecting recrystallisation* are:

(*a*) The greater the degree of cold work the lower the recrystallisation temperature—see broken curve in Fig. 79—and the smaller the grain size. A minimum deformation is necessary to cause recrystallisation.

(*b*) As the temperature is lowered the annealing period necessary to produce a given crystal condition is lengthened and near the minimum recrystallisation temperature it increases remarkably. For example, to attain a constant grain size in iron 8 minutes was required at 675 °C, but 32 hours at 550 °C.

(*c*) Soluble impurities raise the recrystallisation temperature. Very pure aluminium recrystallises below room temperature, whereas commercial aluminium requires heating to 200 °C.

Strain bands denote highly disturbed regions, consequently the new crystals form in these areas first and grow slightly by the time the less deformed areas have recrystallised. Similarly in a rolled solid solution, such as coinage bronze, the central cores which are purer than the remaining metal recrystallise first. As a result of insufficient time of annealing to reach a stable condition strings of fine and coarse grains form with resultant directional properties. Fig. 70 shows a cast bronze, worked and annealed insufficiently to remove all traces of the coring and also shows variation in grain size.

The electron micrograph, Fig. 75, shows the effect of cold work on iron in producing tangles of dislocations, outlining fairly clear cells. This cell formation depends on stacking fault energy. Brass (low sfe) shows no cells and no recovery, but immediate recrystallisation. Aluminium (high sfe) shows many cells and good recovery with a marked decrease in stored energy so that recrystallisation can be retarded due to insufficient driving energy. Some 5% of the total energy of plastic working is stored in the dislocation tangled structure (the rest is liberated as heat) and this energy is retained so long as the dislocations are confined to their slip planes. Thermal energy allows the dislocations to move out of their slip planes first by the cross slip of screw dislocations and then at higher temperatures by the climb of edge dislocations. During *recovery* this thermally assisted movement of dislocations 'tidies up' or simplifies the tangled worked structure. Dislocations of opposite sign annihilate each other; those of the same sign polygonise into well defined cell walls. Fig. 76 shows recovered cells with low-angle boundaries.

When cold-worked metals are heated to a higher temperature than for recovery ($\sim 0.3$ Tm for pure metals and $\sim 0.5$ Tm for alloys) drastic softening occurs by *recrystallisation* in which high-angle boundaries sweep through the metal and replace the tangled dislocation structure by a new set of perfect grains. The driving force for moving the boundaries is provided by the greater number of dislocations on one side of a grain acting as tension strings pulling the boundary into the worked region at a rate proportional to the difference in dislocation density and which increases exponentially with temperature.

The actual mechanism for nucleation of these grains is the subject of much research; some indicating that nucleation of grains occurs by bulging of the original high angle boundaries into high energy regions, others consider deformation bands, inhomogeneity of dislocation distribution, and specially oriented grains initiate recrystallisation. Fig. 77 shows partial recrystallisation, and Fig. 78 shows complete recrystallisation with high-angle boundaries and only a few dislocations pinned by carbides.

Dispersed phases play an important role in recovery and recrystallisation processes; not only do they affect the dislocation distribution, but they also affect the migration of boundaries (Fig. 61). With large particles, high dislocation densities may be produced, often at the particle surface, with severe misorientations, and it is from these regions that new grains are thought to nucleate. Small particles (e.g. 1 μm) may inhibit cell formation during cold work and the migration of boundaries during recrystallisation.

The initial orientation of the grains and the size, shape, deformability

75

76

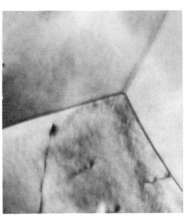

77

78

Fig. 75 Cell structure in cold-worked iron consisting of tangles of dislocations with regions of lower dislocation density ($\times$ 40 000)

Fig. 77 Cold-worked iron after annealing at a higher temperature, showing a recrystallised grain (upper part of photograph) and sub-grains. This is the partially recrystallised condition ($\times$ 40 000)

Fig. 76 Cold-worked iron after a low temperature anneal, showing sub-grains formed from cells in cold-worked structure. This is the recovered state ($\times$ 40 000)

Fig. 78 Fully recrystallised grains, one containing a few dislocations pinned by carbide particles ($\times$ 40 000)

and separation of the particles are all found to be critical in retarding or accelerating recrystallisation.

By allowing precipitation of second phases to occur after cold work, the size, shape and texture of the recrystallised grains can be controlled.

*Factors affecting grain growth* are:

(*a*) The new crystallites grow in size by absorbing each other, at first rapidly and then rather slowly, as shown by the curves (Fig. 80).

(*b*) As the temperature increases the grain size produced by annealing increases (Fig. 80).

(*c*) Insoluble impurities tend to inhibit grain growth, e.g. thorium oxide in tungsten filaments (also p. 167).

(*d*) The degree of cold work previous to the anneal greatly affects the subsequent grain size. Annealing produces a small crystal size in severely

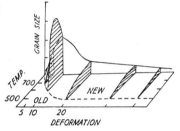

Fig. 79 The relation of grain size, annealing temperature °C and cold work (%) for a 0·18% carbon steel (constant annealing time)

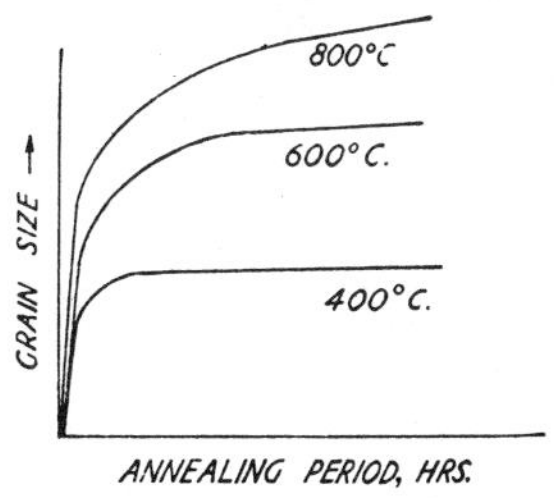

Fig. 80 Effect of duration and temperature of annealing on grain size

cold-worked metal even after grain growth has finished, but slight deformations frequently cause exaggerated grain growth. This critical degree of cold working is about 10% elongation for iron, 2 for aluminium and 1 for lead. Certain crystals become 'germinated' and absorb neighbouring grains not so affected. To produce these conditions a high annealing temperature is required, and hence this trouble may be reduced by annealing the metal at a lower temperature for a longer time. These relations of time, previous deformation and grain size are illustrated in Fig. 79, and an illustration of the wide variation in grain size produced by same annealing treatment of specimens deformed different amounts is seen in Fig. 81. It is interesting to note the recrystallisation formed from the scratch occurring down the centre of the specimen which had been extended 4%.

A sheet with very large crystals would tend to give a rumpled surface (orange-peel effect) after a pressing operation, which is undesirable.

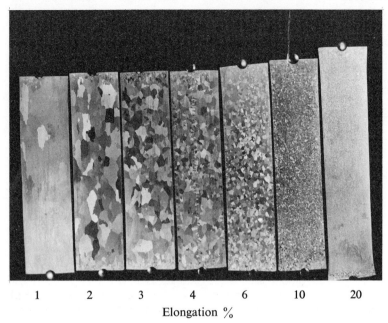

1   2   3   4   6   10   20
Elongation %

FIG. 81 Effect of annealing 5 hours at 640°C on aluminium, cold worked to various amounts

It will be seen, therefore, that the amount of cold deformation, the time and temperature of annealing are of prime importance with reference to grain size produced in material. Typical annealing data are:

| Metal | Mild Steel | Brass | Monel | |
|---|---|---|---|---|
| Temperature, °C | 600–700 | 650 | 760 | 930 |
| Time at temp, hours | 8–24 | $\frac{1}{2}$–1 | 1 | 3 min |

*Annealing castings*

In unstrained metal, such as a casting, no change in crystal size occurs after annealing. One effect of annealing on a casting is that diffusion takes place and the coring is reduced (Fig. 66), or is entirely removed if sufficient time is allowed (Fig. 69). The time, however, varies according to the alloy. Coring is far more persistent in bronze and cupro-nickels, than in brass; while in steel the carbon diffuses at about 100 times the rate of the phosphorus.

## Hot working

Hot working may be regarded as the simultaneous occurrence of deformation and recovery processes—unlike cold working where no appreciable recovery occurs during deformation. If the temperature of working is gradually raised there is a progressive increase in the rate at which the recovery takes place and an associated fall in the stress necessary to cause deformation.

In all normal hot-working operations, however, the rate of hardening due to deformation is greater than the rate of softening associated with the recovery, and the metal is in a more or less strain-hardened condition when the working process is completed. Since the metal is still at an elevated temperature further softening takes place either by a continuation of recovery or by recrystallisation. The structure, when subsequently examined at room temperature, depends on the conditions of working (particularly the working temperature) and on the conditions of cooling, and may be slightly or fully recovered, partially or fully recrystallised. If fully recrystallised, there appears to be no difference between this structure and the structure obtained by cold working followed by full annealing.

Lead can recrystallise at room temperature. Light rolling of lead will cause the hardness to increase, but as the reduction increases the hardness begins to fall, due to the more rapid recrystallisation of heavily deformed material. The heavy working of lead at room temperature constitutes 'hot' working, but deformation of iron at 400 °C is similar to cold working, since rapid recrystallisation does not occur until the temperature is raised above 600 °C. Such material may be referred to as *warm-worked* material.

*Importance of finishing temperature*

If working is finished while the object is at a temperature high above that necessary for recrystallisation, grain growth occurs during the undisturbed cooling. The large grains produced cause low reduction of area and Izod values. The finishing temperature should, therefore, be just above the recrystallisation temperature (or lower critical range in steel), resulting in the production of fine crystals. A lower temperature than this will lead to cold-working effects, which are indicated by a high value of the ratio of yield stress to ultimate tensile stress.

The finishing temperature is sometimes affected by the presence of brittle ranges in a material. Owing to the fact that Monel metal is brittle in the range 870 to 650 °C, the finishing temperature (usually 1040 °C) is much higher than the minimum recrystallisation temperature.

Typical hot forging ranges in °C are:

| Duralumin | Mild Steel | High Carbon Steel | Monel | Zinc |
|---|---|---|---|---|
| 450 to 350 | 1200 to 900 | 900 to 725 | 1150 to 1040 | 150 to 110 |

The amount of deformation should be great enough to break down the original coarse grains and also to prevent grain growth from critical amounts of strain. In large steel forgings it is often specified that the cross-section be reduced by at least 80%.

*Reasons for hot working*

One of the important reasons of hot working is that the metal is usually soft and plastic at elevated temperatures and the power required for the shaping process is reduced. Some alloys, such as 60/40 brass, cannot be cold rolled economically but are very satisfactory worked above 500 °C. Cast zinc cannot be rolled at room temperature without cracking, but can be readily shaped at 150 °C. The delay caused by, and the cost of, annealing are eliminated.

*Effects of hot working* are:

- (*a*) refinement of crystal structure;
- (*b*) elimination of the oriented cast structure (ingotism);
- (*c*) promotes uniformity of material by facilitating diffusion of alloy constituents and breaks up brittle films of hard constituent or impurity, e.g. cementite in steel;
- (*d*) cracks and unoxidised blowholes are sometimes welded up; alternatively, serious cracks or faults are usefully shown up at an early stage;
- (*e*) mechanical properties, especially elongation, reduction of area and Izod values are improved, but fibre and directional properties are produced.

The effect of hot working at 0·2% carbon steel is as follows:

|  | TS $N/mm^2$ | Elongation % on 50 mm | Reduction of Area % | Izod Impact, ft lbf | Joules |
|---|---|---|---|---|---|
| As cast | 417 | 15 | 10 | 5 | 7 |
| Hot worked * | 463 | 35 | 52 | 30 | 40 |
| Hot worked † | 452 | 19 | 25 | 12 | 16 |

\*In direction of rolling.  †Across rolling axis.

## Working processes

*Hot-working processes*

For hot-working steel two methods, forging and rolling, are extensively used. Most non-ferrous metals such as nickel, nickel–chromium alloys, 60/40 brass are also hot rolled, but a few (e.g. leaded brass) are broken down cold. Forging may be carried out by steam hammer, hydraulic press, drop forging or upsetting machine.

The hammer consists of a tup of substantial weight which is lifted and dropped or forced down (double acting) on the metal supported on an anvil, weighing about twenty times that of the tup, and cushioned by timber baulks (Fig. 82). The rating of the hammer is usually indicated by the weight of the tup. The hammer exerts compressive pressure on a relatively small area, for a very short time. The deformation is localised to the exterior, and a high degree of refinement is obtained provided the section worked is a small one. Wrought iron, high carbon and high-speed steels are usually forged initially. Round bars should never be 'drawn down' to a smaller diameter by hammering between flat dies because stresses are set up at the outside edge causing tension at the centre, which sometimes produces a central cavity. By means of tapered rolls this mechanism is utilised in the Mannesman process (Fig. 82), in order to pierce a solid billet as a first stage in the production of seamless tubing. A square bar hammered alternately on its sides has a compressive force at the centre.

For mass-production work, involving large numbers of articles, billets are forged to shape in dies, fixed to the tup and anvil. This is called 'drop forging'. Two possible ways for making a gear blank are shown in Fig. 82(A). In the up-ended sample each tooth will have a similar arrangement of fibre (Fig. 82(B)), while in forging (Fig. 82(A)) the strength of the teeth varies according to their position. Stampings of brass and aluminium alloys are also made in dies.

In order to work the centre of large ingots, used for large shafts, guns, steam drums, etc., the large hydraulic press, capable of exerting a direct squeeze of 2000–3000 tons, is necessary.

Any metal having good malleability can be forged, since compressive forces are involved.

Several machines are now available for high rate forging permitting greater deformation in a single blow because stock temperature drop is insignificant. The motive power is generally nitrogen gas, which has been compressed by a hydraulic system and the energy of decompression being used to drive the dies and platten together, e.g. Dynapak. In the Petroforge the energy is derived from the combustion of a fuel/air mixture in the cylinder of an internal combustion engine. The process can reduce preforming, blank weight and time, but more expensive die material is often required.

Cold forging of steel has evolved from the impact extrusion of lead and aluminium tubes and cans, and eliminates forging scale, de-carburisation and dimensional variation. Efficient lubrication of the stack is essential, and on steel is usually provided by a zinc phosphate coat, together with stearate soap. Most cold forged parts are symmetrical about their axis and regular in section, e.g. bevel pinion valves, bolts, sparking plug bodies.

*Extrusion* is an ideal process for obtaining rods from metal having poor ductility (e.g. lead), but it is also economical to form tubes, and complicated sections from many non-ferrous metals, particularly copper, brass, aluminium and its alloys and also steel and nickel alloys. A round heated billet of metal placed in a container, is forced through a die by a plunger operated by a 750–3000 ton hydraulic press (Fig. 82(D)).

The flow of metal during extrusion can be quite complex and the amount of deformation imposed on the product can vary considerably from point to point. The temperature of the metal emerging from the die can also vary during the course of extrusion due to chilling (if the billet is initially hotter than the container) and due to heating effects (arising from

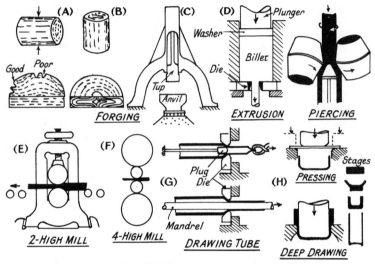

FIG. 82 Shaping processes

plastic working). The structure of an extruded bar often, therefore, varies across its section and along its length. The mechanical properties can also vary, but the most marked effects occur at the extreme front and back ends, and these regions are normally cropped off and discarded. At the back end the oxide film on the outside of the billet is extruded down the centre of the rod, giving rise to the so-called 'extrusion defect'.

*Rolls* exert a squeeze for a short interval and are used extensively for producing billets, plates and sections, and are noted for speed of production. The ingot is usually reduced to a slab or bloom in a cogging mill, normally of a two-high reversing type (Fig. 82(E)). For billets and sections the rolls elongate the metal in one direction only, except for slight side spread. Plates or sheets are frequently rolled in two directions at right angles. Grooved rolls are used for rods.

*Cold-working processes*

Whenever close sizes, good finish and close control of properties are demanded in bars, sheets, wire tubes and sections, cold working is a necessity. This frequently involves the removal of mill or annealing scale prior to cold-working operations.

*Preparation.* Such descaling is accomplished by mechanical means or by *pickling*, which consists of the immersion of the material in an acid solution, e.g. 5–10% sulphuric acid at 60–80°C. A small amount of organic 're-strainer' is added to the solution to diminish the attack of the acid on the metal areas from which the scale has been removed. In the case of most scales formed above 600°C, the acid gains access to the interlayer of ferrous-oxide via pores and cracks in outer ferric-oxide layers; the ferrous oxide is dissolved, the outer layers of scale become detached and fall to the bottom of the tank.

Hydrogen is evolved during pickling and is absorbed by the steel, which is thereby embrittled. This hydrogen embrittlement gives trouble if cold-working operations follow too soon after pickling, but it can be obviated by ageing or by a short treatment in boiling water. The hydrogen diffuses through the steel in the atomic state but whenever it passes into a contraction cavity or blowhole it becomes converted into molecular hydrogen ($H_2$), in which form it cannot diffuse, and will develop very high pressures in the metal sufficient to cause *blisters*.

*Rolling*

Strip and sheets are usually cold rolled to get surface finish and accuracy of dimensions or increased strength. To prevent roll deflection, with consequent variation in thickness of metal, the working rolls may be backed by two or four large rolls, giving rise to four-high and cluster mills (Fig. 82(F)).

For large-scale production (e.g. wide strip mild steel) the metal is handled in coils and a single production unit is used consisting of three four-high mills in tandem and a recoiler. The speeds are arranged so that the strip is under tension, allowing large deformations to be imposed and strip of high dimensional accuracy to be produced. Lubricants are normally sprayed on to the rolls and strip during cold working to reduce the loads involved, to cool the material and to control the surface finish.

With the softer metals very large deformations can be imposed by using a succession of cold-rolling passes. With many materials, however, the rolling sequence has to be interrupted and the metal softened by *intermediate annealing* (process annealing) in order to avoid cracking and to reduce the load and power requirements. With some aluminium alloys only about 30% reduction in thickness can be allowed between anneals.

## Drawing

The drawing of rod, wire and seamless tubes depends essentially on the material possessing high ductility in the cold state. The material is pulled through a die having a conical hole with the result that a reduction in diameter takes place. In tube drawing three modifications may be noted:

- (a) *Sinking* (20–35% reduction) consists in pulling the tube through the die without a mandrel. The wall section is thickened, the inside surface is poor.
- (b) *Plug drawing* (30–40% reduction). The tube is drawn over a plug held relative to the die. Floating plugs are now used, enabling long lengths of tube to be drawn. Both inside and outside diameters are reduced and the wall thickness can be thickened if wished.
- (c) *Mandrel drawing* (45–50% reduction). A hardened mandrel is inserted in the tube and travels forward with tube, but must be removed subsequently. This method is very suitable for thin-walled tubes (Fig. 82(G)).

*Pressing and deep drawing* are special cases of the drawing of sheet or strip (Fig. 82). In pressing, a sheet of metal is deformed between two suitably shaped dies usually to produce a cup- or dish-shaped component. One of the dies may be replaced by a thick pad of rubber—a process giving reduced tooling costs and allowing larger deformations to be imposed.

For deep drawing the starting flat sheet of metal is larger than the area of the punch and the outer parts of the sheet are drawn in towards the die as the operation proceeds. A pressure plate, fixed to the machine, prevents wrinkling of the edges during 'drawing in'.

A deep drawn component may show regions near the edges which are thicker than the original sheet (the thickening being caused by the 'drawing in' process) and regions which are thinner than originally due to local extension particularly near to any small radii in the base of the component.

The process is limited by the possibility of fracture occurring during drawing: the maximum blank diameter is rarely more than about twice the cup diameter. Cups may be *redrawn*, however, to make them deeper. This may be achieved by *ironing* (Fig. 82(H)) which is similar to tube drawing. Annealing may be necessary.

## Machining

All machining operations—turning, drilling, boring, milling and broaching—have in common a wedge shaped tool which is forced asymmetrically into the work material to remove a thin layer in the form of broken chips or a continuous ribbon of swarf from a larger body. The ease of machining depends on the design of tools, method of lubrication and also on the micro-structure and properties of the metal. The term machinability applied to the material being cut is ill-defined and it cannot be measured by any single laboratory test but the following are some of the more well established examples of the way in which machinability is affected by metallurgical qualities. Fig. 83(a) shows some of the essential features of the

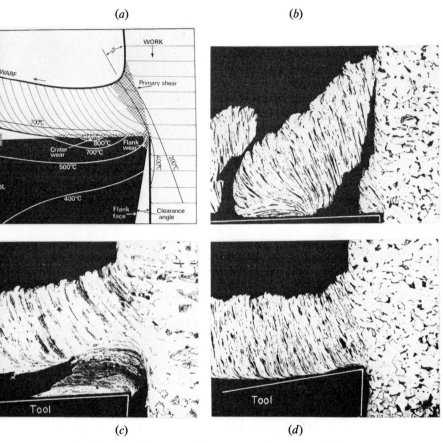

FIG. 83 (a) Diagrammatic representation of machining
(b) Discontinuous chip
(c) Built-up edge on tool
(d) Continuous chip

cutting process, including the primary shear zone, in which the chip is formed by deformation of the structure and also its drag or shear along the rake surface of the tool.

If the area of contact is high, as when cutting ductile metals, the drag force is high and the swarf tends to become very thick and slow moving.

This imposes very high forces on the tool, the swarf is strong and difficult to dispose of and the surface left on the work is torn and rough. Thus high ductility materials such as pure annealed copper, aluminium or iron have poor machinability.

Machining of soft metals can be facilitated by the following means:

(a) *presence of brittle or weak constituent*, for example, graphite in cast iron, slag in wrought iron, manganese sulphide or lead in free-cutting steel (p. 173), selenium in stainless steels (p. 248) and lead in brass (p. 309);

(b) *cold-working*;

(c) *presence of hardening elements* (copper in aluminium, carbon in iron). Low-carbon steels are readily machined in the normalised condition in which the continuity of the ferrite is broken by numerous islands of hard pearlite. The machining of 'fine grained steels' (p. 167) is poorer than that of the coarse-grained steel of similar carbon content. Machining of such inherent fine-grained steel is improved by heating to a temperature high enough to coarsen the austenite grains.

Very hard materials, such as high-carbon steels, become more difficult to machine, because the primary shear force increases and the tool tip is deformed. Such steels can be machined, however, if the cementite is 'balled-up' into small isolated masses, dispersed throughout the soft ferrite (p. 181). The work done in shearing the material at the tool face gives rise to high temperatures, which increase as the speed of cutting is raised, and softening of the tool limits the rates of cutting of materials, such as steels, and Nimonic alloys (Fig. 83(a)) shows temperature isothermals.

When cutting many multi-phase materials, layers of the work material, much hardened by deformation, adhere to the tool surface to form a 'built-up edge' (Fig. 83(d)). This may protect the tool from wear or by breaking away, it may chip the tool, but in most cases it gives a rough finish on the cut surface. It can often be avoided by increasing the cutting speed above approximately 100 m/min giving rise to a continuous chip (Fig. 83(d)).

Non-metallic inclusions in steels and other alloys can greatly modify the machinability. Manganese sulphide and certain plastic silicate inclusions in steel can form planes of easy shear and may extrude on to the tool's rake surface to form a lubricating layer reducing drag. The MnS should be in a massive form; short thin inclusions are ineffective and this necessitates special deoxidation of the steel.

In machine shop practice there are many differences in detail which influence the conditions of cutting and the efficiency of the cutting operation. These often tend to obscure the basic metallurgical factors which limit the rates at which metal can be removed by machining. However, Table 8 shows typical cutting speeds for uninterrupted turning at one selected rate of feed. The influence both of the tool material and of the work material are clear in this case.

TABLE 8

| Work Material | Cutting Speed in plain turning—Metres per minute | |
|---|---|---|
| | High Speed Steel Tools | Cemented Carbide Tools |
| Low carbon steel | 18–46 | 46–152 |
| High carbon or low alloy steels | 15–30 | 46–122 |
| Austenitic stainless steel | 12–27 | 46–90 |
| Nimonic alloys | 4·5–7 | 10–24 |
| Titanium | 12–18 | 30–61 |
| Graphic cast-iron | 15–30 | 46–76 |
| Chilled cast-iron | — | 46–15 |
| Copper | 46–90 | 90–150 |
| Free cutting brass | 76–120 | 150–304 |
| Aluminium alloys | 46–90 | 180–460 |

## Powder metallurgy

The term 'powder metallurgy' covers the art of producing objects from metal powders, with or without the addition of non-metallic constituents and without completely melting the material. A number of different processes are used to make metal powders, and the characteristics of the powder depend on the production route Fig. 84 shows examples. Brittle metals or alloys can be crushed and milled mechanically to give angular or rounded particles, e.g. tungsten carbide, but mechanical comminution of ductile powders produces thin flakes. Some oxides can be reduced to give metal powders by heating in hydrogen or with carbon, for example, iron or tungsten. An atomisation process in which a stream of molten metal is disintegrated by jets of gas or liquid may produce either spherical or irregular particles, for example, tin, aluminium or iron. Electrolysis can be adapted to give deposits of a powdery form on the cathode, for example, copper. Nickel is produced by precipitation from the carbonyl gas. There are many other possible ways of producing metal powders and they are selected for particular applications in terms of particle size, size distribution, shape, surface area and surface condition. The powders may be pressed to a desired form in a suitable die and subsequently or simultaneously heated to produce a welded, alloyed, or coalesced mass. The material may then be suitable for immediate use or may be further worked by conventional methods, as, for example, by drawing down to fine wire, e.g. tungsten and molybdenum. The chief advantages of this technique are that:

(1) It is possible to produce articles having unique properties unobtainable by other methods, e.g. controlled porosity in bearings and filters.

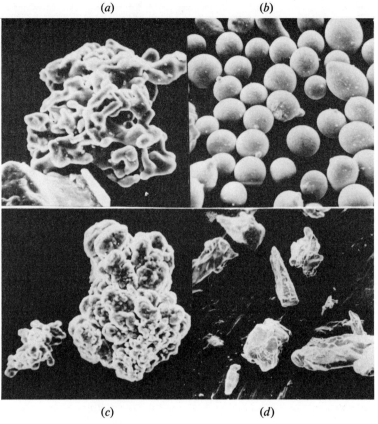

FIG. 84 (a) Iron powder—reduction of oxide ($\times$ 3800)
(b) Monel powder—atomised ($\times$ 370)
(c) Copper powder—electrolytic ($\times$ 4000)
(d) Iron powder—fragmented from brittle electrolyte deposit ($\times$ 1000)
(scanning electron micrographs)

(2) It is suitable for mixtures of metals and non-metals, e.g. copper and graphite (or 'Fluon') in self-lubricating bearings, dynamo brushes; metal-ceramic mixtures for high temperature service.
(3) It offers a method of producing parts in metals which cannot be melted commercially or only with considerable difficulty, and do not lend themselves to casting, e.g. tungsten, molybdenum, tantalum.

(4) It can be used for metals which do not alloy; or have widely separated melting points or greatly different densities, e.g. copper–molybdenum; copper–tungsten; used for switch contacts, conductive brake linings.
(5) It offers a mass-production method of producing large quantities of parts identical in size and quality, perhaps of complex shape, without further operations; and is specially competitive where final machining constitutes a large proportion of the finishing costs by conventional methods, e.g. automobile oil pump gears, very small magnets.

The limitations are:

(a) Restriction on the size of parts that can be produced on the presses available.
(b) The restriction on the design of parts that lend themselves to pressing.
(c) The fact that large quantities are required to justify the cost of dies and tools.
(d) The cost of metal powders.
(e) The fact that the volume of the uncompressed powder is 2–3 times that of the finished piece.

An example of the use of powder metals is 'hard metal' used for cutting tools, drawing dies, etc. Fused tungsten carbide is too brittle for this purpose but some degree of toughness is achieved by the use of a cobalt binder.

The main stages in the manufacture of 'hard metal' are:

(a) Preparation of the individual or combined carbides of controlled grain size by reduction and carburisation of the oxides.
(b) Dispersion of the constituents by ball milling in a liquid.
(c) Pressing the dried, sieved and lubricated powder.
(d) Presintering at about 900 °C in hydrogen atmosphere.
(e) Machining to required form.
(f) Sintering at 1400–1570 °C in hydrogen (see p. 242).

# 8 Corrosion and Oxidation Phenomena

*Mechanism of corrosion of metals*

Corrosion and its prevention is still a problem of great importance to the engineering industry.

Extensive researches which have been devoted to the elucidation of the causes of corrosion have now fairly well established the main facts that the actual attack is electrochemical in character although the various hypotheses differ largely as to the factors by which the electrolytic action is brought about. The presence of moisture is a vital factor and oxygen also plays an extremely important part in the process. Very pure water can attack iron or steel in the presence of oxygen, but the corrosion is slow and usually tends to stifle itself owing to the low solubility of the oxides of iron which are formed on the surface. With a salt solution (such as sodium chloride) and an electric current, generated by methods to be discussed later, the corrosion can be greatly accelerated.

*Electrode potentials*

The theory of electromotive force is discussed fully in text-books on physical chemistry and only a brief outline is given here of the way in which an electrolytic cell is set up. When a salt such as zinc sulphate is dissolved in water it splits up—partially dissociates—into zinc particles having a positive charge and sulphate particles negatively charged, expressed thus:

$$ZnSO_4 \rightarrow Zn^{++} + SO_4^{--}$$

In this case two units of charge (electrons) are involved, and the charged particles are given the name of *ions*. If we immerse a piece of zinc in such a solution of zinc sulphate, containing a definite concentration of ions, we find there is only one electric potential (voltage) at which equilibrium can exist between the metal and the solution. For by dissolving, the zinc tends to form zinc ions until an equilibrium condition is reached between the liquid and the metal. This removal of ions, each with two units of positive electricity, leaves the zinc electrode with an excess negative charge, which is mainly characteristic of the metal. The greater the negative potential the greater is the tendency of a metal to dissolve. If a suitable scale is chosen the single electrode potential can be expressed in volts, as is frequently done by connecting the zinc electrode to a 'hydrogen' electrode, consisting of blackened platinum immersed in a certain concentration of hydrogen ions; its potential being taken as zero. In this way metals can be arranged in the order shown in Table 9, according to their potentials giving a list known as the *electrochemical series*.

TABLE 9   GALVANIC AND ELECTROCHEMICAL SERIES AND VALUES OF OVERVOLTAGE

| Metal | Electrode Potential (volts) | Hydrogen Overvoltage (volts) | Galvanic Series in Sea Water |
|---|---|---|---|
| Platinum | +1·20 | 0·01 | Titanium |
| Silver | +0·80 | 0·1 | Monel |
| Copper | +0·35 | 0·25 | Passive 18/8 |
| Hydrogen | 0·00 | — | Silver |
| Lead | −0·13 | 0·6 | Nickel |
| Tin | −0·14 | 0·5 | Cupro nickel |
| Nickel | −0·25 | 0·15 | Aluminium bronze |
| Cadmium | −0·42 | 0·5 | Copper |
| Iron (ferrous) | −0·44 | — | Brass |
| Zinc | −0·77 | 0·7 | Active 18/8 |
| Aluminium | −1·34 | — | Cast Iron |
| Magnesium | −1·80 | — | Steel |
| Sodium | −2·72 | — | Aluminium |
| | | | Zinc |
| | | | Magnesium |

The ability of metals to resist corrosion is to some extent dependent upon their position in the electrochemical series. On account of their proximity to hydrogen, nickel and copper are not dissolved readily by sulphuric acid, which may be regarded as a salt of hydrogen. The farther two metals are separated from one another in this series, the more powerful is the electric current produced by their contact in the presence of an electrolyte (i.e. a solution having good electrical conductivity). Also the more rapidly the lower metal is attacked the more will the higher metal be protected. It must be remembered, however, that the order in the above series may vary slightly under special corrosive conditions, and the galvanic series in service media (e.g. sea water) are often more useful from the corrosion aspect.

*Pourbaix diagrams*

If the values of electrode potential are combined with solubilities of oxides and hydroxides and with equilibrium data, a diagram can be plotted showing the nature of the stable phase as a function of electrode potential and pH. This is known as a Pourbaix diagram. The full diagram is a very complex figure, and the simplified version is sufficient for most purposes. Fig. 85 shows the simplified form of the Pourbaix diagram for iron. In the regions on the diagram where solid compounds are formed, it is possible that the metal may be protected from further attack by a coating. If this

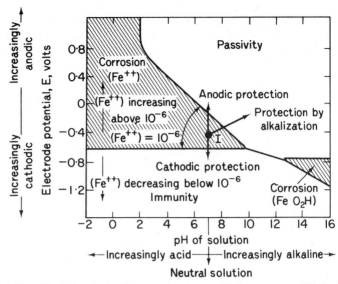

FIG. 85 Simplified Pourbaix diagram for the Fe–H$_2$O system at 25 °C showing the domains of corrosion behaviour. The symbol [Fe$^{++}$] represents the equilibrium concentration of Fe$^{++}$ ions in units of moles per litre of solution. Point I represents iron in ordinary water and the arrows indicate three ways of reducing corrosion

occurs it is known as passivation. On this diagram, passivation is provided by Fe$_3$O$_4$, formed close to the passivity/immunity line, and Fe$_2$O$_3$ forms in the rest of the passive region. The area marked 'Immunity' shows at that potential and pH, metallic iron is the stable state, so that corrosion will not occur. The large area marked 'Corrosion' is where the ions Fe$^{++}$ or Fe$^{+++}$ are stable, while the small area at extremely alkaline pH is corrosion due to the formation of the ion FeO$_2^-$. Numbers of Pourbaix diagrams are available in the literature, and they can provide a useful guide to corrosion behaviour of metals. However, they are based mainly on thermodynamic data rather than on kinetics, and as such, they tend to be *subject to the same types of limitation as are equilibrium diagrams.* Point I represents iron in ordinary water and corrosion can be prevented in three ways shown by arrows: (i) cathodic protection, e.g. by connecting iron to zinc (ii) apply potential of opposite sign to move into passivity (iii) increase alkalinity, e.g. inhibitor.

*Attack by acids*

An example of a short circuited cell is provided by an imperfect coating of copper on steel immersed in dilute sulphuric acid, shown diagrammatically in Fig. 86. The current generated passes from the copper to the steel by the path of lowest resistance (broken line) and returns to the copper through the solution by the passage of ions. The iron, which has the greatest negative potential, dissolves and is called the *anode*; while the copper is called the *cathode*. In such acid attack the hydrogen, which is freed as the iron dissolves, is deposited on the surface of the cathode and as it increases in amount two things may occur. The corrosion of the iron is either brought

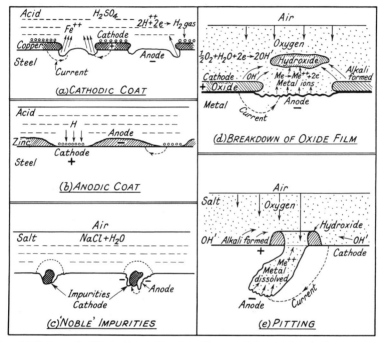

FIG. 86 Schematic illustration of various forms of electrochemical corrosion
*e*—represents one unit of negative electricity
*Me*—represents any metal

to a standstill because of the formation of an opposing hydrogen electrode, i.e. the cell is *polarised*; or the hydrogen may be evolved as bubbles which stream away, with the result that the corrosion will occur continuously. In the first case the corrosion can be accelerated by the use of oxidising agents (e.g. air) which remove the hydrogen from the cathode.

Also, a small cathode, e.g. copper rivet in a steel plate, is quickly polarised and corrosion of the plate is small. On the other hand, a large cathode coupled to a small anode is dangerous.

Thus metals used for coating iron can be divided into two classes:

(1) Metals like nickel or copper, having a positive potential towards iron, protect *mechanically*, by excluding corrosive influences. If the coat-

ing is porous, attack occurs, which is occasionally more intense than that on uncoated steel. This accelerated attack only takes place if the potential between the metals is high and the liquid a good conductor.

(2) Metals such as zinc or aluminium having a negative potential towards iron, are attacked preferentially to the ferrous base and can sometimes give electrochemical protection of the iron exposed at a gap, provided that the covering is attacked sufficiently rapidly to give the necessary current density. The iron remains bright until the exposed area is so large that the central portion is beyond the protective zone. This is an example of *sacrificial* protection (Fig. 86(*b*)). Zinc protectors are frequently used in marine boilers, and near propellers and rudders.

In the choice of coatings the properties of the various metals must therefore be carefully compared and considered. The same mechanism of corrosion is also involved in the corrosion of sheets of commercial metal containing impurities, soldered and welded joints which set up local cells, as illustrated in Fig. 86(*c*). A requirement of good resistance to corrosion is *structural* and *chemical uniformity* of the material. A second constituent or impurity only increases the rate of corrosion if it has a smaller *hydrogen overvoltage* than has the parent metal. By overvoltage we mean the difference in potential between that required to produce gaseous hydrogen on a rough platinum surface and the more negative value on any other metal, the surface of which has an important influence. Thus the corrosion of zinc is increased by additions of platinum, nickel, copper, cobalt, gold, antimony, silver and bismuth, in decreasing order of activity; whereas thallium, cadmium, tin, lead and mercury have no effect owing to their high overvoltage (see Table 9).

Unless the initial voltage of the cell is greater than the overvoltage, all action will eventually cease. With zinc and copper the voltage of the cell is $0.35 - (-0.77) = 1.12$; and the overvoltage of the copper is $0.25$ volts; therefore the reaction proceeds. On the other hand, the voltage of a cadmium–zinc cell is only about $0.35$ volts and the overvoltage of cadmium is $0.5$; hence the reaction ceases.

The attack by acids, which has been discussed, depends on the *strength* of the acids (hydrogen ion concentrations) and most acids also exhibit increased attack when *movement* occurs and when *oxygen* is present, except where passive or inert films are produced on the surface of the metal by oxygen and oxidising acids, such as nitric acid. For this reason mild steel is not attacked by concentrated nitric acid but is rapidly dissolved in the dilute acid. Other very reactive metals such as chromium and aluminium can also be rendered passive by certain oxidising agents due to the formation of invisible oxide films.

## Effect of oxide films

With 'passive' metals the immunity to corrosion depends upon the properties of the film such as:

(a) Thickness and impermeability to media.
(b) Adherence to the base metal.
(c) Resistance to chemical attack.
(d) Mechanical strength.
(e) The ability to repair defects which may arise in the film.

Such oxide films are produced even on exposure to air, and, although invisible while in contact with the base metal, have been separated and studied. The film produced on iron in many environments does not fulfil the above conditions, consequently corrosion occurs. Chloride and sulphate ions are especially liable to cause film breakdown, but chromates and phosphates promote repair. On the other hand, aluminium and chromium form films of the impervious type and this property is usually conferred on their alloys, so long as reducing conditions do not prevail. This dependence for resistance on the presence of oxygen is well illustrated by the corrosion rates of monel metal and stainless 18/8 steel in 3% sulphuric acid.

TABLE 10

| Metal | Loss, mg/dm²/day | |
|---|---|---|
| | Reducing Conditions | Oxidising Conditions |
| 18/8 Steel | 450 | 5 |
| Monel | 10 | 620 |

A tank made of 18/8 steel would be immune from attack if the dilute sulphuric acid was aerated, but stagnant reducing conditions may set up attack at the bottom.

In acid solutions oxide films may dissolve or be reduced to a soluble oxide by the hydrogen formed in the pores.

In the corrosion of some metals there is a danger of *break-away corrosion*, which means that after a period of a low rate of corrosion the film of oxide becomes discontinuous and flakes away from the surface and the corrosion rate accelerates, e.g. zirconium in water at 400 °C.

## Attack by natural waters

Aerated salt solutions usually attack metal surfaces at the weakest points in the oxide film, such as scratches, cut edges and points of high strain. Once attack has commenced, undermining of the primary oxide and consequent flaking off causes the attack to spread out from the initial points of attack, with the exception of small weak points which often heal up owing to the precipitation of corrosion products *in situ*. The exposed parts of the metal become anodic towards the parts still covered with an oxide film and current flows between them. Hydroxyl ($OH^-$) are formed near the oxide film and, if sufficient concentration is produced, they will keep it in good repair. Diffusion of the hydroxyl and metal ions occurs and when they meet a precipitate of hydroxide is formed *away* from the seat of attack. This corrosion product does not therefore stifle corrosion, and the anodic areas spread and the cathodic diminish until a certain current density is reached, called 'minimum protective cathodic current density' (see Fig. 86(*d*)).

As a general rule the presence of oxygen is necessary for the continuance of corrosion of metals in neutral salt solutions. Natural waters are usually saturated with air, but the supply of oxygen may vary according to the depth below the surface of the solution. Evans has shown that such variations in *oxygen concentration* set up electrolytic cells, producing a current between two portions of the same metal to which oxygen is supplied at two different rates. The portion supplied with the smallest quantity of oxygen becomes the anode and is dissolved, while at the aerated part is formed a film with some protective properties. This mechanism explains both 'pitting' and the attack of plates immersed vertically in salt solutions. Various contaminants, some of which are organic sulphur compounds due to bacterial activity in the water, have a great effect on the corrosion rate.

## *Inhibitors*

Pickling inhibitors are organic substances which appear to form a film over the descaled portions of a metal. Cathodic inhibitors are salts, such as zinc sulphate and calcium bicarbonate which form sparingly soluble products on the cathode in natural waters. Sodium phosphate and chromates form sparingly soluble anodic products and are known as anodic inhibitors. Cathodic inhibitors are usually *safe but inefficient* since the corrosion rate will be reduced only when a thin film of cathodic product has been formed over the cathodic area. Anodic inhibitors are usually *efficient but dangerous* since they diminish the rate of attack, but to an even greater extent diminish the area on which attack is concentrated, and thus lead to pitting if an insufficient quantity is used.

## *Pitting*

Cavities in a metal surface are particularly inaccessible to oxygen and therefore become anodic to the surrounding metal to which atmospheric oxygen can diffuse. This effect is illustrated in Fig. 86(*e*). The metal ions formed at the bottom of the cavity migrate upwards and react with the hydroxide ions to form hydroxide of the metal at the mouth of the crevice. This position of the corrosion product accentuates the corrosion by making the diffusion of oxygen to the anode more difficult, and if the cathodic area is large severe pitting may occur. Foreign matter such as sand in condenser tubes causes this type of failure.

Pitting of brass condenser tubes sometimes occurs due to *dezincification*. This consists in the solution of the brass followed by the precipitation of the copper by zinc in the brass; the net result is the selective removal of zinc. A red copper deposit, underneath a white layer, is usually seen near the perforation. It is now the practice to add arsenic to brass to minimise this trouble at low water velocities. A localised attack frequently occurs near the inlet ends of condenser tube, due to *impingement* of air bubbles which carry away the corrosion products. Under impingement conditions copper is also poor but an improvement is obtained by adding nickel and iron (p. 312). Pumps and ship propellers are liable to an attack known as *cavitation* erosion due to the impacts caused by the collapse of vapour bubbles.

*Atmospheric attack*

Corrosion products formed by the atmosphere are more or less adherent and the rate of attack depends largely on their nature, especially their hygroscopic properties. Some surfaces appear wet when there is no visible moisture in the air, that is at humidities below the saturation point. It appears that materials have critical humidities, such as 85% for nickel and 65% for iron corrosion products, above which attack occurs at a destructive rate.

Chlorine compounds are important near the coast, and so are sulphur compounds in industrial areas. Solid particles help to form fog which produces concentration cells when deposited on the metal. The washing effects of rain have an influence in removing such dirt and also corrosion products.

*Corrosion in soils*

The type of attack changes from one soil to another; marshy soils have high acid concentrations, while loose well drained soils have low acidity. Dry soils frequently exhibit alkaline reactions, but have low rates of corrosion owing to lack of moisture. Pitting may occur due to concentration cells caused by the shielding effects of masses of soil and electrochemical attack may arise where the metal is situated in two different soils. Stray currents from electric generators also cause trouble. Certain bacteria are capable of reducing sulphates in clay districts even in the absence of oxygen and so cause intense attack on iron pipes. To minimise this biochemical corrosion the pipe may be zinc coated or wrapped with fabric impregnated with acriflavine dyes or the trench filled with sand or chalk.

*Minimising corrosion*

(1) *Design features.* Obvious points to avoid are (*a*) dissimilar metals in juxtaposition, (*b*) impingement and turbulence, (*c*) crevices which allow collection of moisture and dirt to form stagnant areas, (*d*) structures designed so that cleaning and maintenance of the scheme is difficult or impossible, (*e*) difficult replacement of parts, (*f*) badly ventilated spaces, (*g*) stray electric current, (*h*) inadequate cover of steel in reinforced concrete.

(2) *Change composition of metal.* Copper may be helpful in ferritic steel (p. 228); stainless steels or non-ferrous metal desirable in other applications.

(3) *Alter corrosive environment.* Examples are (*a*) dehumidisation in ship-holds and air conditioning in needle factories, (*b*) deaeration of boiler water, (*c*) inhibitors in radiators.

(4) *Cathodic protection* of pipes, tanks, ships, by zinc or magnesium anodes or external dc current applied to graphite or platinum-clad tantalum anodes.

(5) *Protective coatings.*

## Metallic protection of steel

The characteristics of a few metals commonly used for protecting iron and steel will now be discussed in brief outline (p. 141).

The methods of applying a thin coating of these metals to a steel base are galvanising, sherardising, tinning, electrodeposition, metal-spraying and rolling of composite billets such as nickel-clad mild steel.

*Zinc*

The use of zinc is largely confined to the protection of steel from the action of the atmosphere and natural waters. It protects iron even when the coat is scratched or porous. The different types of zinc contain from 0·01 to 1·5% impurity, but where the resistance of the zinc is due to the accumulation of protective corrosion products, as in atmospheric corrosion, impurities are usually an unimportant factor.

Zinc has a useful resistance only in a narrow neutral range of aqueous solution. The protective hydroxide film is soluble in both acid and alkaline solutions with $p_H$ values less than 6 and greater than 11.

Acids attack zinc very readily with evolution of hydrogen; soft waters are more corrosive than those carrying calcium salts, while potassium bichromate is a film-forming substance.

Atmospheric attack of zinc is slow; bright zinc rapidly tarnishes on exposure, forming a smooth adherent film of zinc oxide, carbonate and hydroxide. In industrial areas where the atmosphere is high in sulphur dioxide and carbon monoxide, a large portion of the film is washed off by rain. In resistance there is little to choose between the various zinc coatings their life being mainly governed by the *thickness*. The corrosion rate is usually below 1 mg per $dm^2$ per day (0·005 mm per year). The 'passivation' of zinc depends on the formation of a chromate film which is protective. On sprayed zinc red-lead base paints should be avoided.

The uniformity of zinc coatings are indicated by the Preece Test (BS 729, 1961), consisting of 1-minute dips at 15°C in neutral copper sulphate solution, sp gr 1·170 (33 gm copper sulphate, 100 cc distilled water neutralised with copper hydroxide) until bright red copper adheres to specimen in places where the solution of the zinc has exposed the iron base. The number of dips is sometimes specified.

*Tin*

The use of tin coatings depends on the facts that:

(*a*) It is physiologically inactive.
(*b*) It is not corroded by foodstuffs in the *absence of oxidisers*.
(*c*) Iron is cathodic to tin under certain conditions.

Block tin is only moderately resistant to acid solutions in the presence of air and tin coatings on steel corrode rapidly. In the absence of air the coatings become resistant due to their high overvoltage; the hydrogen which develops on their surfaces increases the resistance to the flow of current and stops corrosion. If this hydrogen develops very rapidly by immersion in strong acids, evolution of the gas may occur and the polarisation effects no longer prevent corrosion. Tin coatings are frequently used on vessels carrying or processing milk. Even when strongly aerated, milk seems to have only a slight action on tin, whereas fruit juices are much more corrosive, especially when hot. Tin resists the attack by distilled water and also by the atmosphere to a high degree.

*Lead*

The resistance to corrosion of lead is determined by the character of the film which forms on the surface. A protective layer is formed in the neutral range of about $p_H$ 3–11, which is dissolved by certain acids and alkaline solutions. Lead is thus resistant to sulphuric, sulphurous, chromic, phosphoric acids and to the atmosphere; is attacked by hydrochloric and hydrofluoric acids; and strongly corroded by nitric, acetic and formic acids and nitrate solutions.

*Aluminium* is inherently a very reactive metal and its resistance to attack by certain media is due entirely to the formation of an adherent impervious film of aluminium oxide, which is formed by oxidising substances, but is dissolved by strong alkaline and certain strong acid solutions.

Aluminium coatings are finding extending use for resisting sulphurous atmospheres and also for covering the steel or duralumin parts of aeroplanes. On steel work in the open 150–215 g per m$^2$ of aluminium is recommended by BS. Aluminium–iron galvanic couples are found to behave in various ways depending on the nature of aluminium and the exposure conditions. Evans has shown that steel specimens sprayed with aluminium have rusted in hard tap-water at cracks in the coating, yet in sodium chloride solution the aluminium protected the steel (i.e. the necessary current density was obtained). In general, aluminium in contact with copper alloys, nickel, lead and chromium is rapidly attacked when exposed to sea-water, but accelerates the corrosion of magnesium. In sea-water aluminium alloys which contain copper and zinc have poor resistance to corrosion and pitting is common in both aluminium and its alloys.

Chromate additions reduce the corrosion of aluminium in refrigerating and air conditioning plants employing calcium chloride brine, and the metal has a good resistance to carbonates, chromates, acetates, nitrates and sulphates in the range $p_H$ 6·4–7·2.

In industrial atmospheres the rate of attack of aluminium is in the region of 0·5 mg per dm$^2$ per day and usually decreases with time of exposure. The attack by food products is usually 0–5 mg per dm$^2$ per day, although hot fruit juices frequently give higher figures. Concentrated nitric and acetic acids can be handled satisfactorily in pure aluminium vessels, but dilute nitric, sulphuric, phosphoric and hydrochloric acids are corrosive. The attack by alkalis, even in the form of soaps, is severe, and inhibitors such as sodium silicate, chromates and dichromates are used to protect laundry machines.

The resistance of aluminium to attack is vastly improved by the anodic treatment, a process for artificially increasing the thickness of the protective oxide film (p. 328).

## *Paints*

Oil paints consist of a vehicle (linseed oil), drier, and pigment. Since few coats are entirely watertight, the properties of the pigment should:

(1) Resist mechanical abrasion, rain and acids.
(2) Inhibit attack of the metal by creating passivity.

Zinc dust, chromates and red lead possess good inhibiting action and are useful constituents in the priming coat. Iron oxide resists abrasion but provides no protection of the steel at bare places. Attack of the steel is increased by:

(a) Salt shut in below a paint coat.
(b) Invisible moisture film present prior to painting.
(c) Graphite in the priming coat.
(d) Loose rust and partially removed mill scale.

Immersion in hot baths containing acid phosphates of manganese and zinc forms a phosphate film on iron which is a useful base for paint (Bonderizing).

## Oxidation and scaling

With the increasing use of alloys at high temperatures and the use of fuels containing impurities such as vanadium, sodium and sulphur the problem of scaling has become increasingly important. Oxidation is a diffusion process and usually continued oxidation proceeds by the diffusion of metal ions and electrons through the oxide layer.

The resistance to oxidation of any material at elevated temperatures is dependent on the nature of the oxide scale which is formed. If the scale is loose and porous then oxidation will continue and the scale will thicken until ultimately the complete section of metal will be oxidised. If, on the other hand, the oxide scale is adherent and non-porous, then the thin film first formed will act as a protection to the underlying metal.

The oxides formed on the surface of a metal may be of different kinds:

(1) If the oxide's dissociation pressure is less than the applied partial pressure of oxygen *no* oxide forms, e.g. gold, silver at 300–500 °C.
(2) Volatile oxides may reach an equilibrium thickness at which evaporation rate and growth rate are equal, e.g. Mo, Os, Ir. Below the melting point of the $MoO_2$–$MoO_3$ eutectic oxidation rate increases linearly with time; while above it rapid volatisation occurs associated with a sharp rise in oxidation rate which is thereafter relatively independent of temperature.
(3) Oxides with a smaller volume than the metal form a porous layer which allows continuous oxidation, e.g. sodium, calcium, magnesium.
(4) Oxides with a greater volume than the metal form a continuous film which tends to stifle the reaction according to a decreasing parabolic growth law.

As the scale becomes more voluminous cracking and spalling often occurs with consequent increase in rate of oxidation. When an oxide adheres tightly to a metal there is usually a well-defined atomic relationship between the oxide and the underlying metal grain. The scale is often multi-layered, containing different oxides, for example iron oxidised at 1000 °C has scale layers: Fe:95% FeO:4% $Fe_3O_3$:<1% $Fe_2O$ (Fig. 87).

Cobalt forms CoO, $Co_3O_4$ below 900 °C. Zirconium and titanium first form a dense black scale, but later a powdery white scale forms associated with a rapid increase in oxidation rate, i.e. a breakaway.

*Alloying elements* may have different effects such as

(a) an alloy element (e.g. Cr) may form a protective film in one metal (e.g. Fe) but fail to do so in another (e.g. Cu);

(b) traces may produce disproportionate improvements, e.g. 0·1% cerium in Ni–Cr resistance wire increases life ten times. This is probably due to an interlocking effect of its oxide and scale. Traces of beryllium and calcium in Magnox prevent burning;

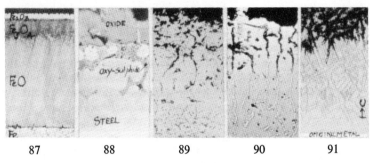

87  88  89  90  91

FIG. 87 Scale layers of iron in range 700–1250 °C
FIG. 88 Intercrystalline penetration of oxysulphide complex into surface of metal  (× 150)
FIG. 89 $V_2O_5$ attack of 60–20–20 Ni–Cr–Co alloy at 1000° C  (× 300)
FIG. 90 Preferential attack along the paths of carbides in cast Ni–Cr–Fe alloy  (× 100)
FIG. 91 Structure of Ni–Cr–Fe alloy attacked by Green rot, i.e. attack into carburised layer  (× 1000)

*(Figs. 89, 90 and 91 by courtesy of the Mond Nickel Co.)*

(c) an alloy element may dissolve in oxide phase of parent metal with little or no improvement;

(d) it may form a new stable oxide which does not dissolve parent metal atoms, e.g. Al, Be, Si in copper form oxides which prevent copper atom reaching the outside and minimise scaling;

(e) it may promote low melting point oxides.

Residual oils in gas turbines contain vanadium and sodium which form low melting compounds, $V_2O_5$ (660–690 °C) and sodium sulphate which flux the protective oxide film and give intercrystalline oxide penetration (Fig. 89). This often results in a catastrophic attack. The presence of sulphur is not very troublesome in the presence of oxygen (e.g. 4%), but

# OXIDATION AND SCALING 151

in oxygen-free gases sulphur can penetrate a scale, and at a sufficiently high temperature (900 °C for iron), it may concentrate at the oxide-metal interface as a molten oxysulphide complex (MPt 970 °C) often in intercrystalline form (Fig. 88).

(*f*) alloy may form carbides with changes in matrix. In Ni–Cr alloys precipitation of chromium carbides impoverishes the matrix in chromium which suffers oxidation mainly along the path of the carbide (Fig. 90).

'Green rot' is another form of attack of Ni–Cr alloys in a reducing-oxidising furnace atmosphere. Carburisation occurs first with precipitation of chromium carbide particles followed by oxidation through the chromium depleted matrix (Fig. 91).

*Internal oxidation* is the formation within the matrix of one metal of particles of a more stable oxide of a second metal that was originally in solid solution (e.g. Al or Si in Cu) by inward diffusion of oxygen from the surface. The nucleated oxide particles grow by diffusion of the solute to them. This is one means of producing dispersion hardening.

*Fretting Corrosion* is the damage caused when two closely fitting metal surfaces are subjected to vibration. The initiation of the fretting process consists of 'molecular plucking' of the surface particles which oxidise to form a debris which is like red rust in the case of steel. The oxide debris can subsequently act as an abrasive, causing serious damage, especially fatigue, ruins dimensional accuracy and jams clearances.

Fretting corrosion is usually reduced at high humidities and increased in the presence of oxygen. Fretting can be reduced by using a bonded coating of $MoS_2$ reducing vibration and load and using surfaces as dissimilar as possible in respect of finish and hardness.

# 9  Irons and Carbon Steels

It is difficult to give a brief outline of the processes of making the various classes of irons and steels but the following generalisations, limited as they are, serve to show how certain characteristic properties of the materials are related to the particular mode of manufacture.

*Pig iron* is a material of great importance in the foundry and in many steel-making processes. The raw materials used in its manufacture are:

*Ore*, consisting of iron oxide or carbonate associated with earthy impurities.

*Limestone*, a flux to combine with the non-metallic portion of the ore in order to form a fluid slag, mainly calcium silicate.

*Coke*, to reduce the iron oxides to iron and to provide heat for melting the metal and slag.

These materials are charged into a tall shaft furnace, sealed at the top by a bell and hopper arrangement. Hot air is blown in through tuyères near the bottom (Fig. 92). The molten iron accumulates in the hearth with the slag floating on top of it, both being tapped periodically. The molten iron is transferred to the steel furnaces or cast into sand or chill moulds to give 'pigs'.

While in the blast furnace the iron absorbs carbon, silicon, manganese and sulphur, and the whole of the phosphorus in the charge. The iron produced is classed according to this phosphorus content, the low phosphorus (0·03%) iron being made from red hematite ore. Each class of pig iron can be graded according to the fracture, grey, mottled and white, which, for a constant rate of cooling, is largely dependent on the silicon content (see Chap. 14). Production has been increased by building larger blast furnaces, improved burden preparation, increasing use of self-fluxing sinters and higher blast temperatures. There is also a tendency to inject alternative cheaper fuels than coke through the tuyères.

*Iron castings*

The crude iron as tapped from the blast furnace is rarely suitable for making castings without remelting, consequently a mixture of pig irons, foundry scrap, steel scrap, limestone and coke is melted in a cupola, tapped and run into sand moulds to give cast iron. The improved structures and properties of these castings are discussed in Chap. 14.

## Wrought iron

In this country wrought iron is made by the old puddling process, which consists in melting grey pig iron and millscale in a small coal-fixed reverberatory furnace, the hearth being lined with iron oxides. The impurities in the pig iron react with the iron oxide to form a slag, largely iron silicate. The removal of silicon, manganese, phosphorus, and finally carbon, causes

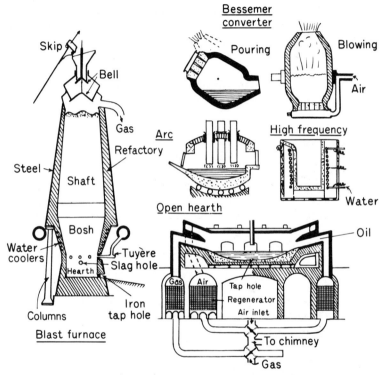

FIG. 92 Furnaces used for making pig iron and steels. RH side of open hearth furnace shows use of oil instead of gas

the freezing-point of the metal in the furnace to rise, until it is actually higher than the furnace temperature, and hence the metal solidifies into a pasty mass of metal closely intermixed with considerable quantities of slag. The pasty balls are withdrawn, hammered and rolled into crude bars. These are cut into lengths, 'piled' together, heated to a welding temperature and rolled into suitable sections. This rolling serves to weld the layers together, to elongate the slag into fibres and so promotes greater uniformity throughout the mass. The structure is shown in Fig. 106.

The *characteristic properties* of wrought iron are due to the presence of slag fibres in a soft matrix. A crack is diverted along the weaker slag fibres, producing a fracture as in Fig. 54, which serves as a warning of possible failure. In forge-welding the slag acts as a flux and also tends to prevent grain growth with consequent embrittlement. Machinability is good, and the Staffordshire iron resists atmospheric corrosion about 20% better than mild steel, but the purer Swedish iron does not show this improvement.

Wrought iron also possesses the property of recovering rapidly from overstrain, which enables it to accommodate sudden and excessive shocks without permanent injury. On the other hand, wrought-iron chains are apt to deteriorate in use owing to the effect of work hardening of the outer skin. Sudden failures can be prevented by annealing the metal.

Wrought iron is used for chains, anchors, railway couplings and crane hooks. Various qualities are available of varying analysis and properties, as Table 11 shows. To cheapen the iron, pieces of mild steel, high in man-

TABLE 11  COMPOSITION AND PROPERTIES OF IRONS

| % Analysis | Pig Iron | Swedish,* Lancashire Wrought Iron | Staffordshire Wrought Iron | Ingot Iron |
|---|---|---|---|---|
| Carbon | 3·5 | 0·03 | 0·02 | 0·03 |
| Silicon | 1·9 | 0·02 | 0·12 | 0·03 |
| Sulphur | 0·06 | 0·008 | 0·018 | 0·038 |
| Phosphorus | 1·0 | 0·052 | 0·228 | 0·009 |
| Manganese | 0·7 | nil | 0·02 | — |
| Slag | | 0·07 to 1·0 | | |
| Yield point N/mm² | — | 213 | 230 | 178 |
| TS | 124 | 355 | 355 | 309 |
| El % | — | 38 | 25 | 43 |

*Made from low sulphur and silicon iron

ganese, are sometimes introduced in the 'piles', producing an inferior 'fagotted' iron. High phosphorus content in the iron renders it cold short and liable to excessive grain-growth, which cannot be remedied by simple annealing. Some commercial irons have a brittle transition temperature near room temperature (p. 15) and are very sensitive to the speed of testing. In the American Aston process molten Bessemer iron (p. 156) is poured into a synthetic slag to form pasty balls, which are rolled into bars.

## Other irons

Other commercial forms of pure iron are ingot iron (Armco) and electrolytic iron. The latter is exceptionally pure except for occluded hydrogen which embrittles it. Annealing removes this brittleness and the material is used for transformer cores.

The ingot iron is made in the basic open-hearth furnace, in which oxidation is carried so far as to remove practically all the impurities. Excess iron oxide remains in the metal. It differs from wrought iron in that the metal is produced in a molten state, separated from the slag, and cast into ingots.

**Steel-making processes**

Steel is made by the Bessemer, Siemens Open Hearth, basic oxygen furnace, electric arc, electric high-frequency and crucible processes.

*Crucible and high-frequency methods*

The Huntsman crucible process has been superseded by the high frequency induction furnace in which the heat is generated in the metal itself by eddy currents induced by a magnetic field set up by an alternating current, which passes round water-cooled coils surrounding the crucible (Fig. 92). The eddy currents increase with the square of the frequency, and an input current which alternates from 500 to 2000 hertz is necessary. As the frequency increases, the eddy currents tend to travel nearer and nearer the surface of a charge (i.e. shallow penetration). The heat developed in the charge depends on the cross-sectional area which carries current, and large furnaces use frequencies low enough to get adequate current penetration. Automatic circulation of the melt in a vertical direction, due to eddy currents, promotes uniformity of analysis. Contamination by furnace gases is obviated and charges from $\frac{1}{4}$ to 5 tonnes can be melted with resultant economy. Consequently, these electric furnaces are being used to produce high quality steels, such as ball bearing, stainless, magnet, die and tool steels.

*Acid and basic steels*

The remaining methods for making steel do so by removing impurities from pig iron or a mixture of pig iron and steel scrap. The impurities removed, however, depend on whether an acid (siliceous) or basic (limey) slag is used. An acid slag necessitates the use of an acid furnace lining (silica); a basic slag, a basic lining of magnesite or dolomite, with line in the charge.

With an acid slag silicon, manganese and carbon only are removed by oxidation, consequently the raw material must not contain phosphorus and sulphur in amounts exceeding those permissible in the finished steel.

In the basic processes, silicon, manganese, carbon, phosphorus and sulphur can be removed from the charge, but normally the raw material contains low silicon and high phosphorus contents. To remove the phosphorus the bath of metal must be oxidised to a greater extent than in the corresponding acid process, and the final quality of the steel depends very largely on the degree of this oxidation, before *deoxidisers*—ferro-manganese, ferro-silicon, aluminium—remove the soluble iron oxide and form other insoluble oxides, which produce non-metallic inclusions if they are not removed from the melt:

$$2Al + 3FeO \text{ (soluble)} \rightleftharpoons 3Fe + Al_2O_3 \text{ (solid)}$$

In the acid processes, deoxidation can take place in the furnaces, leaving a reasonable time for the inclusions to rise into the slag and so be removed before casting. Whereas in the basic furnaces, deoxidation is rarely carried out in the presence of the slag, otherwise phosphorus would return to the metal. Deoxidation of the metal frequently takes place in the ladle, leaving only a short time for the deoxidation products to be removed. For these reasons acid steel is considered better than basic for certain purposes, such as large forging ingots and ball bearing steel. The introduction of vacuum degassing hastened the decline of the acid processes.

*Bessemer steel*

In both the Acid Bessemer and Basic Bessemer (or Thomas) processes molten pig iron is refined by blowing *air* through it in an egg-shaped vessel, known as a converter, of 15–25 tonnes capacity (Fig. 92). The oxidation of the impurities raises the charge to a suitable temperature, which is therefore dependent on the composition of the raw material for its heat: 2% silicon in the acid and 1·5–2% phosphorus in the basic process is normally necessary to supply the heat. The 'blowing' of the charge, which causes an intense flame at the mouth of the converter, takes about 25 minutes and such a short interval makes exact control of the process a little difficult.

The *Acid Bessemer* suffered a decline in favour of the Acid Open Hearth steel process, mainly due to economic factors which in turn has been ousted by the basic electric arc furnace coupled with vacuum degassing.

The *Basic Bessemer* process is used a great deal on the Continent for making, from a very suitable pig iron, a cheap class of steel, e.g. ship plates, structural sections.

For making steel castings a modification known as a Tropenas converter is used, in which the air impinges on the surface of the metal from side tuyères instead of from the bottom. The raw material is usually melted in a cupola and weighed amounts charged into the converter.

*Open-hearth processes*

In the Siemens process, both acid and basic, the necessary heat for melting and working the charge is supplied by oil or gas. But the gas and air are preheated by regenerators, two on each side of the furnace, alternatively heated by the waste gases. The regenerators are chambers filled with checker brickwork, brick and space alternating. The furnaces have a saucer-like hearth, with a capacity which varies from 600 tonnes for fixed, to 200 tonnes for tilting furnaces (Fig. 92). The raw materials consist essentially of pig iron (cold or molten) and scrap, together with lime in the basic process. To promote the oxidation of the impurities iron ore is charged into the melt although increasing use is being made of oxygen lancing. The time for working a charge varies from about 6 to 14 hours, and control is therefore much easier than in the case of the Bessemer process.

The Basic Open Hearth process was used for the bulk of the cheaper grades of steel, but there is a growing tendency to replace the OH furnace by large arc furnaces using a single slag process especially for melting scrap and coupled with vacuum degassing in some cases.

*Electric arc process*

The heat required in this process is generated by electric arcs struck between carbon electrodes and the metal bath (Fig. 92). Usually, a charge of graded steel scrap is melted under an oxidising basic slag to remove the phosphorus. The impure slag is removed by tilting the furnace. A second limey slag is used to remove sulphur and to deoxidise the metal in the furnace. This results in a high degree of purification and high quality steel can be made, so long as gas absorption due to excessively high temperatures is avoided. This process is used extensively for making highly alloyed steel such as stainless, heat-resisting and high-speed steels. Oxygen lancing is often used for removing carbon in the presence of chromium and enables scrap stainless steel to be used.

The nitrogen content of steels made by the Bessemer and electric arc processes is about 0·01–0·25% compared with about 0·002–0·008% in open hearth steels.

*Oxygen processes*

The high nitrogen content of Bessemer steel is a disadvantage for certain cold forming applications and continental works have, in recent years, developed modified processes in which oxygen replaces air. In Austria the *L/D process* (Linz–Donawitz) converts low phosphorus pig iron into steel by top blowing with an oxygen lance using a basic lined vessel (Fig. 93(b)). To avoid excessive heat scrap or ore is added. High quality steel is produced with low hydrogen and nitrogen (0·002%). A further modification of the process is to add lime powder to the oxygen jet (*OLP process*) when higher phosphorus pig is used.

The *Kaldo* (Swedish) process uses top blowing with oxygen together with a basic lined rotating (30 rev/min) furnace to get efficient mixing (Fig. 93(a)). The use of oxygen allows the simultaneous removal of carbon and phosphorus from the (P, 1·85%) pig iron. Lime and ore are added. The German *Rotor process* uses a rotary furnace with two oxygen nozzles, one in the metal and one above it (Fig. 93(c)).

The use of oxygen with steam (to reduce the temperature) in the traditional basic Bessemer process is also now widely used to produce low nitrogen steel. These new techniques produce steel with low percentages of N, S, P, which are quite competitive with open hearth quality.

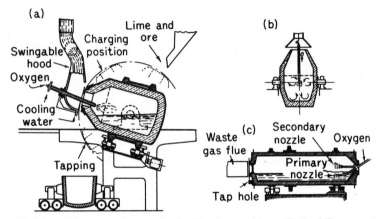

FIG. 93 (a) Kaldo rotary furnace in the blowing position; (b) the LD converter; (c) the principle of the Rotor furnace

Other processes which are developing are the Fuel-oxygen-scrap, FOS process, and spray steelmaking which consists in pouring iron through a ring, the periphery of which is provided with jets through which oxygen and fluxes are blown in such a way as to 'atomise' the iron, the large surface to mass ratio provided in this way giving extremely rapid chemical refining and conversion to steel.

*Vacuum degassing* is also gaining ground for special alloys Some 14 processes can be grouped as stream, ladle, mould and circulation (e.g. DH and RH) degassing methods, Fig. 94. The vacuum largely removes hydrogen, atmospheric and volatile impurities (Sn, Cu, Pb, Sb), reduces metal oxides by the $C - O$ reaction and eliminates the oxides from normal deoxidisers and allows control of alloy composition to close limits. The clean metal produced is of a consistent high quality, with good properties in the transverse direction of rolled products. Bearing steels have greatly improved fatigue life and stainless steels can be made to lower carbon contents.

## STEEL-MAKING PROCESSES

*Vacuum melting and ESR.* The aircraft designer has continually called for new alloy steels of greater uniformity and reproducibility of properties with lower oxygen and sulphur contents. Complex alloy steels have a greater tendency to macro-segregation, and considerable difficulty exists in minimising the non-metallic inclusions and in accurately controlling the analysis of reactive elements such as Ti, Al, B. This problem led to the use of three processes of melting.

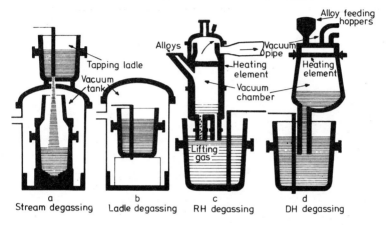

FIG. 94 Methods of degassing molten steel (*after C. Holden*)

(a) *Vacuum induction melting* within a tank for producing super alloys (Ni and Co base), in some cases for further remelting for investment casting. Pure materials are used and volatile tramp elements can be removed.

(b) *Consumable electrode vacuum arc* re-melting process (Fig. 95) originally used for titanium, was found to eliminate hydrogen, the Λ and V segregates and also the large silicate inclusions. This is due to the mode of solidification. The moving parts in aircraft engines are made by this process, due to the need for high strength cleanness, uniformity of properties, toughness and freedom from hydrogen and tramp elements.

(c) *Electroslag refining (ESR)* This process, which is a larger form of the original welding process, re-melts a preformed electrode of alloy into a water-cooled crucible, utilising the electrical resistance heating in a molten slag pool for the heat source (Fig. 96). The layer of slag around the ingot maintains vertical unidirectional freezing from the base. Tramp elements are not removed and lead may be picked up from the slag.

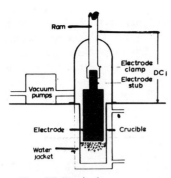
Fig. 95 Typical vacuum arc remelting furnace

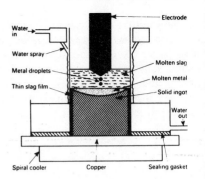
Fig. 96 Electroslag remelting furnace

## Structure of plain steels

The essential difference between ordinary steel and pure iron is the amount of carbon in the former, which reduces the ductility but increases the strength and the susceptibility to hardening when rapidly cooled from elevated temperatures. On account of the various micro-structures which may be obtained by different heat-treatments, it is necessary to emphasise the fact that the following structures are for 'normal' steels, i.e. slowly cooled from 760–900 °C depending on the carbon contents.

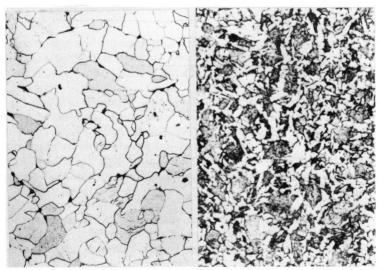

Fig. 97 Armco iron: ferrite grains ($\times$ 200)

Fig. 98 0·4% carbon steel. Ferrite and pearlite ($\times$ 200)

# STRUCTURE OF PLAIN STEELS

The appearance of pure iron is illustrated in Fig. 97. It is only pure in the sense that it contains no carbon, but contains very small quantities of impurities such as phosphorus, silicon, manganese, oxygen, nitrogen, dissolved in the solid metal. In other words, the structure is typical of pure metals and solid solutions in the annealed condition. It is built up of a number of crystals of the same composition, given the name *ferrite* in metallography (Brinell hardness 80).

The addition of carbon to the pure iron results in a considerable difference in the structure (Fig. 98), which now consists of two constituents, the white one being the ferrite, and the dark parts representing the constituent containing the carbon, the amount of which is therefore an index of the quantity of carbon in the steel. Carbon is present as a compound of iron

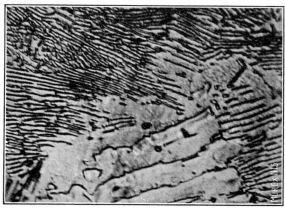

FIG. 99  0·87% carbon steel  (× 1000)
Pearlite resolved into ferrite and cementite bands

and carbon (6·67%) called *cementite*, having the chemical formula $Fe_3C$. This cementite is hard (Brinell hardness 600 +), brittle and brilliantly white. On examination the dark parts will be seen to consist of two components occurring as wavy or parallel plates alternately dark and light (Fig. 99). These two phases are ferrite and cementite which form a eutectoid mixture, containing 0·87% carbon and known as Pearlite. The appearance of this pearlite depends largely upon the objective employed in the examination (p. 55), and also on the rate of cooling from the elevated temperature.

Since the pearlite contains 0·87% carbon, it is evident from Fig. 100 that a steel with 0·17% carbon contains about 80% ferrite and 20% pearlite; while 0·5% and 0·87% carbon steels contain 60 and 100% pearlite respectively. Any further increase in carbon gives rise to free cementite at the grain boundaries or as needles. The highest strength is obtained when the structure consists entirely of pearlite. The presence of free cementite masses still increases the hardness but reduces the strength.

*Allotropy of iron*

Certain substances can exist in two or more crystalline forms; for example charcoal, graphite and diamonds are allotropic modifications of carbon. Allotropy is characterised by a change in atomic structure which occurs at a definite transformation temperature.

Four changes occur in iron, which give rise to forms known as alpha, beta, gamma and delta. Of these, α β and δ forms have the same atomic structure (body centred cubic) while γ-iron has a face centred cubic structure. Iron can, therefore, be considered to have *two* allotropic modifications.

Details of the changes are:

TABLE 12

| Change | Temperature, °C | | Name | Structural Change |
|---|---|---|---|---|
| | Heating | Cooling | | |
| α–β | 769 | 769 | $A_2$ | intra-atomic |
| β–γ | 937 | 910 | $A_3$ | b.c.c.–f.c.c. |
| γ–δ | 1400 | 1399 | $A_4$ | f.c.c.–b.c.c. |
| Austenite–pearlite | 730 | 695 | $A_1$ | |

The $A_2$ change at 769 °C, at which the α-iron loses its magnetism, can be ignored from a heat-treatment point of view. These changes in structure are accompanied by thermal changes, together with discontinuities in other physical properties such as electrical, thermo-electric potential, magnetic, *expansion* and tenacity. The $A_3$ change from a b.c.c. to an f.c.c. atomic structure at 937 °C is accompanied by a marked contraction while the reverse occurs at 1400 °C. These changes in structure are accompanied by recrystallisation, followed by grain growth.

*Critical points*

The addition of carbon to iron, however, produces another change at 695 °C, known as $A_1$, and associated with the formation of pearlite.

These structural changes, which occur during cooling, give rise to evolutions of heat, which cause *arrests* on a cooling curve. The temperatures of these arrests are known as *critical points* or 'A' points. These arrests occur at slightly higher temperatures on heating, as compared with cooling, and this lag effect, increased by rapid cooling, is known as thermal hysteresis. To differentiate between the arrests obtained during heating and cooling, the letters *c* and *r* respectively are added to the symbol A (from *c*hauffage and *r*efroidissement). In a steel containing about 0·8–0·9% carbon the

# STRUCTURE OF PLAIN STEELS

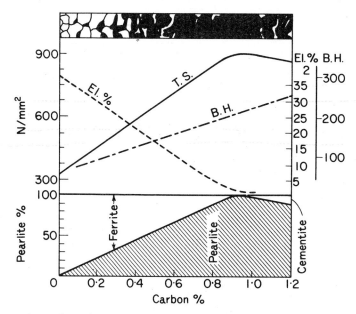

FIG. 100 Effect of carbon content on mechanical properties and microconstituents of iron

evolution of the heat at $Ar_1$ is sufficient to cause the material to become visibly hotter and the phenomenon is called 'recalescence'.

*Iron-cementite equilibrium diagram*

The addition of carbon to iron not only gives rise to the $A_1$ point but also influences the critical points in pure iron. The $A_4$ point is raised; and the $A_3$ point lowered until it coincides with $A_1$. The $\alpha$, $\beta$ and $\delta$ modifications, which may be called *ferrite*, have only slight solubility for carbon, but up to 1·7% of carbon dissolves in $\gamma$-iron to form a solid solution called *Austenite*. These effects are summarised in the iron–$Fe_3C$ equilibrium diagram (Fig. 101), which is of much importance in the study of steels. The iron–iron carbide system is not in true equilibrium, the stable system is iron–graphite, but special conditions are necessary to nucleate graphite.

From what has been discussed in Chap. 5 it will be seen that the complicated Fe—$Fe_3C$ diagram can be divided into several simple diagrams:

Peritectic transformation    CDB  — $\delta$-iron transforms to austenite.
Eutectic at E                             — austenite and cementite.
Solid solution D to F            — primary dendrites of austenite form.
Eutectoid point at               P — formation of pearlite.

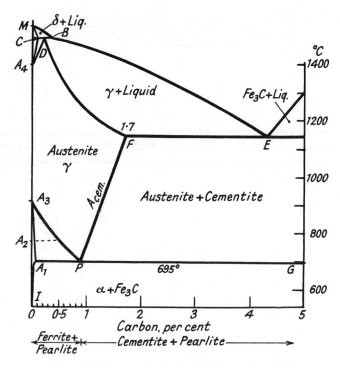

FIG. 101 Iron-cementite equilibrium diagram

*The ferrite solubility line*, $A_3P$, denotes the commencement of precipitation of ferrite from austenite.

*The cementite solubility line*, FP, indicates the primary deposition of cementite from austenite.

*The pearlite line*, $A_1PG$, indicates the formation of the eutectoid at a constant temperature.

Let us consider the freezing of alloys of various carbon contents.

*0·3% carbon*

Dendrites of δ-iron form, the composition of which is represented eventually by C (0·07%), and the liquid, enriched in carbon, by B. The solid crystals then react with the liquid to form austenite of composition D. Diffusion of carbon occurs as the solid alloy cools to line $A_3P$. Here α-ferrite commences to be ejected from the austenite, consequently the remaining solid solution is enriched in carbon, until point P is reached at which cementite can be also precipitated. The alternate formation of ferrite and cementite at 695 °C gives rise to pearlite. The structure finally consists of masses of pearlite embedded in the ferrite.

## 0·6% carbon

When line BE is reached dendrites of austenite form, and finally the alloy completely freezes as a cored solid solution, which, on cooling through the critical range (750–695 °C), decomposes into ferrite and pearlite.

## 1·4% carbon

Again the alloy solidifies as a cored solid solution, but on reaching line FP, cementite starts to be ejected and the residual alloy becomes increasingly poorer in carbon until point P is reached, when both cementite and ferrite form in juxtaposition. The structure now consists of free cementite and pearlite.

## Cast steel

The equilibrium diagram does not, however, tell us what form is taken by the ferrite or cementite ejected from the austenite on cooling.

Without going too deeply into the matter, it may be considered that the ferrite has a choice of three different positions, which, in order of degree of supercooling or ease of forming nuclei, are:

(1) boundaries of the austenite crystals (Fig. 102);
(2) certain crystal planes (octahedral) (Fig. 103);
(3) about inclusions as shown in (Fig. 104).

Thus, ferrite starts to form at the grain boundaries, and if sufficient time is allowed for the diffusion phenomena a ferrite *network structure* is formed, while the pearlite occupies the centre, as in Fig. 102. The size of the austenite grains existing above $A_3$ is thereby betrayed.

If the rate of cooling is faster, the complete separation of the ferrite at the boundaries of large austenite grains is not possible, and ejection takes place within the crystal along certain planes (octahedral, Fig. 30), forming a mesh-like arrangement known as a *Widmanstätten Structure*, shown in Fig. 103. The appearance will vary with the angle of cut through the grain, as shown in Fig. 105.

In steels containing more than 0·9% carbon, cementite can separate in a similar way and Widmanstätten structures are also found in other alloy systems (p. 103).

*Steel with Widmanstätten structures* are characterised by (1) low impact value, (2) low percentage elongation since the strong pearlite is isolated in ineffective patches by either weak ferrite or brittle cementite, along which cracks can be readily propagated. This structure is found in overheated steels and cast steel, but the high silicon used in steel castings modifies it into a 'feathery' structure, shown in Fig. 114.

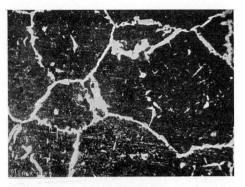

FIG. 102  0·5% carbon steel, slow cooled. Shows network of ferrite  (× 100)

FIG. 103  0·5% carbon steel, slow cooled from 1450°C. Shows Widmanstätten structure—ferrite precipitated at boundaries and on planes of austenite grain (× 200)

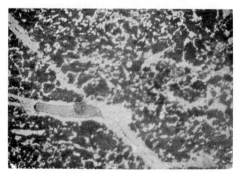

FIG. 104  0·5% carbon steel. Shows ferrite formed round inclusion of manganese sulphide, even after annealing  (× 200)

It is highly desirable that Widmanstätten and coarse network structures generally be avoided, and as these partly depend upon the size of the original austenite grain, the methods of securing small grains are of importance.

Large austenite grains may be refined by (*a*) hot working, (*b*) normalising (p. 177).

Such refined austenite grains are liable to coarsen when the steel is heated well above the $Ac_3$ temperature, in such operations as welding, forging and carburising unless the grain growth is restrained. This restraint can be brought about by a suitable mode of manufacture of the steel.

FIG. 105 Planes revealed by different sections through cube and equivalent micro-structure: ferrite, white

*Controlled grain size*

It is now possible to produce two steels of practically identical analysis with inherently different grain growth characteristics, so that at a given temperature each steel has an 'inherent austenite grain size', one being fine relative to the other. The so-called 'fine-grained' steel increases its size on heating above $Ac_3$, but the temperature at which the grain size becomes relatively coarse is definitely higher than that at which a 'coarse-grained' steel develops a similar size.

The fine-grained steels are 'killed' with silicon together with a slight excess of aluminium which forms aluminium nitride as submicroscopic particles that obstruct austenite grain growth and is an example of a general phenomenon (p. 125). At the coarsening temperature the AlN goes into solution rapidly above 1200 °C and virtually completely at 1350 °C. The austenite grain size is frequently estimated by the following tests:

(1) *McQuaid–Ehn Test.* Micro-sections of structural steels carburised for not less than 8 hours at 925 °C and slowly cooled to show cementite networks are photographed at a magnification of 100. Comparison is made with a grain-size chart issued by the American Society for Testing Materials (E 19·46).* This test is also valuable in detecting 'abnormality' of pearlite (p. 208).

(2) *The Quench and Fracture test* consists in heating normalised sections of the steel, above $Ac_3$, quenching them at intervals of 30 °C. An examination of the fractured surface enables the depth of hardness and grain size to be estimated by comparison with standard fractures.

*Properties of fine and coarse-grained steels*

Fine-grained steel (3–16 grains per cm$^2$) has the following characteristics compared with a coarse grain (0·2–2 grains per cm$^2$):

(*a*) Enhanced notch-bar toughness in plain steels, but increase is less in alloy steels.
(*b*) Shallower hardening; less tendency to warp and crack.
(*c*) Carburises less deeply; absence of grain growth in core permits single quench from the carburising temperature.
(*d*) Less tendency for cracks to originate in flash trimming of forgings.
(*e*) Mass production machining is more difficult.
(*f*) Deterioration by quench and strain ageing is reduced. Greater standardisation can therefore be obtained by using a steel with an inherent grain size suitable for a particular purpose.

## Impurities in steel

As a result of deoxidation, and the incomplete removal of impurities, the molten steel cast into a mould has (*a*) dissolved impurities (Mn, Si, P and gases); (*b*) suspended non-metallic inclusions. When solid we get, therefore, a cored solid solution with segregations of inclusions as discussed in Chap. 4. On cooling through the transformation range the original austenite crystals are changed into ferrite and pearlite grains, but the non-metallic inclusions and also soluble impurities which diffuse very slowly, such as phosphorus, persist in their original positions. Thus the primary structure can be revealed by etching reagents which attack these impurities (Fig. 113). After rolling the steel exhibits a fibrous structure.

It should be noted that the amount of an element reported in an analysis

*ASTM number $n$ is derived from: Number of grains per in.$^2$ $\times$ 100 = $2^{n-1}$.

# IMPURITIES IN STEEL

can be misleading as to its effect on the properties of a steel. This is because many elements exist in different forms, with characteristic effects, e.g. manganese can dissolve in ferrite (solid solution strengthen) and form $Mn_3C$, MnS, MnO and manganese silicates.

*Manganese* confers depth of hardening, but also a particular liability to crack in quenching, for which reason high carbon steels, intended to be quenched in water, should contain less than 0·5%. Manganese raises the yield point and Izod values.

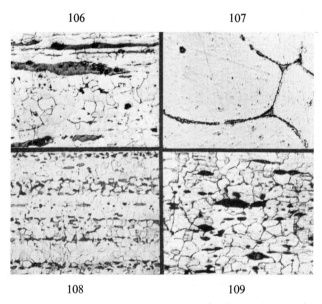

FIG. 106 Longitudinal sections of wrought iron, ferrite + slag streaks. Etched nitric acid ($\times$ 80)
FIG. 107 FeS films in iron ($\times$ 200)
FIG. 108 Ghost bands in mild steel, etched ($\times$ 80)
FIG. 109 Free-cutting steel: MnS streaks in ferrite, etched ($\times$ 80)

*Silicon* is present in most steels and is beneficial. The amount is usually less than 0·2% especially in tool steels owing to the tendency for graphite to form.

*Sulphur* exists as either manganese sulphide (MnS) or ferrous sulphide (FeS). The ferrous sulphide forms brittle, low melting point yellowish-brown films round the solid steel crystals and causes the metal to split when forged (Fig. 107).

The high melting point dove-grey manganese sulphide is only slightly soluble in iron and collects into large globules irregularly distributed through the steel. It is plastic at high temperatures, being elongated into threads by rolling without seriously impairing the properties of the material (Figs. 108 and 109). The manganese should be about five times that theoretically required to combine with sulphur present. Modern free cutting steels use 0·2% sulphur and 1·5% manganese with low phosphorus (Fig. 109). But in tool steels the sulphur should not exceed 0·035%, except where free machining is vital.

Higher sulphur levels such as MnS seem to improve fatigue life possibly by envelopment of oxides and also raise overheating temperatures. The form of MnS is affected by oxygen content of the steel. The form of the MnS at the freezing point of the ingot changes from liquid globules to eutectic networks to solid angular particles as the oxygen is reduced by Si and Al additions.

*Phosphorus*

Both ferrous sulphide and phosphorus have a powerful tendency to segregate, as would be expected from the wide separation of the liquidus and solidus in the Fe—S, Fe—P equilibrium diagrams. Hence in steels with an average phosphorus content of 0·05% a few areas many contain 0·1%, which then become dangerous. Every endeavour is made to keep the phosphorus to 0·02–0·05% or render it beneficial (p. 228).

Phosphorus forms a compound $Fe_3P$ which dissolves in iron up to 1·7%, but in the presence of 3·5% carbon its solubility is reduced to 0·3% and the excess forms the brittle eutectic found in cast iron.

In rolled steel the areas containing high phosphorus are elongated into bands which are characterised by the absence of pearlite. Such light streaks are termed *ghost bands* and are weak owing to the presence of phosphorus and the absence of pearlite. Ghosts are due to local segregation of impurities during the solidification of the ingot; consequently inclusions are frequently present in the bands (Fig. 108) and dissolved oxygen also causes similar bands. Although the phosphorus must be strictly controlled, for many applications it is quite illogical to assume that very low phosphorus contents always confer desirable qualities on steels. Very low percentages in basic steels are associated with increased amounts of non-metallic inclusions, due to the high degree of oxidation of the bath.

Positive advantages accrue from the presence of phosphorus in certain cases; with 0·08% the tendency of tin plates to stick together is prevented and the fire welding of spades is improved, while with 0·07% in the presence of 0·3% copper, increased resistance to atmospheric corrosion is obtained.

*Effects of dissolved carbon and nitrogen*

After cold rolling the properties of mild steel change with time and conditions of temperature during storage. The phenomenon is known as *strain-ageing*, the cause of which seems to be associated with dissolved carbon and nitrogen. The Izod value is particularly affected as shown by Table 13, which also indicates that a steel of 'fine-grained' characteristics is comparatively free from this trouble. Toughness can be restored by heating to 500–650 °C.

TABLE 13   EFFECT OF STRAIN-AGEING ON IZOD VALUES

| Grain Size | Treatment of 0·095% Carbon Steel | Izod, Impact, Joules Average | % Decrease |
|---|---|---|---|
| Coarse | Normalised 920 °C | 125 | — |
|  | Normalised 920 °C Strained 15% | 48 | 61 |
|  | Normalised 920 °C Strained ht 250 °C ½ hour | 8 | 93 |
| Fine | Normalised 920 °C | 134 | — |
|  | Normalised 920 °C Strained 15% | 115 | 13 |
|  | Normalised 920 °C Strained ht 250 °C ½ hour | 109 | 17 |

*Quench-ageing* relates to the hardening, usually accompanied by a reduction in toughness, which develops on resting at 0–200 °C after quenching. It appears to be due to 'precipitation hardening' (see p. 96) and is most pronounced in a mild steel with a carbon content of 0·04% quenched from 700 °C, corresponding to point $A_1$ in Fig. 101. Toughness can be restored by reheating the steel to 300 °C.

*Packaging and cold forming steels*

These steels are produced in the form of thin sheet or strip and fall into two groups:
(i) Steel where cold formability is the most important property, e.g. *Extra deep-drawing* (EDD) steel for car bodies.

The formability requirements of different cold-pressing operations differ widely. Failure in simple bending occurs by fracture, starting at the surface of maximum tensile strain, but the operation may also be limited by springback. In deep-drawing, failure occurs when the load required to draw-in the outer part of the pressing is sufficient to cause fracture in the central area which is stretched over the punch. Finally, in stretching operations performance is limited by tensile instability (necking) or by fracture.

The material properties, of primary importance in cold-forming, are the elastic modulus, work-hardening, fracture behaviour and flow strength, particularly in the through-thickness direction of the sheet, as given by $\bar{r}$ values (see p. 118).

The principal steels are rimmed and Al-stabilised steels produced by hot rolling, coiling, cold rolling, followed by batch annealing in coil form, and final temper rolling. Strain ageing after pressing can increase final strength and rigidity.

For high r̄ values in Al-stabilised steels the aluminium must be closely controlled (0·02–0·05%) and a high coiling temperature avoided because AlN is then precipitated at too early a stage for it to influence texture development. For cold forming applications the yield stress should be low and therefore a medium grain size (6–8) is required and which can only be obtained by prolonged batch annealing. To reduce quench ageing, decarburisation and denitriding are carried out in *open coil annealing* (i.e. recoiled loosely) so that surfaces are exposed to the gas atmosphere. Such steels are very soft and find extensive use for enamelling, galvanising and electrical steels.

(ii) *High strength packaging steel*, e.g. Tin plate 0·1 C, 0·3–0·4 Mn. The conditions for EDD can be relaxed and continuous annealing at speeds up to 700 M/min can be used which give little chance for grain growth (10·5–12) and therefore produce yield strength of 400 N/mm$^2$, double that of EDD steel. Steel *quenched* from above Acl to form lath martensite can give TS up to 900 N/mm$^2$, with less directionality than *double reduced* steel formed by 30–50% reduction after annealing, i.e. work hardened.

## Typical compositions and applications of carbon steels

The sulphur and phosphorus are kept below 0·05% unless otherwise stated.

*Rimming steels* (see p. 66)

1. *Carbon less than 0·07%* The lowest carbon steels are used for wire and rails for electrical conductors.

2. *Carbon 0·07–0·15%* These steels constitute a large bulk of the steel made and their applications include:

   (*a*) Rod and wire for nails, rivets, fencing, binding, cable armouring, ferro-concrete bars and mattress wire.
   (*b*) Hot rolled strip and bars for general purposes.
   (*c*) Cold rolled and annealed strip for pressings, car bodies, tin plate.
   (*d*) Tubes.

*Balanced steels*

   (*a*) *Carbon 0·06–0·1%*, Si 0·01% used for bright drawing and machining types of rod.
   (*b*) *Carbon 0·25–0·3* (43A BS 4360) general structural steel TS = 432–510 N/mm$^2$.

*Killed steels*

3. *Dead mild steel* (C, 0·07–0·15; Si, 0·1; Mn, 0·5%) will withstand a remarkable amount of cold work such as flanging, pressing and is used for solid-drawn tubes.

4. *Mild steel* (C, 0·10–0·25%; Mn, 0·6/1%) is used for:

   (a) *Drop forging* and stamping purposes.
   (b) *Section steel* for joists, channels and angles, known as 28/32 tensile.
   (c) *Ship and boiler* plate to requirements of Lloyd's.
   (d) *Case hardening*, in which case the manganese may be increased up to 1·2%, especially if the percentage of sulphur is high.
   (e) *Free-cutting*. The tendency of mild steels to tear when being machined is reduced by either quenching or cold drawing, or adding sulphide. The addition of lead (0·18%) improves machinability (e.g. Ledloy steels) without ill effects.

5. *Medium carbon steel* used for:

   (a) Forgings (C, 0·2–0·5) for general engineering purposes, boiler drums, drop forgings, and in particular for agricultural tools, such as hoes, spades, forks.
   (b) (C, 0·2–0·4) Bright drawn material, with an TS of 618 N/mm$^2$, and 15% El, for machined details.
   (c) (C, 0·3–0·4) Shafts, high tensile tubes, wire, fish plates.
   (d) (C, 0·4–0·5) Turbo-electric discs, shafts, rotors, die blocks, gears and tyres.

6. *High carbon steels* These are used for withstanding wear; hardness is obtained at the expense of ductility and toughness.

*Carbon %*

| | |
|---|---|
| 0·5 –0·65 | Railway rails, tyres, large forging dies. Laminated springs for automobiles and for railway purposes. Wire ropes, wheel spokes, hammers for pneumatic riveters, snaps, sets. |
| 0·65–0·75 | Saws, mandrels, hollow drills, caulking tools, diesel engine liners, hammers, hot setts, keys, torsion bars. |
| 0·75–0·85 | Laminated springs, car bumpers, pneumatic cold chisels, small forging dies, large dies for cold-presses, shear blades, cold setts. |
| 0·85–0·95 | Small cold chisels, punches, shear blades. |
| 0·95–1·1 | Screwing dies, mint dies, axes, picks, milling cutters. |
| 1·1 –1·4 | Razors, broaches, gauges, reamers, drills, woodworking tools, turning and planning tools. |

The carbon content varies according to the size, shape and duty of the tool and the details given above are necessarily only a rough guide.

Typical mechanical properties of carbon steels are given in Table 14.

TABLE 14  MECHANICAL PROPERTIES OF CARBON STEELS

| C | Si | Mn | S | P | Condition | YP N/mm² | TS N/mm² | El | RA | BH | Izod ft lbf | Izod Joules |
|---|---|---|---|---|---|---|---|---|---|---|---|---|
| 0·14 | 0·20 | 0·56 | 0·03 | 0·03 | N, 900°C | 216 | 417 | 44 | 61 | 117 | 90 | 123 |
| 0·24 | 0·16 | 0·55 | 0·04 | 0·04 | N, 870°C (3½ in.) | 247 | 478 | 34 | 56 | 134 | 60 | 82 |
| | | | | | OQ, 860°C; T, 650°C | 309 | 494 | 36 | 62 | 144 | 90 | 123 |
| 0·15 | 0·09 | 1·1 | 0·23 | 0·08 | Hot Rolled, Free cutting | 262 | 417 | 40 | 44 | 118 | 60 | 80 |
| | | | | | Cold drawn, 11% | 525 | 556 | 24 | 42 | 155 | 26 | 35 |
| 0·2 | 0·25 | 1·0 | 0·03 | 0·01 | Casting | 278 | 478 | 18 | 20 | 149 | 13 | 17 |
| | | | | | Casting A, 880°C 6 hours | 278 | 463 | 30 | 43 | 147 | 34 | 45 |
| 0·35 | 0·15 | 0·70 | | | Forging | 309 | 571 | 23 | 36 | 164 | 14 | 18 |
| | | | | | OQ, 850°C; T, 600°C | 417 | 664 | 27 | 49 | 190 | 54 | 73 |
| 0·58 | 0·37 | 0·79 | 0·04 | 0·03 | N, 850°C | 463 | 772 | 18 | 33 | 220 | 10 | 13 |
| | | | | | OQ, 830°C; T, 600°C | 540 | 880 | 20 | 42 | 250 | 20 | 27 |
| 0·65 | 0·28 | 0·76 | | | As rolled | 463 | 772 | 22 | 40 | | | |
| | | | | | OQ, 820°C; T, 600°C | 571 | 957 | 17 | 36 | 275 | 5 | 7 |
| 1·15 | 0·02 | 0·32 | 0·03 | 0·03 | As rolled | 587 | 865 | 7 | 6 | | 2 | 3 |
| 1·56 | 0·03 | 0·18 | | | As rolled | 556 | 741 | 2 | 2 | | 1 | 1 |

# 10 Heat-treatment of Steels

Heat-treatment is an operation involving the heating of the solid metal to definite temperatures, followed by cooling at suitable rates in order to obtain certain physical properties, which are associated with changes in the nature, form, size and distribution of the micro-constituents.

**Annealing**

The purpose of annealing may involve one or more of the following aims:

(1) To soften the steel; improve machinability.
(2) To relieve internal stresses induced by some previous treatment (rolling, forging, uneven cooling).
(3) To remove coarseness of grain.*

The treatment is applied to forgings, cold-worked sheets and wire, and castings. The operation consists in (*a*) heating the steel to a certain temperature, (*b*) 'soaking' at this temperature for a time sufficient to allow the necessary changes to occur, (*c*) cooling at a predetermined rate.

*Sub-critical anneal*

It is not always necessary to heat the steel into the critical range and mild steel products which have to be repeatedly cold worked in the processes of manufacture are softened by annealing at 500° to 650 °C for several hours. This is known as 'process' or 'close' annealing, and is commonly employed for wire and sheets. The recrystallisation temperature of pure iron is in the region of 500 °C (see p. 125) consequently the higher temperature of 650 °C brings about rapid recrystallisation of the distorted ferrite, and since mild steel contains only a small volume of strained pearlite a high degree of softening is induced. As shown, Fig. 110B illustrates the structure formed consisting of the polyhedral ferrite with elongated pearlite (see also Fig. 111).

Prolonged annealing induces greater ductility at the expense of strength,

*This may depend on (1) recrystallisation of cold-worked steel or (2) the critical ranges in steel.

owing to the tendency of the cementite in the strained pearlite to 'ball-up' or spheroidise, as illustrated in Fig. 110c. This is known as 'divorced pearlite'. The ferrite grains also become larger, particularly if the metal has been cold worked a critical amount (see p. 125). A serious embrittlement sometimes arises after prolonged treatment owing to the formation of cementitic films at the ferrite boundaries. With severe forming operations, cracks are liable to start at these cementite membranes.

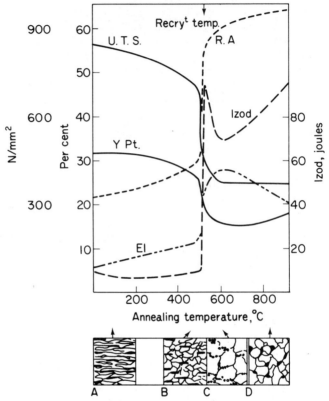

FIG. 110 Effect of annealing cold-worked mild steel

The modern tendency is to use batch or continuous annealing furnaces with an inert purging gas. Batch annealing usually consists of 24–30 hrs 670°C, soak 12 hrs, slow cool 4–5 days. Open coil annealing consists in recoiling loosely with controlled space between wraps and it reduces stickers and discoloration. Continuous annealing is used for thin strip (85% Red) running at about 400 m/min. The cycle is approximately up to 660°C 20 sec, soak and cool 30–40 sec. There is little chance for grain growth and it produces harder and stiffer strip; useful for cans and panelling.

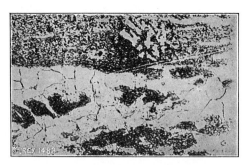

FIG. 111 Effect of annealing at 650°C on worked steel. Ferrite recrystallised. Pearlite remains elongated (× 600)

'Double reduced' steel is formed by heavy reduction (~ 50%) after annealing but it suffers from directionality. This can be eliminated by heating between 700–920°C and rapidly quenching.

*Full anneal and normalising treatments*

For steels with less than 0·9% carbon both treatments consist in heating to about 25–50°C above the upper critical point indicated by the Fe–Fe$_3$C equilibrium diagram (Fig. 112). For higher carbon steel the temperature is 50°C above the lower critical point (see p. 181).

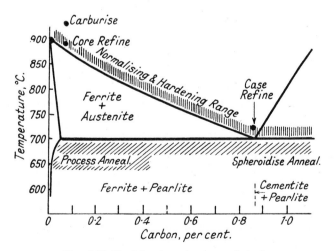

FIG. 112 Heat-treatment ranges of steels

Average annealing and hardening temperatures are:

| Carbon % | 0·1 | 0·2 | 0·3 | 0·5 | 0·7 | 0·9 to 1·3 |
|---|---|---|---|---|---|---|
| Av. temp. °C | 910 | 860 | 830 | 810 | 770 | 760 |

These temperatures allow for the effects of slight variations in the impurities present and also the thermal lag associated with the critical changes.

After soaking at the temperature for a time dependent on the thickness of the article, the steel is very *slowly* cooled. This treatment is known as *full annealing*, and is used for removing strains from forgings and castings, improving machinability and also when softening and refinement of structure are both required.

*Normalising* differs from full-annealing in that the metal is allowed to cool in *still air*, the structure and properties produced, however, varying with the thickness of metal treated. The tensile strength, yield point, reduction of area and impact value are higher than the figures obtained by annealing.

*Changes on annealing*

Consider the heating of a 0·3% carbon steel. At the lower critical point ($Ac_1$) each 'grain' of pearlite changes to several minute austenite crystals and as the temperature is raised the excess ferrite is dissolved, finally disappearing at the upper critical point ($Ac_3$), still with the production of fine austenite crystals. Time is necessary for the carbon to become uniformly distributed in this austenite.

The properties obtained subsequently depend on the coarseness of the pearlite and ferrite and their relative distribution. These depend on:

(a) *the size of the austenite grains*; the smaller their size the better the distribution of the ferrite and pearlite.
(b) *the rate of cooling through the critical range*, which affects both the ferrite and the pearlite.

As the temperature is raised above $Ac_3$ the crystals increase in size and at any temperature the growth, which is rapid at first, diminishes (see Fig. 80). Treatment just above the upper critical point should be aimed at since the austenite crystals are then small.

On cooling slowly through the critical range ferrite commences to deposit on a few nuclei at the austenite boundaries; large rounded ferrite crystals are formed, evenly distributed among the relatively coarse pearlite. With a higher rate of cooling, many ferrite crystals are formed at the austenite boundaries and a network structure of small ferrite crystals is produced with fine pearlite in the centre.

## Overheated, burnt and underannealed structures

When the steel is heated well above the upper critical temperature large austenite crystals form and slow cooling gives rise to the Widmanstätten type of structure, with its characteristic lack of both ductility and resistance to shock (see p. 165). This is known as an *overheated structure*, and it can be refined by reheating the steel to just above the upper critical point. Surface decarburisation usually occurs during overheating.

During the Second World War, aircraft engine makers were troubled with overheating (above 1250°C) in drop-stampings made from *alloy* steels.* In the hardened and tempered condition the fractured surface shows dull facets. The minimum overheating temperature depends on the 'purity' of the steel and is substantially lower in general for electric steel than for open-hearth steel. The overheated structure in these alloy steels occurs when they are cooled at an *intermediate* rate from the high temperature—at faster or slower rates the overheated structure may be eliminated. This together with the fact that the overheating temperature is very materially raised in the presence of high contents of MnS and inclusions suggests that this overheating is concerned in some way with a diffusion and precipitation process, *involving MnS*. Such overheating can occur in an atmosphere free from oxygen, thus emphasising the difference between overheating and burning.

As the steel approaches the solidus temperature, incipient fusion and oxidation take place at the grain boundaries. Such a steel is said to be *burnt* and it is characterised by the presence of brittle iron oxide films which render the steel unfit for service, except as scrap for remelting. For *Underannealing*, see p. 402.

## Annealing of castings

In the case of steel castings full annealing is the only way of completely effacing the coarse grains and Widmanstätten structure, with its associated brittleness.

The following properties of a 0·32% carbon steel illustrate the improvement brought about by annealing:

TABLE 15

|  | BH | YP $N/mm^2$ | TS $N/mm^2$ | El | RA | Cold Bend | Izod J |
|---|---|---|---|---|---|---|---|
| Cast | 160 | 309 | 448 | 6 | 10 | 45° | 12 |
| Annealed 880°C, 6 hr, furnace cooled | 153 | 247 | 479 | 24 | 28 | 165° | 32 |

*J. *Iron and Steel Inst.*, March 1946, January and March 1950.

The Widmanstätten structure can be modified into a 'feathery' arrangement of the ferrite by the influence of silicon. This is shown in Fig. 114 which consists of a portion of the boundary-ferrite, Widmanstätten and feathery structures. Fig. 113 shows the macroform of the primary crystals,

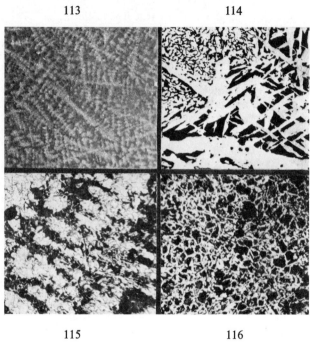

113      114

115      116

0·35% cast steel—etched—2% nitric acid.

FIG. 113 Macro-structure of cast steel revealing large primary austenite crystals due to presence of impurities ( × 4)

FIG. 114 Micro-structure of same steel showing part of ferrite network, Widmanstätten and feathery structure. Ferrite—white. Pearlite—dark ( × 80)

FIG. 115 Same steel imperfectly annealed: ferrite formed in masses outlining original cast structure ( × 80)

FIG. 116 Same steel properly annealed: ferrite and pearlite uniform and fine ( × 80)

revealed by the segregation of the impurities. The effects of this segregation have to be effaced as much as possible by annealing and this necessitates temperatures higher than those used for worked steels. An imperfect anneal is illustrated by Fig. 115 in which the original cast structure is still outlined by the deposition of the ferrite in the old positions, especially around inclusions. They can be prevented by double annealing.

*Double annealing* consists in heating the steel to a temperature considerably over the $A_3$ point, cooling rapidly to a temperature below the lower critical range and then immediately reheating to a point just over the upper critical point ($Ac_3$), followed by slow cooling.

This method is particularly useful for *castings*. The high temperature treatment effaces the strains, coalesces the sulphide films in the ferrite which embrittle the steel and produces homogeneity by rapid diffusion. The quick cooling prevents the coarse deposition of ferrite in the large grains, but tends to harden the metal. The second heating refines the coarse grains and leaves the steel in a softened condition. A typical structure is shown in Fig. 116 which should be compared with Fig. 115.

*Softening tool and air-hardening steels*

To soften high carbon and air-hardening steels, in order to allow machining operation to be carried out, they are heated just below the lower change point (650–700 °C) so that the cementite balls-up into rounded masses (i.e. *spheroidising anneal*). When the cementite is in this condition high carbon steels can be cold drawn; but too high a temperature causes pearlite to be reformed, with consequent high resistance to deformation. It should be remembered, however, that coarse laminated cementite spheroidises extremely slowly, and the above treatment is therefore carried out on 'hardened' material, obtained by suitable cooling from above $A_3$ or after cold working. A short cycle anneal is to heat just above $Ac_1$, cool below $Ar_1$ and raise temperature to just below $Ac_1$ for 8 hours. Although the softest condition is obtained when large globules of cementite are embedded in ferrite, a smooth machined surface is difficult to obtain due to tearing. Groups of large globules cause failure of sharp-edged cutting tools.

During subsequent hardening operations the time required to dissolve fine spheroidised cementite is less than for the lamellar type and this property is made use of in hardening thin sections, such as safety razor blades and needles, in order to reduce decarburisation.

*Annealing and hardening temperatures for tool steels*

The annealing or hardening temperatures of steels containing more than 0.9% carbon is just above the lower critical point (730–790 °C) instead of above the upper range. The reason for this is as follows:

Fig. 117 shows the appearance of a 1.3% carbon steel as cast, in which the cementite exists as brittle networks and plates. This type of structure must be replaced by rounded particles of cementite in fine pearlite before hardening, otherwise cracks will propagate through the continuous masses of brittle cementite.

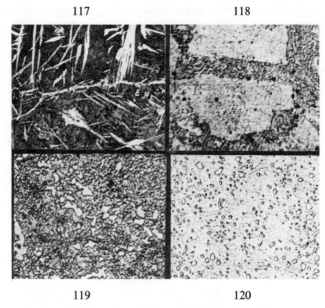

1·3% carbon steel—etched—picric acid

FIG. 117 As cast: cementite network and plates in pearlite (× 100)
FIG. 118 Heated to 1050 °C and quenched in water. Large grains (× 100)
FIG. 119 Cementite globules in properly hot-worked steel (× 200)
FIG. 120 Cementite globules in martensite, in hardened steel (× 200)

The upper critical line ($A_{cem}$, in Fig. 101) rises steeply with increasing carbon content above 0·87% and an excessively high temperature is required to dissolve all the free cementite. This tends to develop coarse austenite crystals which cause the steel to become brittle and cracks to form on quenching. This grain growth is shown by Fig. 118, which is the structure of steel quenched from 1050 °C. On the other hand, particles of cementite restrain grain growth.

Forging is, therefore, carried out through the critical range in order to disperse the free cementite. This is followed by annealing at about 760 °C (just above $Ac_1$), to ball-up the free cementite and to remove strains. Fig. 119 shows the structure formed. Even a fine cementite network structure would cause trouble when drastic quenching is used, such as for files.

Fig. 120 shows the structure of a steel hardened from 760 °C, consisting of particles of cementite dispersed in a matrix of martensite.

*Products of quenching: constituents of hardened steel*

The equilibrium diagram (Fig. 101) indicate the changes which occur under the slow cooling conditions of annealing. The rapid rates of cooling

# HARDENED STEEL

necessary to harden a steel cause the austenite to persist to a lower temperature and to transform into a variety of micro-constituents discussed below.

*Austenite* is sometimes present with martensite, in drastically quenched steels (Fig. 122). It cannot be completely retained in carbon steels by even drastic quenching, but suitable additions of alloying elements allow the retention of this constituent, for example 18/8 austenitic stainless steel. This austenite consists of polyhedral grains, showing twins (Figs. 121 and 122). It is non-magnetic and soft.

*Martensite* is the hardest constituent obtained in a given steel, but the hardness increases with the carbon content of the steel up to 0·7%. The micro-structure exhibits a fine needle-like structure, which becomes more pronounced when the steel is quenched from high temperatures. See Figs. 121 and 124.

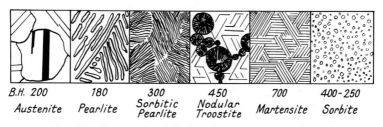

FIG. 121 Forms of carbide in micro-constituents in steel

The nature of martensite* has not been definitely agreed upon, but for the present purpose it might be considered to be highly stressed α-iron which is supersaturated with carbon.

*Bainite*, which occurs in alloy steels, has a rapidly etched needle-like structure, somewhat resembling tempered martensite. As the temperature of its formation becomes higher the acicular nature becomes less accentuated, the needles increasing in size. Figs. 130 and 131 show upper bainite and lower bainite formed at 400 °C and 325 °C respectively (see also p. 194).

*Troostite* (nodular) rapidly etches black and is practically unresolvable under the ordinary microscope. Special microscopic technique has shown that nodular troostite is a mixture of *radial* lamellae of ferrite and cementite and therefore differs from pearlite only in degree of fineness and carbon content which is the same as that in the austenite from which it is formed. Figs. 121 and 125 show a typical martensite–troostite structure; nodules

*The f.c.c. lattice of γ-iron is equivalent to a body centred tetragonal lattice with ratio 1·414. The tetragonal lattice of martensite is formed from this by compressing its height and increasing its cross-section. A slight further compression to give a ratio of 1 gives rise to α-iron.

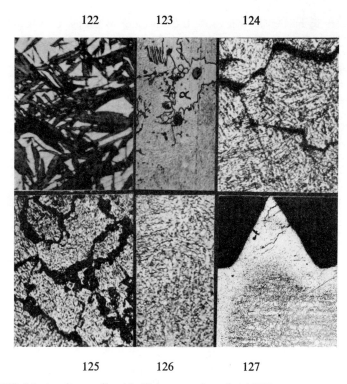

FIG. 122 Martensite needles (dark) in austenite ($\times$ 1200)
FIG. 123 Steel (C, 0·4) quenched from between $A_1$ and $A_3$. Undissolved ferrite around inclusion in martensite ($\times$ 100)
FIG. 124 Martensite and quench crack. Steel (C, 0·5) quenched in water from 900°C ($\times$ 400)
FIG. 125 Nodular troostite in martensite ($\times$ 400)
FIG. 126 Sorbite in quenched and tempered (600°C) steel (C, 0·5) ($\times$ 500)
FIG. 127 Case-hardened screw. Cracked martensitic case (white), martensite and ferrite core ($\times$ 30)

outline the boundaries of the original austenite grains. Troostite* is softer than martensite and small amounts in the steel lessen the risks of cracking and distortion.

*Some confusion arises as to the nomenclature of micro-constituents found in hardened and tempered steels, since the terms troostite and sorbite are frequently used to indicate constituents formed during quenching and also during tempering. In the former case (quenching) the cementite always occurs in a laminated form, while in the latter (tempering) it has a granular form. Hence, the term troostite has been adopted in this book for constituents possessing a laminated structure. Sorbite is used to denote the granular structures.

## Quench hardening of steel

Hardening of steel is obtained by a suitable quench from within or above the critical range. The temperatures are the same as those given for full annealing (p. 177). The soaking time in air furnaces should be 1·2 min for each mm of cross-section or 0·6 min in salt or lead baths. Uneven heating, overheating and excessive scaling should be avoided. The quenching is necessary to suppress the normal breakdown of austenite into ferrite and cementite, and to cause a partial decomposition at such a low temperature that martensite is produced. For this to occur each steel requires a *critical cooling velocity* which is greatly reduced by the presence of alloying elements, which therefore cause hardening with mild quenching (e.g. oil and hardening steels). Steels with less than 0·3% carbon cannot be hardened effectively, while the maximum effect is obtained at about 0·7% due to an increased tendency to retain austenite in high carbon steels Fig. 128.

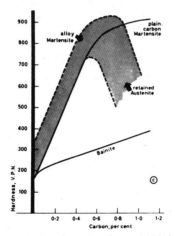

FIG. 128 Variation of hardness of martensite and bainite with carbon content

Water is one of the most efficient quenching media in commercial use where maximum hardness is required, but it is liable to cause distortion and cracking of the article. Where hardness can be sacrificed, oils such as whale, cotton seed and mineral are used. These tend to oxidise and form sludge with consequent lowering of efficiency.

The quenching velocity of oil is much less than water; ferrite and troostite are formed even in small sections. Intermediate rates between water and oil can be obtained with water containing 10–30% Ucon, a substance with an inverse solubility which therefore deposits on the article to slow rate of cooling. To minimise distortion long cylindrical articles should be quenched vertically, flat sections edgeways and thick sections should enter the bath first. To prevent steam bubbles forming soft spots, a water quenching bath should be agitated.

*Fully* hardened and tempered steels develop the best combination of strength and notch-ductility.

*Tempering and toughening*

The martensite of a quenched tool steel is exceedingly brittle, highly stressed, consequently cracking and distortion of the article are liable to occur after quenching. Retained austenite is unstable and as it changes with time dimensions may alter, e.g. dies may alter 0·012 mm. It is necessary, therefore, to warm the steel below the critical range in order to relieve stresses and to allow the arrested reaction of cementite precipitation to take place. This is known as *tempering*.

150–250 °C. The article is heated in an oil bath, immediately after quenching, to prevent belated cracking, to relieve internal stress and to decompose austenite without much softening.

200–450 °C. Used to toughen the steel at the expense of hardness. Brinell hardness is 350–450.

450–700 °C. The precipitated cementite coalesces into larger masses, and the steel becomes softer, weaker and more ductile as shown by Fig. 126. The structure is known as *sorbite*, which at the higher temperatures becomes coarsely spheroidised (Fig. 121). It etches more slowly than troostite and has a Brinell hardness of 220–350.

Sorbite is commonly found in heat-treated constructional steels, such as axles, shafts and crankshafts, subjected to dynamic stresses. A treatment of quenching and tempering in this temperature range is frequently referred to as *toughening*, and it produces an increase in the ratio of the elastic limit to the ultimate tensile strength.

The reactions in tempering occur slowly and time as well as temperature of heating is important. Tempering is carried out to an increasing extent under pyrometric control in oil, salt,* lead baths and also in furnaces in which the air is circulated by fans. After tempering, the articles may be cooled either rapidly or slowly except in the case of steels susceptible to temper brittleness (see p. 221).

*Temper colours* formed on a cleaned surface are still used occasionally as a guide to temperature. They are due to the interference effects of thin films of oxide formed during tempering, and they act similarly to oil films on water. Alloys such as stainless steel form thinner films than do carbon steels for a given temperature and hence produce a colour lower in the series; for example pale straw corresponds to 300 °C, instead of 230 °C (Table 16).

*E.g. equal parts sodium and potassium nitrates for 200–600 °C. (See Home Office form 1988 re explosion risk.)

TABLE 16

| Temper Colour | Temperature, °C | Articles |
|---|---|---|
| Pale straw | 230 | Planing and slotting tools |
| Dark straw | 240 | Milling cutters, drills |
| Brown | 250 | Taps, shear blades for metals |
| Brownish-purple | 260 | Punches, cups, snaps, twist drills, reamers |
| Purple | 270 | Press tools, axes |
| Dark purple | 280 | Cold chisels, setts for steel |
| Blue | 300 | Saws for wood, springs |
|  | 450–650 | Toughening for constructional steels |

With turning, planing, shaping tools and chisels, only the cutting parts need hardening, and this is frequently carried out in engineering works by heating the tool to 730 °C, followed by quenching the cutting end vertically. When this end is cold it is cleaned with stone and the heat from the shank of the tool is allowed to temper the cutting edge to the correct colour, then the whole tool is quenched. Oxidation can be reduced by coating the tool with charcoal and oil.

*Changes during tempering*

The principles underlying the tempering of quenched steels have a close similarity to those of precipitation hardening (see p. 186) discussed and are conveniently considered at this stage. The overlapping changes which occur when high carbon martensite is tempered are shown in Fig. 49(*d*) and as follows:

*Stage 1.* 50–200 °C  Martensite breaks down to a transition precipitate known as ε-carbide ($Fe_{2.4}C$) across twins and a low carbon martensite which results in slight dispersion hardening, decrease in volume and electrical resistivity.

*Stage 2.* 205–305 °C  Decomposition of retained austenite to bainite and decrease in hardness.

*Stage 3.* 250–500 °C  Conversion of the aggregate of low carbon martensite and ε-carbide into ferrite and cementite precipitated along twins which gradually coarsens to give visible particles and rapid softening, Fig. 129.

*Stage 4.* Carbide changes in alloy steel. 400–700 °C. In steels containing one alloying addition cementite forms first and the alloy diffuses to it. When sufficiently enriched the $Fe_3C$ transforms to an alloy carbide. After further enrichment this carbide may be superseded by another and this formation of transition carbides may be repeated several times before the equilibrium carbide forms. In chromium steel, changes are: $Fe_3C \rightarrow Cr_7C_3 \rightarrow Cr_{23}C_6$. In steels containing several carbide-forming elements the reactions are often more complex, and the carbides which decompose are not necessarily followed by carbides based on the same alloy elements. The transformation can also occur *in situ* by gradual exchange of atoms without any appreciable hardening; or by resolution of existing iron carbides and fresh nucleation of a coherent carbide with considerable hardening which counteracts the normal softening that occurs during tempering. In some

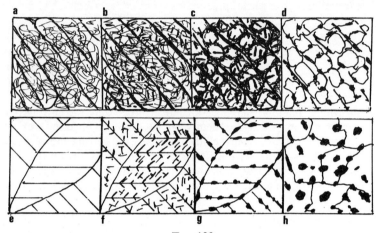

FIG. 129

Lath M.
(a) As quenched
Laths with high density of dislocation
(b) Tempered 300 °C
Widmanstätten precipitation of carbides within laths
(c) Tempered 500 °C
Recovery of dislocation structure into cells with laths
(d) Tempered 600 °C
Recrystallisation cementite re-nucleated equioxed ferrite boundaries

Twinned M.
(e) High C twinned martensite
(f) Tempered 100 °C
fine ε-carbides across twins
(g) Tempered 200 °C
Coherent cementite along twins. ε-carbides dissolve
(h) Tempered 400 °C
Breakdown of twinned structure. Carbides grow and spheroidise

alloy steels, therefore, the hardness is maintained constant up to about 500 ° C or in some cases it rises to a peak followed by a gradual drop due to breakdown of coherence and coalescence of the carbide particles. This age-hardening process is known as *secondary hardening* and it enhances high temperature creep properties of steel (e.g. steel E in Fig. 49(d)). Chromium, for an example, seems to stabilise the size of the cementite particles over a range 200–500 °C. Vanadium and molybdenum form a fine dispersion of coherent precipitates ($V_4C_3, Mo_2C$) in a ferrite matrix with considerable hardening. When over-ageing starts the $V_4C_3$ grows in the grain boundaries and also forms a Widmanstätten pattern of plates within the grain.

These changes occurring over long periods, are illustrated in Fig. 49(e), which show the relation of temperature and time.

Low carbon lath martensites have a high Ms temperature and some tempering often occurs on cooling, i.e. *autotempering*.

Tempering at 300 °C causes precipitation of carbides within the laths in Widmanstätten form (Fig. 129). Tempering at 500 °C, promotes the recovery of the dislocation tangle into cells within the laths with carbides precipitated along boundaries. Tempering at 600 °C gives rise to recrystallisation into equioxed ferrite with carbides re-nucleated at the boundaries.

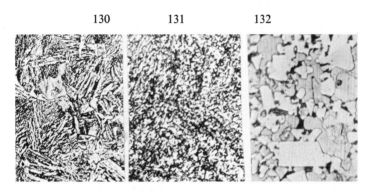

FIG. 130 Upper bainite  ( × 500). C, 0·35; Ni, 1·34; Cr, 0·85 . Transformed at 400 °C
FIG. 131 Lower bainite  ( × 1000). C, 0·33; Ni, 3; Transformed at 325 °C
FIG. 132 Tungsten–titanium carbide in hard-metal  ( × 1500)

*Quench cracks*

The volume changes which occur when austenite is cooled are:

(a) expansion when gamma iron transforms to ferrite;
(b) contraction when cementite is precipitated;
(c) normal thermal contraction.

When a steel is quenched these volume changes occur very rapidly and unevenly throughout the specimen. The outside cools most quickly, and is mainly martensitic, in which contraction (b) has not occurred. The centre may be troostitic and contraction (b) started. Stresses are set up which may cause the metal either to distort or to crack if the ductility is insufficient for plastic flow to occur. Such cracks may occur some time after quenching or in the early stages of tempering.

Quench cracks are liable to occur:

(*a*) due to presence of non-metallic inclusions, cementite masses, etc.;
(*b*) when austenite is coarse grained due to high quenching temperature;
(*c*) owing to uneven quenching (see Martempering, p. 199);
(*d*) in pieces of irregular section and when sharp re-entrant angles are present in the design.

*The relation of design to heat-treatment* is very important. Articles of irregular section need special care. When a steel has been chosen which needs a water-quench, then the designer must use generous fillets in the corners and a uniform section should be aimed at. This can sometimes be

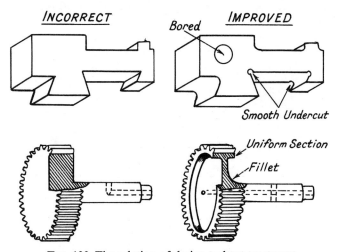

FIG. 133 The relation of design to heat-treatment

obtained by boring out metal from bulky parts without materially affecting the design; examples are given in Fig. 133. A hole drilled from the side to meet a central hole may cause cracking and it should be drilled right through and temporarily stopped up with asbestos wool during heat-treatment. A crack would also form at the junction of the solid gear with the shaft and there would be a serious danger of cracks at the roots of the teeth owing to the great change in size of section. This design could be improved by machining the metal away under the rim to make a cross-section of uniform mass.

*Fundamentals of heat-treatment*

Heat-treatment of steel involves the change of austenite, a face-centred cubic iron lattice containing carbon atoms in the interstices, into a body-centred cubic ferrite with a low solubility for carbon. The carbon atoms segregate into areas to form cementite. This involves mobility or diffusion of the carbon atoms and both time and temperature are important (p. 95). Atomic movements are rapid at high temperatures but increasingly sluggish as the temperature decreases. As the rate of cooling of an austenitised steel increases the time allowed for the changes is shortened and the reactions are incomplete at 600–700 °C. Residual austenite, therefore, transforms at lower temperatures, with shorter movements of atoms and finer structures. At temperatures below about 250 °C diffusion is so slow that another transition structure is formed.

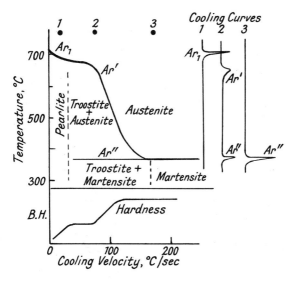

Fig. 134 Effect of cooling rate on the transformation of austenite

*The effect of rapid cooling on the critical points* is complex (Fig. 134). Increase in the rate of cooling has the following effects:

(1) Arrest temperatures are depressed.
(2) $Ar_3$ merges with $Ar_1$ producing a single depressed point known as $Ar'$. Fine laminated troostite is formed.
(3) Accelerated cooling causes another arrest to appear at 350–150 °C, known as $Ar''$. Troostite and martensite are formed.
(4) Rapid quenching causes $Ar'$ to merge into $Ar''$. Martensite is formed.
(5) The arrest due to the formation of bainite at 500–250 °C does not usually appear with carbon steel, but is present with many alloy steels.

## Constant temperature transformation—TTT-curves

The structures formed during continuous cooling of steel from above $Ac_3$ can be understood best by studying the constant-temperature (isothermal) transformation of austenite, thus separating the two variables—time and temperature. One method of doing this consists in heating small specimens above $Ac_3$ to form austenite, then quenching into a suitable bath (e.g. liquid tin) at some constant sub-critical temperature. After holding for selected periods of time, the specimens are withdrawn from the bath and rapidly quenched in cold water which converts any untransformed austenite into martensite the volume of which can be estimated microscopically. Another method consists in following length changes caused by the decomposition of austenite at the constant temperature by means of a dilatometer.

When a carbon steel is quenched in baths held at *constant* temperatures the austenite is found to transform at velocities characteristic of each temperature. The time for the beginning and completion of the transformation of austenite are plotted against temperature to give the Bain 'S-curve', shown in Fig. 135 now called TTT-curve, time-temperature-transformation. The logarithmic scale of time is used to condense results into a small space. $Ae_1$ and $Ae_3$ lines represent the *equilibrium* transformation temperatures. Austenite is completely stable above $Ae_3$, partially unstable between $Ae_3$ and $Ae_1$, and below $Ae_1$ it is completely unstable and transforms in time. Two regions of rapid transformation occur about 550° and 250°C. The form of each of the curves and their positions with respect to the time axis depend on the composition and grain size of the austenite which is transforming.

The TTT-curve is most useful in presenting an overall picture of the transformation behaviour of austenite which enables the metallurgist to interpret the response of a steel to any specified heat-treatment and to plan practical heat-treatment operations to get desirable microstructures; to control limited hardening or softening and the time of soaking.

The decomposition of austenite occurs according to three separate but sometimes overlapping mechanisms and results in three different reaction products: pearlitic, bainitic, martensitic.

*Pearlitic*

The upper dotted curve in Fig. 135 represents the beginning of the formation of ferrite; the curve just below it indicates the beginnings of the breakdown of the austenite remnant into a ferrite–carbide aggregate. In the high-temperature pearlitic range in Fig. 135 the process resembles the solidification of crystals from a liquid by the formation and growth of nuclei of carbide followed by ferrite by side nucleation with side and edge growth, Fig. 136(a) and (b). At 700°C the formation of nuclei *is slow* (i.e. incubation period), then growth proceeds rapidly to form large pearlite colonies covering several austenite grains in some cases.

As the transformation temperature is lowered to 500°C the incubation

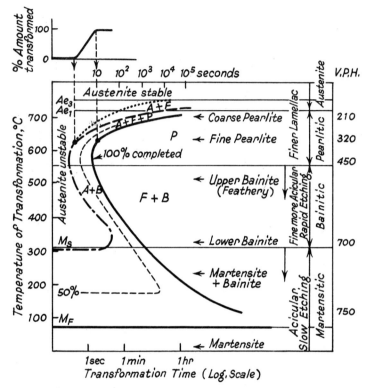

Fig. 135 Idealised TTT-curve for 0·6% carbon steel depicting time interval required for beginning, 50% and 100% transformation of austenite at a constant temperature

A = Austenite    F = Ferrite    P = Pearlite    B = Bainite

period decreases and the pearlite becomes increasingly fine. Large numbers of nuclei form in the austenite boundaries, but growth is slower and this produces nodular troostite, Figs. 125, 136(a). In the case of medium carbon steels the excess ferrite decreases in volume and begins to show an acicular or Widmanstätten type of distribution. The relative amounts of free ferrite to be expected after a given heat-treatment is indicated by the size of the 'austenite and ferrite' field and by the temperature interval between $Ae_1$ and $Ae_3$.

## Bainitic

Between about 500° and 350°C initial nuclei are ferrite which is coherent with the austenite matrix. Cementite then precipitates from the carbon-enriched layer of austenite, allowing further growth of the ferrite as shown in Fig. 136(c). The carbides tend to lie parallel to the long axis of the bainite

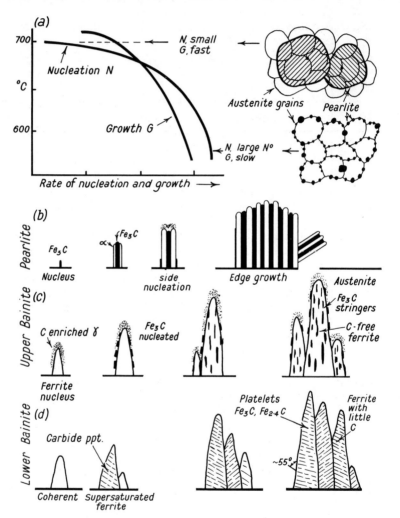

FIG. 136 (a) Effect of different speeds of nucleation and growth on formation of pearlite colonies; (b), (c), (d) diagrammatic representation of formation of pearlite, upper bainite and lower bainite

needle to form the typical open feathery structure of *upper* bainite (Fig. 130). Below 350°C coherent ferrite, supersaturated with carbon, forms first and is then followed by the precipitation of carbide *within* the ferrite needle, transversely at an angle of 55°. A proportion of the carbide is $Fe_{2.4}C$ and the ferrite contains a little dissolved carbon. This *lower* bainite structure is somewhat similar to lightly tempered martensite (Figs. 131, 136(*d*)). It is possible to predict the strengths available in steels which can be air cooled to form bainite structures from their chemical composition up to 0·25% C, e.g. Tensile strength (tsi) = 16 + 125 (% C) + 15 (% Mn + % Cr) + 12 (% Mo) + 6 (% W) + 8 (% Ni) + 4 (% Cu) + 25 (% V + % Ti).

*Martensitic*

In quenching down to about 250°C the temperature drops rapidly through the interval in which 'nucleation' could take place at an observable rate, to a temperature so low that the molecular mobility, i.e. diffusion, becomes too small for the formation of nuclei. In the third stage, therefore, the austenite changes incompletely into a distorted body-centred structure, with little or no diffusion of the carbon into particles of cementite, to form martensite the plates of which are formed at a high speed (less than 0·002 sec). This suggests that the mechanism of formation of this structure is not nucleation and growth but a shearing process resembling the process of mechanical twinning and involving *very little atomic movement* but considerable internal stress due to the shear and to the position of the carbon atoms. As the temperature falls the elastic energy increases and eventually causes a shear in a part of the matrix which stabilises the rest. Further shear can only occur when the temperature is lowered and more energy gained. The amount of martensite formed, therefore, is practically *independent of time* and *depends principally* on the *temperatures* at which the steel is held. Hence a proportion of austenite is usually retained in a quenched steel which can be reduced in amount by a decrease in temperature. This fact is used in sub-zero quenching (p. 250). The temperature at which martensite begins to form (Ms) is progressively lowered as the carbon content of the steel increases, e.g.

| C% | 0·02 | 0·2 | 0·4 | 0·8 | 1·2 |
|---|---|---|---|---|---|
| Ms°C | 520 | 490 | 420 | 250 | 150. |

It is also affected by the alloy content, but the following empirical formula (Steven and Haynes) can be used for calculating Ms from the chemical analyses, provided all carbides have been dissolved in the austenite:

Ms in °C = 561 − 474 (% C) − 33 (% Mn) − 17 (% Ni) − 17(% Cr) − 21 (% Mo).

Mf is about 215°C below the Ms.

Plastic and elastic stresses promote the formation of martensite but it is retarded when cooling is interrupted. When cooling is resumed after such a *stabilisation* arrest martensite only begins to form again after cooling to a lower temperature. The rate and extent of stabilisation (depression) depend on the temperature and time of holding, amount of prior transformation and alloy content.* Two forms of martensite have been identified depending on carbon content. In low carbon steels laths containing many dislocations are found while in high carbon steels the plates are heavily twinned, Fig. 137(a) and (b).

FIG. 137 (a) Lathe martensite formed in 0·08 °C steel quenched in brine from 100 °C ( × 20 000)

FIG. 137 (b) Twinned martensite in Fe 30 % Ni ( × 110 000)

(Courtesy P. M. Kelly)

*Effect of alloys on TTT-curves*

TTT-curves are built up from overlapping curves for ferrite, pearlite and bainite and these are affected in different ways by alloys as shown in Fig. 138. Carbon, nickel, manganese, silicon and copper move both pearlite and bainite C-curves to the right but do not separate them appreciably on the temperature scale. Molybdenum, chromium and vanadium move the pearlite C-curve markedly to the right and also displace it upwards to higher temperatures; the bainite curve is not moved so much to the right but is depressed to lower temperatures. Thus there may be two 'noses' to the curve (Fig. 138(b)) or the two reactions may be completely separated and the bainite may become a region of fast and controlling reaction. The bainite range also tends to overlap the martensitic range and often the incubation period for the formation of lower bainite suddenly begins to get shorter once martensite begins to form. In other words, the C-curve begins

*Two groups of phase transformation are now given the name *civilian*, in which atoms move in a random manner (e.g. pearlite) and *military* because of its orderly disciplined manner, e.g. martensite. Martensite transformations also occur in non-ferrous alloys often differing greatly from the rather special ease in steel.

to change into an S-curve; consequently a steel quenched to a temperature somewhat below $M_s$ may transform partially to martensite followed by bainite.

The TTT-curves are also affected by other factors. Increasing the austenite grain size has an insignificant effect on the rate of formation of bainite structures but shifts the pearlite curve to the right because there is less grain boundary area to provide nucleation sites and growth occurs over a greater

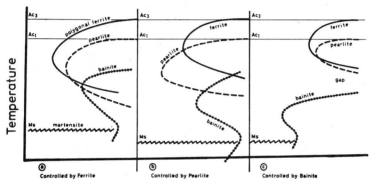

FIG. 138 TTT diagrams showing the separation of ferrite, pearlite and bainite curves for start of transformation
(a) Transformation controlled by the ferrite C-curve
(b) Transformation controlled by the pearlite C-curve
(c) Transformation controlled by the bainite C-curve and stable region separates pearlite and bainite

distance. Compositional change in the austenite may arise due to the presence of undissolved carbides, which impoverish the austenite and also nucleate the pearlite rapidly; and heterogeneity which gives rise to a less precise diagram because the start of the reaction will correspond to the lean portion and the end of the transformation to the richest alloy region.

*Transformation on continuous cooling*

The TTT-curve refers specifically to transformation at a *constant* temperature, whereas most hardening operations involve continuous cooling. The effects of continuous cooling on the transformation behaviour can be derived from Fig. 139, which shows superimposed on the TTT-curve a number of cooling curves, each representing a different rate of continuous cooling. Curve 1 indicates that a slow cool, such as in annealing, allows transformation to pearlite to start at (*a*) and to be complete at (*b*). From lower temperatures than (*b*) the rate of cooling can be rapid without affecting hardness of the steel. Curve 2 represents the fastest cooling rate at which austenite can transform entirely to pearlite (*c*) which, however, is much finer than that indicated by curve 1. With curve 3 partial transformation to pearlite will occur between (*d*) and (*e*); no change occurs between (*e*) and (*f*) (due to the long incubation period); then some martensite forms between (*f*) and (*g*). Curve 4 represents the critical cooling rate since it just misses the nose *n* of the TTT-curve and transformation to martensite occurs between (*h*) and (*i*).

In most cases the actual TTT-curve obtained under continuous cooling conditions occupies a position downward and to the right with respect to

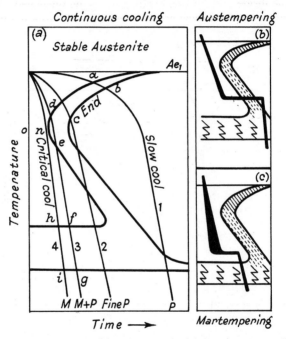

Fig. 139 (*a*) TTT-curve for 0·8% carbon steel with continuous cooling curves, 1, 2, 3, 4, superimposed. M = martensite; P = pearlite; (*b*) and (*c*) the cooling curves for austempering and martempering in relation to the TTT-curve

the TTT diagram. Several methods of deriving and portraying the transformations under continuous cooling conditions have been used but the one illustrated in Fig. 140 (after Steven and Mayer, *J. Iron & Steel Inst.*, May 1953) gives the data for various positions in oil quenched cylindrical bars of 25 to 150 mm dia covering the ruling sections in BS 970 (Table 29) and is therefore of practical significance for directly assessing the transformation behaviour of simple shapes quenched in oil. The abscissa scales cover the conditions existing at the surface, mid-radius and centre of the bar. The lines in the diagram represent the various stages of completion of the transformation, i.e. 10, 50, 90%. The temperature range for the transformation at the centre of a 75 mm oil quenched bar is indicated by the intersection of the curves with the ordinate from the 75 mm position. Thus transformation would begin at 600 °C and finish at 400 °C. Bars up to about 20 mm dia would be fully hardened by oil quenching.

*Martempering* is a method of forming martensite with a minimum of distortion and residual stresses by reducing the difference in temperature between the inside and outside of the piece being quenched. The piece is cooled faster than the critical cooling velocity down to a temperature just

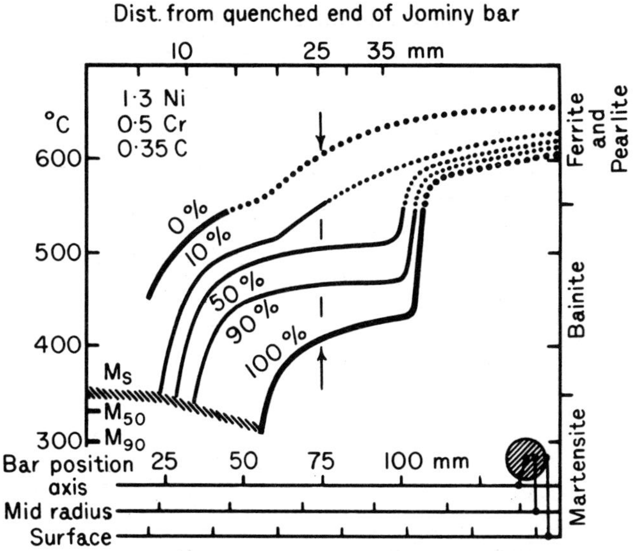

FIG. 140 Continuous cooling transformation curve of low NiCr steel (after Steven and Mayer, *J. Iron and Steel Inst.*, May 1953)

above the martensite transformation. It is held in a salt bath sufficiently long for its temperature to become uniform without transforming into bainite, after which it is cooled in air (Fig. 139).

*Austempering* consists in quenching a steel from above $Ar_3$ at a critical cooling rate into a salt or lead bath maintained at a temperature within the bainitic zone. The piece remains in the bath until the austenite is completely transformed to bainite, after which it is allowed to cool to room temperature, the rate being immaterial (Fig. 139).

Bainitic structures produced in this way are free from cracks and are softer than martensite and possess good impact resistance. Tempered bainite, however, has mechanical properties inferior to those of tempered martensite.

## Heat-treatment of high carbon steel wire-patenting

Wire ropes for haulage purposes are usually made from carbon steel wires ranging from 0·35 to 0·5% carbon, and before drawing the material is subject to a heat-treatment known as *patenting*. This consists in passing the wire through tubes in a furnace at about 970 °C. This high temperature treatment produces a uniform austenite of rather large grain size. The subsequent cooling—in air or molten lead—is rapid since the sections treated are generally small—e.g. wire rods, so that the resulting structure consists of very fine pearlite preferably with no separation of primary ferrite.

The large crystals would give rise to brittleness if the material was left in the heat-treated condition, but this effect is not noticed after a few drawing passes. Variation in hardness—either softer or harder—can be produced by tempering martensite, but such material does not draw so well as patented wire, which is able to withstand reductions of area up to 90%. The strength is explained on the basis of the reduced ferrite cells and the alignment of cementite in fibres.

The capacity for drawing is shown by the tests obtained on drawn steel after different heat-treatments, shown in Table 17.

TABLE 17   0·5% CARBON STEEL

| Treatment | RA % by Drawing | TS $N/mm^2$ | El | RA |
|---|---|---|---|---|
| Annealed and drawn | 56 broke | 1019 | 7 | 1 |
| Patented and drawn | 80 | 1436 | 7 | 11 |

The wire is usually cold-worked to yield a TS of 1530–1840 $N/mm^2$ and the micro-structure exhibits a pronounced fibrous effect (Fig. 153). Decarburisation should be avoided.

## Hardenability, mass effect, ruling section

Hardenability is the measure of the *depth* to which a steel will harden on quenching; the *maximum* hardness is mainly a function of the carbon content. The hardenability of steel depends on:

(1) The quenching medium and method of quenching.
(2) Composition of the steel and method of manufacture.
(3) Section of the steel.

The so-called 'Mass effect' arises from the fact that even with the most severe quench the cooling of a bar is progressively slower, from the outside to the centre due to the low thermal conductivity of the steel. It must be appreciated, therefore, that it is the *rate of cooling* of a piece of steel which determines the properties resulting from a quenching process, and not mass or weight.

The effects of mass are shown by the following results (Table 18) on quenched bars of varying section, from the centre of which standard test specimens were machined.

To prevent misuse of steels it has become desirable to state the limit of thickness or *'ruling section'* up to which the mechanical properties quoted can be obtained. For sections other than cylinders, reference should be made to BS 970 for conversion factors.

*One of the most important functions of alloying elements in high tensile steels is to produce fully hardened structures in large sections.*

TABLE 18 EFFECT OF MASS ON PROPERTIES: CARBON STEEL
C, 0·45; Si 0·32; Mn, 0·78; S and P, 0·02%
Water quenched from 870°C and not tempered

| Dia of Bar, mm | BH | TS | El | RA | Izod ft lbf | Joules |
|---|---|---|---|---|---|---|
| 17 | 550 | 1853 | 3 | 4 | 3 | 4 |
| 29 | 320 | 1035 | 8 | 28 | 14 | 18 |
| 76 | 241 | 803 | 15 | 47 | 24 | 32 |
| Tempered at 600°C | | | | | | |
| 17 | 255 | 850 | 18 | 60 | 76 | 103 |
| 76 | 212 | 741 | 25 | 59 | 44 | 59 |

Shallow hardening steels (C, 1; Mn, 0·3%) are often used for dies in which a tough core and a hard skin is required. The depth of hardening is increased by drastic quenching media, and by manganese and elements such as nickel and chromium, but decreased by fine austenite grains.

The addition of 3% nickel and 1% chromium retards the transformation of austenite so much that oil quenching exceeds the critical cooling velocity and all portions from edge to centre of average sections will harden. The results in Table 19 illustrate this effect and should be compared with the ones in Table 18.

TABLE 19 EFFECT OF MASS ON PROPERTIES OF ALLOY STEEL
C, 0·31; Si, 0·14; Mn, 0·70; Ni, 3·27; Cr, 0·82
Oil quenched from 820°C, tempered at 600°C

| Dia of Bar, mm | BH | TS $N/mm^2$ | El | RA | Izod ft lbf | Joules |
|---|---|---|---|---|---|---|
| 17 | 285 | 958 | 24 | 64 | 62 | 85 |
| 76 | 285 | 927 | 22 | 60 | 60 | 82 |

It has now been found that small additions of a number of alloying elements are more effective than an equivalent total addition of one element, and enables appropriate use to be made of alloy scrap.

The quantitative determination of hardenability has been tackled from the theoretical side by Grossman (Amer. Soc. Metals, Hardenability Symposium 1939 and 1940, p. 969) and from the practical aspect by Jominy, and considerable work is reported in the Iron and Steel Institute Special Report No. 36, 1946.

In the Jominy test (BS 4437) a 25 mm dia bar 100 mm long is heated to the normal hardening temperature, then inserted in a standard jig and a

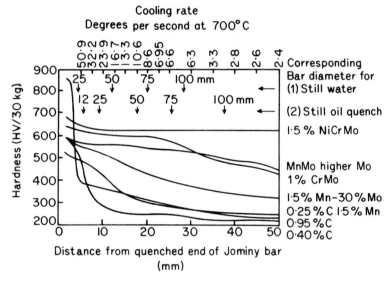

FIG. 141 Hardness distance curves for end quench test samples (Sykes)

12·5 mm dia jet of water at 24 °C is directed against one end of the test piece. The free height of the jet is 62·5 mm. When cold, hardness measurements are made along flats on the bar and these are plotted against distance from the quenched end as in Fig. 141, which also includes the cooling rates at various distances from the quenched end. The positions along the Jominy bar having cooling rates equal to those at the centres of bars of various diameters are marked in Fig. 141 for water and oil quenching.

From Fig. 141 the effectiveness of a given quench on various steels can be judged; for example, bars of $1\frac{1}{2}\%$ Ni Cr Mo steel up to 100 mm dia can be satisfactorily hardened in oil, whereas for the plain 0·95% carbon steel the maximum diameter of the fully hardened bar will be less than 12·5 mm for the same quench. The Jominy end quench test is not only a very reproducible test, but is also an excellent laboratory test for placing steels in order of merit as regards hardenability and it gives a very useful first approximation to the merits of a given steel. As to whether a steel with a particular hardenability will give satisfactory mechanical properties in a particular section must be decided by trial—actually, of course, the method of trial—apart from giving the final answer—is also, in many cases, far simpler than the involved mathematical processes which have been developed in an endeavour to predict the mechanical properties of a steel in any given section from the Jominy curve.

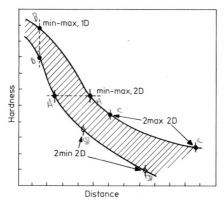

FIG. 142 Designation of hardenability limits

The hardenability of different casts of steel of the same analysis may vary substantially and similarly significant variations in hardenability may occur in samples taken from different parts of the same ingot. As a consequence Jominy results are given as bands not simple curves and BS 4437 uses two points to specify hardenability as shown in Fig. 142 and as follows:

(1) A desired hardness value at 2 desired distances, e.g.     A–A
(2) Minimum and maximum hardness values at desired distance   B–B
(3) Two maximum hardness values at 2 desired distances     C–C
(4) Minimum hardness values at 2 desired distances     D–D
(5) Any minimum and any maximum hardness.

## Surface hardening

### Case-hardening

Some parts of machinery must be very hard to resist surface wear and yet possess adequate ductility to resist breakage. To meet these requirements low carbon steel is employed; the carbon content at the surface is enriched by heating at 900 °C in carbon-rich material and the whole article is heat-treated to produce the desired properties. This is known as case hardening.

*The formation of the case* is obtained by treatment with (*a*) solid, (*b*) liquid, (*c*) gaseous media.

(*a*) Pack carburising consists in packing the roughly machined sections in some material, rich in carbon, in boxes which are covered with lids and luted to exclude air and the escape of gases. The boxes are heated at 900–950°C for a period depending on the depth of case desired. Any portions which are not required to be hard, are protected by a covering of some material which does not absorb carbon such as a mixture of asbestos and fireclay or copper plating (0·075 mm thick).

The carburising material usually consists of wood or bone charcoal, charred leather or petroleum coke, together with an energiser, such as barium carbonate (10–15%) which promotes rapid action on the steel. The absorption of carbon is due to the decomposition of carbon monoxide in the presence of carbon and iron according to the reversible reaction $2CO \rightleftharpoons CO_2 + C$ (nascent carbon is absorbed on surfaces).

The carbon is appreciably soluble only in austenite. The operation must be carried out, therefore, above the upper critical temperature. Once carbon has entered at the surface of the steel it tends to diffuse inwards at a rate which depends upon the composition of the steel and increases with the temperature of operation. This produces a carbon gradient from the outside to the inside. After casing, the articles are cooled down either in the box or quenched if the risk of distortion is not important.

(*b*) Liquid carburising baths consist essentially of 20–50% sodium cyanide with soda ash heated to 900–950°C in calorised iron pots. They are particularly suitable for light cases (0·08–0·25 mm) on mild steel articles subjected to light loads and also on higher carbon materials (0·35–0·50%) such as gears and mainshafts of cars.

The addition of barium chloride produces activated cyanide baths which are used for cases up to 1 mm depth. They give faster penetration, higher carbon (1·0%) and lower nitrogen (0·24%) concentration than normal baths. Cyanide hardening is quick and clean, enabling the article to be quenched from a pot, but the materials used and gases evolved are poisonous.

The chemical reactions that take place in a bath of molten cyanide in the presence of iron are not yet definitely known, but the result may be expressed as:
$$4NaCN + 4O_2 = 2Na_2CO_3 + 2CO + 2N_2.$$
Both the carbon monoxide and the nitrogen produce hardening of the surface of the metal.

(*c*) Gas carburising is carried out in a gas-tight container in an atmosphere consisting of a neutral carrier gas enriched with a hydrocarbon gas such as propane. Heating is more rapid and uniform than pack carburising.

The carbon potential can be closely controlled and can be high initially to give rapid absorption of carbon by the steel followed by a diffusing period at a lower level.

The depths of case vary from 0·25 mm on articles for light work, 0·5–1·0 mm for much automobile work, to 0·37 mm for roller bearing and ball races where compressive stresses are high.

The carbon content of the case has an important influence on the service of the article and a maximum of 0·9% at the outer edge is aimed at, although this can be exceeded if grinding is carried out after casing. With higher carbon contents free cementite net-works are formed which are extremely hard and brittle, consequently during grinding, or in service, cracks start at the cementite masses and cause outer layers to flake off, a phenomenon known as *exfoliation*.

*Heat-treatment after carburising*

As a result of prolonged heating at a high temperature in the carburising operation both the core and the case exhibit overheated structures, which would be unsatisfactory for severe service. The articles are heat-treated therefore, (*a*) to refine the core, (*b*) to refine and harden the case. 'Fine grain' (Al treated) steels are often used now which allow the core refining treatment to be omitted.

*Core refining* is accomplished by heating just above the upper-critical point for the core (870 °C, Fig. 112) at which the coarse ferrite-pearlite structure is replaced by fine grained austenite. After soaking, the article is quenched in water or oil to give a fine dispersion of ferrite in martensite (Fig. 123). At the same time all the cementite in the case will be taken into solution at 870 °C and the rapid cooling prevents any coarse net-works forming again. The case, however, which corresponds to a 0·9% carbon tool steel has coarsened because the temperature was well above its critical range. A coarse martensite is formed which is brittle. The articles are, therefore, given a second heat-treatment.

TABLE 20  EFFECT OF HEAT-TREATMENT ON THE PROPERTIES OF A CASE-HARDENING STEEL

C, 0·17; Si, 0·24; Mn, 0·72; S and P, 0·04
Bars 29 mm diameter. (WQ = water quench)

| Treatment | BH | YP $N/mm^2$ | TS $N/mm^2$ | El | RA | Izod ft lbf | J |
|---|---|---|---|---|---|---|---|
| Normalised at 920 °C | 137 | 308 | 479 | 34 | 64 | 80 | 108 |
| WQ 920 °C | 223 | 525 | 710 | 18 | 51 | 24 | 32 |
| WQ 920 °C reheated and WQ from 760 °C | 183 | 386 | 617 | 28 | 64 | 43 | 58 |

*The refining and hardening of the case* is produced by water quenching from 760 °C, which is just above the critical range for the 0·9 carbon case, but between $A_1$ and $A_3$ for the core as shown by Fig. 112. Fine-grained austenite is formed in the case, without appreciable growth occurring since soaking should not be prolonged, and on quenching a fine martensite is formed which is hard but not excessively brittle. At the same time the core undergoes tempering followed by the solution of the cementite to form islands of fine-grained austenite embedded in a matrix of the excess ferrite. On quenching, this austenite forms small martensite masses possibly fringed with troostite in a ferrite matrix. Tempering at 150 °C is advisable to relieve the quenching stresses.

The effect of these two treatments on the properties of the core are illustrated by the results in Table 20.

*Case-hardening steels*

Since toughness in the core is one of the most important characteristics of case-hardening steels the carbon content is usually below 0·2%, although it may be as high as 0·3% where greater support for the case is desired. For the latter purpose, however, it is advisable to use an alloy steel with high strength, combined with toughness. With alloy steels, also, the desired properties may be obtained by oil quenching, thereby reducing the risk of distortion to which water-quenched carbon steels are liable.

Plain carbon steels are given in Table 21:

TABLE 21

| Carbon % | Mn % | |
|---|---|---|
| 0·06–0·15 | 0·6 | Maximum core toughness, thin sections |
| 0·10–0·18 | 0·8 | General purposes: gears, shafts |
| 0·2–0·3 | 0·5 | High load carrying capacity, roller bearings |

To facilitate machining low-carbon steels are frequently processed in the drawn condition or sulphur with high manganese added intentionally (see free-cutting steel). It is desirable to have a 'blocky ferrite' structure with a fairly coarse grain size.

Manganese aids carburisation, increases depth of hardening but also increases the liability to cracking of the hardened case. Silicon retards carburisation, induces graphitisation and is usually below 0·3%.

*Alloy case-hardening steels*

The most common elements added to case-hardening steels are nickel and chromium. Boron (0·003%) is sometimes used with manganese.

SURFACE HARDENING 207

The main characteristics of these steels as compared with ordinary steels are:

(a) higher yield and maximum strength with almost equal impact strength and ductility;
(b) the critical hardening speed is lowered so that an oil quench enables the hardness to be obtained with less risk of distortion and cracking;
(c) grain growth is very slow at the carburising temperature, consequently for many purposes the preliminary treatment of core-refinement at 900°C may be dispensed with. This method of single quenching is usually restricted to parts having a light case where there is less possibility of the existence of cementite net-works. Certain plain carbon steels of the 'inherent fine grain' type (see p. 167) do not suffer from grain growth during casing and can be rendered fit for service by a single quench;
(d) nickel retards carburisation slightly, and the hardness of the case is lower than that on carbon steels, but the case has a good resistance to wear;
(e) cracking in grinding, exfoliation and spalling of sharp corners, are experienced to a less degree.

The addition of 0·3–0·8% chromium alone is useful where a hard case is essential and core properties are less important.

Chromium added to a nickel steel produces a marked improvement in the wear resisting properties of the case, increases the tensile strength of the core without serious loss of ductility, and also diminishes the mass effect. A popular case-hardening steel is that with 4·0% nickel, 1·1% chromium, and 0·2 to 0·3% molybdenum, which is capable of withstanding the highest surface stresses, even with a comparatively thin case, and is suitable for gears. It is, however, difficult to machine and somewhat prone to over-carburisation. Particulars of typical alloy steels covering a range of core strengths are given in Table 22.

*Selection of steels*

For maximum surface hardness without severe shocks:
Carbon steel (0·1), $1\frac{1}{2}$% manganese (also with addition of sulphur); or chromium 0·3–0·8.
For parts subject to severe shocks and high stresses (automobile gears, steering worm and quadrant, overhead valve gear):

$1\frac{3}{4}$% Ni–Mo, $3\frac{1}{2}$% Ni–Cr–Mo.
1 Mn, 1 Cr used widely in Europe for auto-gears and spline shafts.

As a compromise between surface hardness, load carrying capacity and

TABLE 22  DETAILS OF ALLOY CASE-HARDENING STEELS—BS970

| BS No. | Name | Typical Composition | | | Min. N/mm² | Core Properties E | Izod ft lbf | Izod Joule |
|---|---|---|---|---|---|---|---|---|
| | | Ni | Cr | Mo | | | | |
| 635M15 | ¾% Ni–Cr | 0·9 | 0·6 | | 770 | 15 | 25 | 32 |
| 655M13 | 3¼% Ni–Cr | 3·3 | 0·8 | | 925 | 10 | 35 | 47 |
| 659M15 | 4% Ni–Cr | 4·1 | 1·2 | | 1320 | 8 | 25 | 32 |
| 665M17 | 1¾% Ni–Mo | 1·8 | | 0·25 | 770 | 12 | 35 | 47 |
| 665M23 | 1¾% Ni–Mo > C | 1·8 | | 0·25 | 925 | 10 | 12 | 16 |
| 6805M17 | ½% Ni–Cr–Mo | 0·5 | 0·5 | 0·2 | 770 | 12 | 20 | 27 |
| 815M17 | 1½% Ni–Cr–Mo | 1·5 | 1·0 | 0·15 | 1080 | 8 | 20 | 27 |
| 822M17 | 2% Ni–Cr–Mo | 2·0 | 1·5 | 0·2 | 1320 | 8 | 25 | 32 |
| 832M13 | 3½% Ni–Cr–Mo | 3·5 | 0·8 | 0·2 | 1080 | 12 | 30 | 40 |
| 835M15 | 4% Ni–Cr–Mo | 4·1 | 1·2 | 0·2 | 1320 | 8 | 25 | 32 |

| | Heat-Treatment | |
|---|---|---|
| Steels | 655, 665, 659, 832, 835 | 635, 805, 815, 822 |
| Carburise °C | 880/930 | 880/930 |
| Single Oil Quench °C | 820/840 | 820/840 |
| Refine °C | 850/880 | 850/880 |
| Cooling* | A O W | A O W |
| Harden °C Oil Q | 760/780 | 800/820 |
| Temper °C max. | 150 | 150 |

*A = Air   O = Oil   W = Water
5th and 6th Digits of BS No. is 100 times mean carbon percentage

shock, also for intricate and varying sections (Crown wheels, bevel pins, aero reduction gears):

4% Ni–Cr–Mo, or 2% Ni–Cr–Mo for economy.

*Abnormality*

Certain steels are susceptible to the occurrence of soft spots in spite of correct treatment and carbon content. This defect frequently occurs in American steels, and such steels are called abnormal steels. These show, in the zone containing more than 0·9% carbon, a decided tendency of the pearlite to be very coarse and irregularly lamellar and to have broken down completely at the (austenite) grain boundaries to form massive cementite with free ferrite on either side. The core of the steel is always fine-grained and the effect seems to be due to the use of aluminium as a deoxidiser, a practice far more common in America than in this country. It appears that an abnormal steel has a much higher critical cooling rate than normal steel and therefore requires drastic quenching to retain martensite. In water quenching the steam bubbles formed on the surface of the article retard the cooling and produce soft spots. The McQuaid-Ehn test, described on p. 168, provides a method of detecting abnormal steels.

*Nitrogen hardening*

A process of surface-hardening devised by A. Fry consists in enriching with nitrogen the surface layers of special 'Nitralloy' steels containing 1% aluminium, 1·5% chromium and 0·2% molybdenum. A thin but extremely hard case is produced.

The machine-finished and heat-treated parts are heated at 500°C for 40–90 hours in a gas-tight box through which ammonia gas is circulated, followed by cooling in the box. About 30% of the ammonia dissociates ($NH_3 \rightleftharpoons 3H + N$) and part of the nascent nitrogen is absorbed by the surface layers of the steel. With plain carbon steels iron nitrides (sometimes stated to be $Fe_4N$ and $Fe_2N$) are formed to a greater depth, but the case is less hard and more brittle than in nitralloy steels. The aluminium appears to form stable nitrides which do not diffuse readily, and a shallow (less than 1 mm) but intensely hard (900–1100 VPN) case is formed. The chromium contributes to the hardness and also flattens the hardness gradient below the surface, thus reducing the risk of spalling. The molybdenum also increases hardenability and prevents embrittlement. A nitrided case consists of two layers, an outer brittle white zone 0·01 mm thick, preferably removed by lapping; and an inner zone containing alloy nitrides. With large grain size coarse filaments of nitride form at the boundaries.

The desire for better core properties led to the introduction of steels with lower aluminium content (0·4%). Aluminium free steels containing chromium and molybdenum are now being used where extreme surface hardening is not necessary, such as aero engine crankshafts and air screws. The case is very tough with a diamond hardness of about 750.

The carbon contents of steels used for nitriding vary from 0·2 to 0·5%; the lower content being used for lightly stressed wearing parts, such as spindles and gears, while the higher carbon steels are used to withstand high local pressures such as in die blocks and dies for moulding plastics. The stems of austenitic steel valves for aero work are also nitrided, and this is facilitated by a previous deposition of copper on the surface to be hardened.

By suitable initial hardening and tempering treatments, a wide range of tensile strengths can be obtained even in large sections.

Typical properties are given in Table 23.

The hardness of the core is not appreciably affected by nitriding if tempering has been carried out at 600–650°C and this enables all core treatments to be carried out prior to nitriding. Internal stresses must be removed by heating at 540 °C for 5 hours between rough and final machining and 0·025 mm must be allowed for the expansion which occurs during nitriding. Parts which are required soft are protected by a coat of tin or solder.

The advantages of nitriding are that quench cracks and distortion are prevented, since no subsequent heat-treatments are necessary. The case is in compression and the core in tension and distortion can occur in slender and asymmetrical components. Keyways should be duplicated at diametrically opposite positions and local grinding avoided since distortion

TABLE 23   NITRIDING STEELS

| Composition | | | | Heat-treatment °C | | YP | TS | | | Izod | | Hardness | |
|---|---|---|---|---|---|---|---|---|---|---|---|---|---|
| C | Cr | Mo | | OQ | Te'p'r | N/mm² | N/mm² | El | RA | ft lbf | J | Core | Case |
| 0·39 | 1·6 | 0·2 | Al 1·1 | 900 | 650 | 625 | 880 | 13 | 59 | 40 | 54 | 269 | 1050 |
| 0·29 | 0·88 | 1·1 | Mo 0·9 | 875 | 640 | 925 | 990 | 13 | 57 | 50 | 68 | 293 | 600 |
| 0·39 | 3·2 | 1·0 | V 0·12 | 940 | 625 | 1158 | 1310 | 8 | 63 | 15 | 20 | 380 | 900 |
| 0·5 | 1·3 | 1·0 | Ni 1·8 | 870 | 620 | 1234 | 1450 | 11 | 49 | 25 | 34 | — | 700 |

can occur slowly. Thin walled cylinders can 'bell-mouth' unless supported by removable collars. Resistance to corrosion is good and the extremely hard case retains its hardness up to 500 °C as contrasted with the softening of case-carburised steels at 200 °C and upwards. The treatment is ideal for such parts as link pins, crankshafts, printing dies, pump parts, where resistance to wear and galling is essential.

Cast iron containing about 1·5 % each of chromium and aluminium can be nitrided at 510 °C to give a case 0·37 mm thick and 900 VP hardness. It is principally used for cylinder liners.

*Carbonitriding* is a gaseous process for the simultaneous diffusion of carbon and nitrogen into steel and is carried out in a carburising gas to which is added a nitriding gas usually anhydrous ammonia (2–5 %). Steels employed are similar to those used for carburising and the temperature range is 800–875 °C, thus reducing distortion. Copper plating is used for 'stopping off'.

*The Sulfinuz process* for accelerating running-in and reducing wear involves immersion of preheated parts in a molten bath of cyanide and sulphur compounds at 570 °C for about 2 hours. The first 0·025 mm layer of the surface contains about 5 % S, 2 % N and 2 % C.

## Flame and induction hardening

Both of these processes consist in heating the surface to be hardened above the $Ac_3$ critical point and cooling rapidly. The steel must contain sufficient carbon or alloys to harden by quenching. The gas flame or Shorter process is used for hardening gears, cams, etc. The induction or Tocco hardening is used for hardening crankshaft journals and heating is effected by the eddy currents induced in the surface layers (see p. 155). Heating is very rapid and carbides should be small to facilitate their solution. It is also desirable to use steels with low $Ac_1$ and $Ac_3$ temperatures, little or no free ferrite and medium to deep hardenability.

# 11  Alloy Steels

During the last fifty years engineers have demanded steels with higher and higher tensile strength, together with adequate ductility. This has been particularly so where lightness is desirable, as in the automobile and aircraft industries. An increase in carbon content met this demand in a limited way, but even in the heat-treated condition the maximum strength is about 700 N/mm² above which value a rapid fall in ductility and impact strength occurs and mass effects limit the permissible section. Heat-treated alloy steels provide high strength, high yield point, *combined* with appreciable ductility even in large sections.

The use of plain carbon steels frequently necessitates water quenching accompanied by the danger of distortion and cracking, and even so only thin sections can be hardened throughout. For resisting corrosion and oxidation at elevated temperatures, alloy steels are essential.

The Alloy Steels Research Committee adopted the following definition: 'Carbon steels are regarded as steels containing not more than 0·5% manganese and 0·5% silicon, all other steels being regarded as alloy steels.'

The principal alloying elements added to steel in widely varying amounts either singly or in complex mixtures are nickel, chromium, manganese, molybdenum, vanadium, silicon and cobalt.

*The effect of the alloying element* in the steel may be one or more of the following:

(1) It may go into *solid solution* in the iron, enhancing the strength. The general effectiveness is shown in Fig. 143.
(2) *Hard carbides* associated with $Fe_3C$ may be formed.
(3) It may form intermediate compounds with iron, e.g. FeCr (sigma phase), $Fe_3W_2$.
(4) It may *influence the critical range* in one or more of the following ways:

  (*a*) Alter the temperature. For example, 3% nickel lowers the Ac points some 30°C, while 12% chromium raises the $Ac_1$ temperature to about 800°C and also forms a range of 150/200°C above this in which the pearlite changes to austenite. Fig. 144 shows the effect of alloys on the eutectoid temperature.
  (*b*) Alter the carbon content of the eutectoid (Fig. 144). The carbon content of the pearlite in a 12% chromium steel is 0·33%, as com-

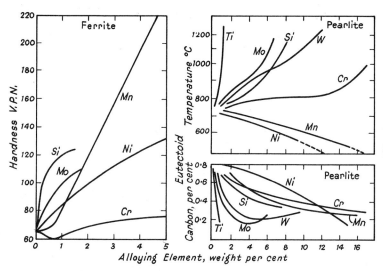

FIG. 143 Hardening effects of alloying elements in solid solution in fully annealed ferrite (*Austin*)

FIG. 144 Effects of alloying elements on the carbon content and temperature of the eutectoid point (*Bain*)

pared with 0·87 in an ordinary steel. Nickel also reduces the amount of carbon in the pearlite and consequently increases the volume of this constituent at the expense of the weaker ferrite. This is illustrated by Figs. 151 and 152, which show the change in structure produced by adding 3% nickel to a 0·3% carbon steel. It has the appearance of a 0·5% carbon steel and it has a similar tensile strength.

(c) *Alter the 'critical cooling velocity'*, which is the minimum cooling speed which will produce bainite or martensite from austenite. Typical critical speeds obtained by quenching from 950 °C are given in Table 24.

TABLE 24   EFFECT OF ALLOYING ON THE CRITICAL COOLING SPEED OF STEEL

| Carbon | Alloying Element | Cooling Speed to form Martensite, |
|---|---|---|
| | % | °C per sec (650 °C) |
| 0·42 | 0·55 Mn | 550 |
| 0·40 | 1·60 Mn | 50 |
| 0·42 | 1·12 Ni | 450 |
| 0·40 | 4·80 Ni | 85 |
| 0·38 | 2·64 Cr | 10 |

The efficiency of the additions of the various alloy elements in reducing the effect of mass during quenching may be judged by the relative reduction of the critical velocity of the steel. Chromium and manganese respectively are far more effective than nickel. See p. 200 for effect on hardenability.

(5) *The volume change* from austenite to martensite may be influenced as shown in Fig. 145. Combinations of elements can be chosen so that the volume change is reduced and also the risk of quench cracking. It may *produce effects characteristic of the alloying element*.

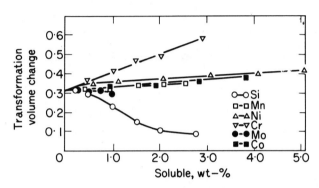

FIG. 145 Effect of alloying elements on the transformation volume change austenite to martensite

ISI Centenary Meeting preprint

(*a*) It may render the alloy *sluggish* to thermal changes, increasing the stability of the hardened condition and so producing tool steels which are capable of being used up to 550 °C without softening and in certain cases may exhibit an increase in hardness.

(*b*) It may have a chemical effect on the impurities. Under suitable slag conditions vanadium, in quite small quantities, 'cleans' the steel and renders it free from slag inclusions. Manganese and zirconium form sulphides.

(*c*) Certain elements such as chromium, aluminium, silicon and copper tend to produce adherent oxide films on the surface of the steel which increase its resistance to corrosion and oxidation at elevated temperatures.

(*d*) Creep strength may be increased by the presence of a dispersion of fine carbides, e.g. molybdenum (p. 233).

## Classification of alloying additions

Classification of alloying metals according to their effect in the steel is difficult, because the influence varies so widely with each addition depending on the quantity used and other elements present. A useful grouping, however, is based upon the effect of the element on (*a*) the stability of the carbides and (*b*) the stability of the austenite.

(1) *Elements which tend to form carbides.* Chromium, tungsten, titanium, columbium, vanadium, molybdenum and manganese.

The mixture of complex carbides is often referred to as cementite.

(2) *Elements which tend to graphitise the carbide.* Silicon, cobalt, aluminium and nickel.

Only a small proportion of these elements can be added to the steel before graphite forms during processing, with attendant ruin of the properties of the steel, unless elements from group 1 are added to counteract the effect.

(3) *Elements which tend to stabilise austenite.* Manganese, nickel, cobalt and copper (e.g. Figs. 157 and 158).

These elements alter the critical points of iron in a similar way to carbon by raising the $A_4$ point and lowering the $A_3$ point, thus increasing the range in which austenite is stable, and they also tend to retard the separation of carbides. They have a crystal lattice (f.c.c.) similar to that of $\gamma$-iron in which they are more soluble than in $\alpha$-iron.

(4) *Elements which tend to stabilise ferrite.* Chromium, tungsten, molybdenum, vanadium and silicon (Figs. 157 and 158).

These elements are more soluble in $\alpha$-iron than in $\gamma$-iron. They diminish the amount of carbon soluble in the austenite and thus tend to increase the volume of free carbide in the steel for a given carbon content. On the binary equilibrium diagram of these elements with pure iron the $A_4$ point is lowered and $A_3$ raised (although it may be lowered initially), until the two points merge to form a '*closed gamma loop*'. Thus with above a certain amount of each of these elements the austenite phase disappears and ferrite exists from the melting-point down to room temperature. No critical points exist and such steels (e.g. 18% chromium irons) are not amenable to normal heat treatment, except recrystallisation after cold work. This effect, however, can be counteracted by adding elements from group 3. For example, 2% of nickel is added to the 18% chromium stainless steel to enable it to be refined by normal heat-treatment; carbon has the same effect. Aluminium has the reverse effect in 12% chromium steel.

The approximate amounts at which the $\gamma$-iron disappears when no carbon is present are:

|            | Cr   | Si  | W | P   | Al  | Ti   | Mo |
|------------|------|-----|---|-----|-----|------|-----|
| Percentage | 12·8 | 2·0 | 6 | 0·5 | 1·1 | 0·75 | 4 |

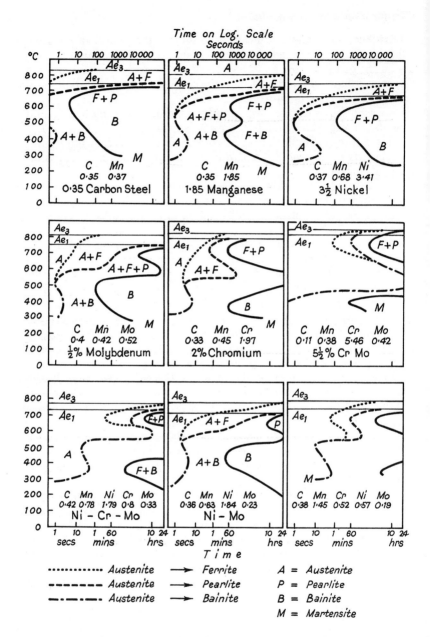

FIG. 146 TTT-curves for alloy steels (after US Steel Corp.)

## Characteristics of alloying elements

### Manganese

All commercial steels contain 0·3–0·8 % manganese, to reduce oxides and to counteract the harmful influence of iron sulphide (see free-cutting steel). Any manganese in excess of these requirements partially dissolves in the iron and partly form $Mn_3C$ which occurs with the $Fe_3C$. Manganese lowers $Ar_3$ and $Ar_1$ and quite small quantities increase hardenability, which is also indicated in Fig. 146. There is a tendency nowadays to increase the manganese content and reduce the carbon content in order to get a steel with an equal tensile strength but improved ductility. This is illustrated by the following results on normalised 29 mm diameter bars.

TABLE 25

| C % | Mn % | TS N/mm² | YP N/mm² | El | RA | Izod ft lbf | J |
|---|---|---|---|---|---|---|---|
| 0·38 | 0·70 | 649 | 417 | 25 | 46 | 38 | 51 |
| 0·30 | 1·34 | 649 | 402 | 30 | 64 | 61 | 83 |

If the manganese is increased above 1·8 % the steel tends to become air-hardened, with resultant impairing of the ductility. Up to this quantity, manganese has a beneficial effect on the mechanical properties of oil hardened and tempered 0·4 % carbon steel.

The manganese content is also increased in certain alloy steels, with a reduction or elimination of expensive nickel, in order to reduce costs. Steels with 0·3–0·4 % carbon, 1·3–1·6% manganese and 0·3 % molybdenum have replaced 3 % nickel steel for some purposes.

Non-shrinking tool steel contains up to 2% manganese, with 0·8–0·9 % carbon.

Steels with 5 to 12% manganese are martensitic after slow cooling and have little commercial importance.

Hadfield's manganese steel contains 12 to 14% of manganese and 1·0% of carbon. It is characterised by a great resistance to wear and is therefore used for railway points, rock drills and stone crushers. Austenite is completely retained by quenching the steel from 1000 °C, in which soft condition it is used, but abrasion raises the hardness of the surface layer from 200 to 600 VPN (with no magnetic change), while the underlying material remains rough. Annealing embrittles the steel by the formation of carbides at the grain boundaries. Nickel is added to electrodes for welding manganese steel and 2% Mo sometimes added, with a prior carbide dispersion treatment at 600 °C, to minimise initial distortion and spreading.

## Nickel

Nickel and manganese are very similar in behaviour and both lower the eutectoid temperature (Fig. 143). This change point on heating is lowered progressively with increase of nickel (approximately 10°C for 1% of nickel), but the lowering of the change on cooling is greater and irregular. The temperature of this change ($Ar_1$) is plotted for different nickel contents for 0·2% carbon steels in Fig. 147, and it will be seen that the curve takes a sudden plunge round about 8% nickel. A steel with 12% nickel begins to transform below 300°C on cooling, but on reheating the reverse change does not occur until about 650°C. Such steels are said to exhibit pronounced lag or hysteresis and are called *irreversible steels*. This characteristic is made use of in maraging steels (p. 232) and 9% Ni cryogenic steel.

TTT-curve (Fig. 146) shows that nickel retards the transformation of austenite at all temperature levels, but less markedly than manganese.

The addition of nickel acts similarly to increasing the rate of cooling of a carbon steel. Compare Figs. 147 and 134 to note the similarity of the curves. Thus with a *constant rate of cooling* the 5–8% nickel steels become troostitic; at 8–10% nickel, where the sudden drop appears, the structure is martensitic, while above 24% nickel the critical point is depressed below room temperature and austenite remains. The lines of demarcation are not so sharp as indicated by Fig. 147, but a gradual transition occurs from one constituent to another.

The mechanical properties change accordingly as shown in the lower part of Fig. 147. Steels with 0–5% nickel are similar to carbon steel, but are stronger, on account of the finer pearlite formed and the presence of nickel in solution in the ferrite (see also p. 195). When 10% nickel is exceeded the steels have a high tensile strength, great hardness, but are brittle, as shown by the Izod and elongation curves. When the nickel is sufficient to produce austenite the steels become non-magnetic, ductile, tough and workable, with a drop in strength and elastic limit.

Carbon intensifies the action of nickel and the change points shown in Fig. 147 will vary according to the carbon content. The influence of carbon and nickel on the structure are shown in the small inset (Guillet) diagram in Fig. 147, for one rate of cooling.

Steels containing 2 to 5% nickel and about 0·1% carbon are used for case hardening; those containing 0·25 to 0·40% carbon are used for crankshafts, axles and connecting rods.

The superior properties of low nickel steels are best brought out by quenching and tempering (550–650°C). Since the $Ac_3$ point is lowered, a lower hardening temperature than for carbon steels is permissible and also a wider range of hardening temperatures above $Ac_3$ without excessive grain growth, which is hindered by the slow rate of diffusion of the nickel.

Martensitic nickel steels are not utilised and the austenitic alloys cannot compete with similar manganese steels owing to the higher cost. Maraging steels have fulfilled a high tensile requirement in aero and space fields. High nickel alloys are used for special purposes, owing to the marked

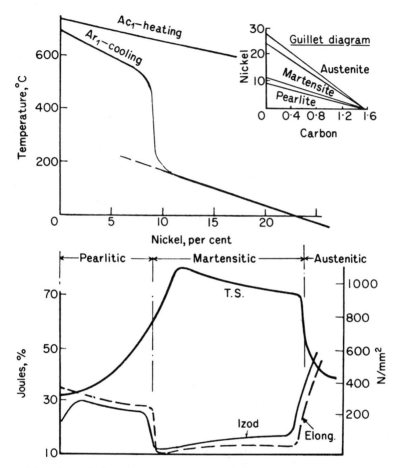

FIG. 147 Effect of nickel on change points and mechanical properties of 0·2% carbon steels cooled at a constant rate

influence of nickel on the coefficient of expansion of the metal. With 36% nickel, 0·2% carbon, 0·5% manganese, the coefficient is practically zero between 0° and 100°C. This alloy ages with time, but this can be minimised by heating at 100°C for several days. The alloy is called *invar* and it is used extensively in clocks, tapes and wire measures, differential expansion regulators, and in aluminium pistons with a split skirt in order to give an expansion approximating to that of cast iron.

A carbon-free alloy containing 78·5% nickel and 21·5% iron has a high permeability in small magnetic fields (p. 299).

*Chromium*

Chromium can dissolve in either $\alpha$- or $\gamma$-iron, but, in the presence of carbon, the carbides formed are cementite $(FeCr)_3C$ in which chromium may rise to more than 15%; chromium carbides $(CrFe)_3C_2$, $(CrFe)_7C_3$, $(CrFe)_4C$, in which chromium may be replaced by a few per cent, by a maximum of 55% and by 25% respectively. Stainless steels contain $Cr_4C$.

The pearlitic chromium steels with, say, 2% chromium are extremely sensitive to rate of cooling and temperature of heating before quenching; for example:

| Temp. of Initial Heating, °C | Critical Hardening Rate (Mins to cool from 836° to 546°C) |
|---|---|
| 836 | 3·5 |
| 1010 | 6·5 |
| 1200 | 13 |

The reason is that the chromium carbides are not readily dissolved in the austenite, but the amount increases with increase of temperature. The effect of the dissolved chromium is to raise the critical points on heating (Ac) and also on cooling (Ar) when the rate is slow. Faster rates of cooling quickly depress the Ar points with consequent hardening of the steel. Chromium imparts a characteristic form of the upper portion of the isothermal transformation curve (Fig. 146).

The percentage of carbon in the pearlite is lowered (Fig. 144). Hence the proportion of free cementite (hardest constituent) is increased in high carbon steel and, when the steel is properly heat-treated, it occurs in the spheroidised form which is more suitable when the steel is used for ball bearings. The pearlite is rendered fine.

When the chromium exceeds 11% in low-carbon steels an inert passive film is formed on the surface which resists attack by oxidising reagents—see Chap. 12. Still higher chromium contents are found in heat-resisting steel.

Chromium steels are easier to machine than nickel steels of similar tensile strength. The steels of higher chromium contents are susceptible to temper brittleness if slowly cooled from the tempering temperature through the range 550/450°C. These steels are also liable to form surface markings, generally referred to as 'chrome lines'.

The chrome steels are used wherever extreme hardness is required, such as in dies, ball bearings, plates for safes, rolls, files and tools. High chromium content is also found in certain permanent magnets.

*Nickel and chromium*

Nickel steels are noted for their strength, ductility and toughness, while chromium steels are characterised by their hardness and resistance to wear. The combination of nickel and chromium produces steels having all these properties, some intensified, without the disadvantages associated with the simple alloys. The depth of hardening is increased, and with 4·5% nickel, 1·25% chromium and 0·35% carbon the steel can be hardened simply by cooling in air.

Low nickel–chromium steels with small carbon content are used for case-hardening (see Table 22), while for most constructional purposes the carbon content is 0·25–0·35%, and the steels are heat-treated to give the desired properties.

Considerable amounts of nickel and chromium are used in steel for resisting corrosion and oxidation at elevated temperatures (see Chap. 12).

*Embrittlement.* The effects of tempering a nickel–chromium steel are shown in Fig. 148, from which it will be noticed that the Izod impact curve No. 1 reaches a dangerous minimum in the range 250–450 °C in common with many other steels. This is known as 350 °C embrittlement. Phosphorus and nitrogen have a significant effect while other impurities (As, Sb, Sn) and manganese in larger quantity may also contribute to the embrittlement.

*Temper brittleness* is usually used to describe the notch impact intergranular brittleness* induced in some steels by *slow cooling* after tempering above about 600 °C and also from prolonged soaking of tough material between about 400° and 550 °C. Temper brittleness seems to be due to grain boundary enrichment with alloying elements—Mn, Cr, Mo—during austenitising which leads to enhanced segregation of embrittling elements—P, Sn, Sb, As—by chemical interaction on slow cooling from 600 °C. The return to the tough condition, obtained by reheating embrittled steel to temperatures above 600 °C and *rapidly cooling*, is due to the redistribution and retention in solution of the embrittling segregation. Antimony (0·001 %), phosphorus (0·008%), arsenic, tin, manganese increase, while *molyb-*

---

*Grain boundaries are revealed in temper brittle samples by etching in 1 gm cetyl trimethyl ammonium bromide; 20 gm picric acid; 100 cc distilled water, 100 cc ether. Shake mixture, allow to stand for 24 hrs; use portion of top layer and return to tube afterwards.

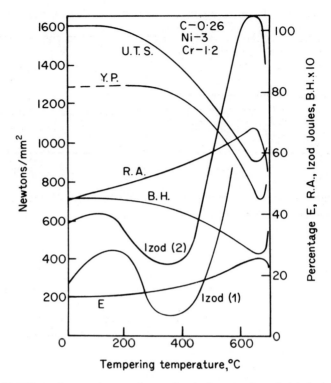

FIG. 148 Effect of tempering on the mechanical properties of nickel–chromium steel, C 0·26, Ni 3, Cr 1·2, 29 mm diam, bars hardened in oil from 830 °C. Izod (2) for steel with 0·25% molybdenum added

*denum decreases the susceptibility of a steel to embrittlement.* 0·25% molybdenum reduces the brittleness as shown by Izod curve No. 2. Table 26 illustrates the effect rate of cooling after tempering and the influence of an addition of 0·45% molybdenum:

TABLE 26   STEEL: 0·3% CARBON, 3·5% NICKEL, 0·7% CHROMIUM, TEMPERED AT 630 °C

| Steel | Cooling Rate | TS N/mm² | El | RA | Izod ft lbf | J |
|---|---|---|---|---|---|---|
| Ni–Cr | Oil | 896 | 18 | 60 | 64 | 87 |
| Ni–Cr | Furnace | 880 | 18 | 60 | 19 | 25 |
| Ni–Cr–Mo | Furnace | 896 | 18 | 61 | 59 | 80 |

## Molybdenum

Molybdenum dissolves in both α- and γ-iron and in the presence of carbon forms complex carbides $(FeMo)_6C$, $Fe_{21}Mo_2C_6$, $Mo_2C$.

Molybdenum is similar to chromium in its effect on the shape of the TTT-curve (Fig. 146) but up to 0·5% appears to be more effective in retarding pearlite and increasing bainite formation.

Additions of 0·5% molybdenum have been made to plain carbon steels to give increased strength at boiler temperatures of 400 °C, but the element is mainly used in combination with other alloying elements.

Ni–Cr–Mo steels are widely used for ordnance, turbine rotors and other large articles, since molybdenum tends to minimise temper brittleness and reduces mass effect.

Molybdenum is also a constituent in some high-speed steels, magnet alloys, heat-resisting and corrosion-resisting steels.

## Vanadium

Vanadium acts as a scavenger for oxides, forms a carbide $V_4C_3$ and has a beneficial effect on the mechanical properties of heat-treated steels, especially in the presence of other elements. It slows up tempering in the range of 500–600 °C and can induce secondary hardening. Chromium–vanadium (0·15%) steels are used for locomotive forgings, automobile axles, coil springs, torsion bars and creep resistance.

## Tungsten

Tungsten dissolves in γ-iron and in α-iron. With carbon it forms WC and $W_2C$, but in the presence of iron it forms $Fe_3W_3C$ or $Fe_4W_2C$. A compound with iron—$Fe_3W_2$—provides an age-hardening system. Tungsten raises the critical points in steel and the carbides dissolve slowly over a range of temperature. When completely dissolved, the tungsten renders transformation sluggish, especially to tempering, and use is made of this in most hot-working tool ('high speed') and die steels. Tungsten refines the grain size and produces less tendency to decarburisation during working.

Tungsten is also used in magnet, corrosion- and heat-resisting steels.

## Silicon

Silicon dissolves in the ferrite, of which it is a fairly effective hardener, and raises the Ac change points and the Ar points when slowly cooled and also reduces the γ–α volume change.

Only three types of silicon steel are in common use—one in conjunction with manganese for springs; the second for electrical purposes, used in sheet form for the construction of transformer cores, and poles of dynamos and motors, that demand high magnetic permeability and electrical resistance; and the third is used for automobile valves.

TABLE 27

|   | C | Si | Mn |
|---|---|---|---|
| 1. Silico–manganese | 0·5 | 1·5 | 0·8 |
| 2. Silicon steel | 0·07 | 4·3 | 0·09 |
| 3. Silichrome | 0·4 | 3·5 | 8 Cr |

It contributes oxidation resistance in heat-resisting steels and is a general purpose deoxidiser.

*Copper* dissolves in the ferrite to a limited extent; not more than 3·5% is soluble in steels at normalising temperatures, while at room temperature the ferrite is saturated at 0·35%. It lowers the critical points, but insufficiently to produce martensite by air cooling. The resistance to atmospheric corrosion is improved and copper steels can be temper hardened (p. 97).

*Cobalt* has a high solubility in α- and γ-iron but a weak carbide-forming tendency. It decreases hardenability but sustains hardness during tempering. It is used in 'Stellite' type alloys, gas turbine steel, magnets and as a bond in hard metal.

*Boron.* In recent years, especially in USA, 0·003–0·005% boron has been added to previously fully killed, fine-grain steel to increase the hardenability of the steel. The yield ratio and impact are definitely improved, provided advantage is taken of the increased hardenability obtained and the steel is fully hardened before tempering. In conjunction with molybdenum boron forms a useful group of high tensile bainitic steels. Boron is used in some hard facing alloys and for nuclear control rods.

**Strengthening mechanisms in alloy steel**

The solid-solution hardening of carbon has a major effect on the strength of martensite, but ductility can only be obtained at low carbon levels.

Although alloying elements affect hardenability (p. 200) they have a minor effect on hardness except to reduce it at high carbon levels by causing austenite to be retained, Fig. 128.

Alternative ways of improving the strength of alloy steels are:

(1) *Grain refinement* which increases strength and *ductility*. This can be developed by severely curtailing the time after the cessation of forging at some low temperature of austenite stability or by rapid heating, coupled with a short austenitising period. Fine grain is produced in 9% Ni steel (p. 216) by tempering fine lath martensite.

(2) *Precipitation hardening* by carbide, nitride or intermetallic compounds.
   (a) by secondary hardening (p. 188), e.g. 12% Cr steel with additions p. 246.
   (b) Age hardening a low carbon Fe–Ni lath martensite supersaturated with substitutional elements, e.g. maraging, p. 232.
   (c) Age hardening of austenite, e.g. stainless steels, see p. 246. Phosphorus and titanium are common additions.
   Stacking faults are often associated with fine carbide precipitates, and strengths could be raised by increasing the number of stacking faults (i.e. lower fault energy).
   (d) Controlled transformation 18/8 austenite steels (p. 246) in which transformation to martensite is induced by refrigeration or by strain.

(3) *Thermomechanical treatments* which may be classified into three main groups:
   (i) *Deformation of austenite prior to transformation.*
   *Ausforming* consists of deforming a steel in a metastable austenitic condition between $Ac_1$ and $M_s$ (e.g. 500°C called LT) followed by transformation to martensite and light tempering (Fig. 149). This results in increased dislocation density in the martensite and a finer carbon precipitation on tempering. Strengths up to 1800 N/mm² can be obtained without impairing the ductility ($\sim$ 6 N/mm²% deformation). Steels must possess a TTT-curve with a large bay of stable austenite, e.g. 826 M40. Typical application is leaf springs.

TABLE 28 IMPROVEMENT IN STRENGTH SHOWN BY AUSFORMED STEEL

| Treatment | 0·4 C, 2·5 Ni, 0·7 Cr, 0·5 Mo | | | | | |
|---|---|---|---|---|---|---|
| | 0·2 PS N/mm² | TS N/mm² | E | RA | Izod ft lbf | J |
| OQ 830°C, T 300°C | 1400 | 1600 | 12 | 42 | 16 | 21 |
| 80% Red at 500°C | 1550 | 2300 | 9 | 24 | 11 | 14 |
| Ditto + T 300°C | 1600 | 1800 | 11 | 45 | 10 | 13 |

Deformation of stable austenite just above $Ac_3$ before cooling (called HT). The properties are somewhat inferior to those produced by ausforming.
*Deformation induced transformation* originally used in Hadfield 13% Mn steel, but can be adapted to metastable austenitic stainless steels. The fully austenitic steel is severely warm-worked

above the lowest temperature at which martensite is produced during straining. The distinctive property is the high rate of straining hardening which increases ductility.

(ii) *Deformation of austenite during transformation*

*Isoforming* is the deformation of a steel (e.g. 1% Cr) during the isothermal transformation to pearlite, which refines the structure and improves fracture toughness (Fig. 149). A somewhat similar thermomechanical process can be used in the bainitic region, producing bainite and martensite.

*Zerolling* consists in forming martensite by deformation at sub-zero temperatures to strengthen 18/8 austenitic steels. The amount of martensite is influenced by alloy composition and increased with deformation and lowering of the temperature.

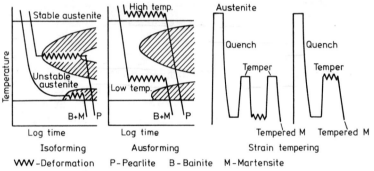

Fig. 149 Methods of thermomechanical treatment

(iii) *Deformation after transformation of austenite*

*Marforming* consists of deforming the maraging steel (p. 232) in the soft martensitic condition—generally cold—and there is a pronounced increase in strength of the subsequent maraged product.

With other steels, considerable increases in strength can be obtained by straining martensite ($\sim 3\%$) either in the untempered or tempered condition. A strengthening effect also occurs on re-tempering, probably due to the resolution and re-precipitation of the carbides in a more finely dispersed form.

*Strain tempering and dynamic strain ageing (Fig. 149)*

Both processes involve about $\frac{1}{2}$–5% deformation at room temperature between two stages of tempering—i.e. strain tempering—while in dynamic strain ageing deformation is concurrent with tempering.

## Applications of alloy steels

Alloy steels may be divided into four classes:

(1) Structural steels, which are subjected to stresses in machine parts.
(2) Tool and die steels.
(3) Magnetic alloys.
(4) Stainless and heat-resisting steels.

### Structural steels

The structural steels can be grouped conveniently on the basis of tensile strength although the dividing lines between the classes are ill defined owing to the wide variation in properties obtained from one steel by varying the heat-treatment, and the ruling section in which the properties are required. The basis of design of machine components generally fall into either static or dynamic loading. For static loading the proof stress should be used with a safety factor of 2 or less. For dynamic loading the working stresses must be related to the fatigue limit which is about 55% of tensile strength for mild steel but only 50% at 770 N/mm² and 40% at 1540 N/mm² tensile strength. Notch sensitivity is also very important. Only a brief survey of the common types is possible and BS 970 should be consulted for more detailed examples.

### Structural steels below 680 N/mm² tensile

The main object of steels in this class is to enable lighter structures to be built by the use of relatively high tensile steels while retaining as far as possible the highly desirable properties of easy workability, adaptability and insensitivity to faulty manipulation possessed by mild steel. These steels may be used in the hot rolled condition.

Four characteristics are important: (1) yield strength for design, (2) notch ductility to avoid brittle fracture (p. 13), (3) weldability, (4) cost. From the point of view of cost, therefore, quench and tempering processes on thick plates and sections are usually ruled out, although the installation of special equipment, such as roller quench presses, may make the technique an alternative to expensive alloy additions. Further, although killed steels are established in European technology, balanced steels remain in favour in UK because hot top practice is not widely used and a greater yield is obtained in balanced steels. Increased use of continuous casting may affect this in the future. The most popular types are those with manganese raised to 1·3–1·7% with carbon 0·2–0·4%, but for welding the carbon is kept low (now covered by BS 4360). The American 'Corten A' Steel has a composition of C, 0·12; Si, 0·5; Cu, 0·5; Cr, 0·8; P, 0·1 and Mn, 0·5%. Although the tensile strength is less than 494 N/mm² the yield is in the region of 371

N/mm². The combination of copper and phosphorus also increases the resistance to atmospheric corrosion which is important when thinner plates are used. The original steel 'A' suffers a decrease in yield strength and notch ductility in thickness over 25 mm, to overcome which 'Corten B' was developed—C 0·14; P 0·04; Mn 1·1; Cr 0·5; Cu 0·4; V 0·1; Bol Al 0·02.

The addition of 0·5% nickel and 0·25% molybdenum to a manganese steel gives a good general purpose steel (785 M 19). Fortiweld steel containing Mo, 0·5; B, 0·003; C, 0·11% has a TS of 618 N/mm² and is readily welded since it transforms in the bainitic region.

*Pearlite Reduced Steel.* Pearlite increases the tensile strength but not the yield stress, and since it raises the brittle–ductility transition temperature, there is a good case for reducing the carbon content. Low carbon steels (< 0·15%) strengthened with Mn + Nb and control rolled have good weldability and toughness and are called Pearlite Reduced Steel (PRS).

*Grain refinement.* Decreasing the ferrite grain size significantly increases both yield strength and notch ductility without increasing the carbon equivalent (p. 273), which affects weldability. The relation of yield strength $\sigma_y$ to structure is given by the Petch equation: $\sigma_y = \sigma_i + K_y d^{-\frac{1}{2}}$

$\sigma_i$ = ferrite strength (internal friction opposing dislocation motion)

$K_y$ = constant (a grain boundary locking term) d = grain size.

Fine grained steels (p. 167) using Al N have to be produced as killed steel with low (production) yield (BS 50D). Niobium and vanadium have a lower affinity for oxygen than aluminium and can be used in semi-killed steels, an economic advantage. Since 1960 about 0·30–0·1% Nb forming $Nb_4 C_3$ has been increasingly used as a grain refiner and precipitation hardening element and is the basis of several weldable steels in BS 4360, replacing 968 7762 3706, and includes 4 tensile ranged (40, 43, 50, 55 h bar) with several sub-grades which are distinguished by increased stringency of yield stress and notch ductility requirements. Both ladle and product analyses and carbon equivalents (p. 273) are included.

*Precipitation Hardening.* In Niobium steels—$Nb_4C_3$ dissolves above 1250 °C, large grains for and subsequent precipitation hardening is pronounced, but brittle transition temperature is high. Normalising at 900–950 °C forms a precipitate, resulting in a grain refined steel, slight precipitation hardening, but low impact transition temperature. By *controlled hot rolling* from 1250 °C to a low finishing temperature (<900 °C) with a substantial amount of deformation in range 950–850 °C, a fine ferrite grain size is obtained in sections up to 25 mm, with a minimum yield of 463 N/mm² due to dispersion hardening occurring in the ferrite during cooling. Thick plates present difficulties in getting the required drop in rolling temperature. Holding at an intermediate temperature produced a partially recrystallised structure of large grains surrounded by small ones. The final structure is of very mixed size with poorer mechanical properties.

It is possible to quench similar steels from 1050 °C to form a low carbon martensite or with lower carbon content, acicular ferrite followed by tempering to give higher properties. Nicuage steel, C, 0·06, Ni, 1, Cu, 1·1, Nb, 0·02, Mn, 0·5 rolled and aged at 500–570 °C has a yield of 600 N/mm$^2$ and a tensile strength of 700 N/mm$^2$, elongation 23% and 34–82 Joules charpy V notch impact at $-20$ °C coupled with good weldability and corrosion resistance.

*Steels above 680 N/mm$^2$ tensile*

Relative to the steels just discussed, those in this group are designed solely for their mechanical properties, which depend on accurately controlled heat-treatment.

In 1941 BS 970 covering bars, billets and forgings in this strength range rationalised steel specifications to conserve essential alloys. The basic principle was the specifying of mechanical properties related to *size* of bar when heat-treated rather than chemical composition. The cheapest steel will develop the requisite properties in the limited ruling section of the component used although other factors may modify the choice such as forging characteristics and die wear, machinability and ease of heat treatment. Suitable compositions within the range have to be chosen in relation to mass. Table 29 gives the limiting sizes up to which the particular steel, when heat-treated within specified limits, can be given certain ranges of mechanical properties. From Table 29 it will also be noted that certain combinations of alloys are far more efficient in producing the desired properties. Thus nickel alone has a low efficiency in spite of its pre-war popularity. A conception new to many engineers is that the *actual steel used for a given tensile strength depends on the size of the article at the time it is heat-treated.*

In 1970 BS 970 was revised and the En designation replaced by a six digit system. The first three digits refer to alloy type; the fifth and sixth digits represent 100 times the mean carbon content. At the fourth digit letters, A, M and H indicate if the steel is supplied to analysis, mechanical property or hardenability requirements, which are the new alternative methods. See p. 202 for Jominy hardenability specification.

*(850–1000 N/mm$^2$) tensile steels*

*530M40.* 1% chromium steel is suitable for gears, connecting-rods and stub axles. Low carbon types are suitable for water hardening. For full toughness tempering at 650 °C is necessary. (Ruling section 30 mm).

*640M40.* Low nickel–chromium steel, suitable for water-hardening in large masses. (Ruling section 63 mm.)

*605M36 and 608M38.* Manganese–molybdenum oil-hardening steels alternative to 3% nickel and nickel-chromium steels in 772–927 N/mm$^2$

TABLE 29  STEELS IN 700—1850 N/mm² RANGE BS 970, 1972

| Condition | R | S | T | U | V | W | X | Y | Z |
|---|---|---|---|---|---|---|---|---|---|
| TS N/mm² min.* | 700 | 770 | 850 | 930 | 1000 | 1080 | 1150 | 1240 | 1540 |
| (TS tonf/in.²) | (45) | (50) | (55) | (60) | (65) | (70) | (75) | (80) | (100) |
| YP min. N/mm.² | 520 | 590 | 680 | 760 | 850 | 940 | 1020 | 1100 | 1240 |
| El mm.% | 17 | 15 | 13 | 12 | 12 | 11 | 10 | 10 | 7 |
| Izod Joules | 54 | 54 | 54 | 47 | 47 | 40 | 34 | 34 | 13 |
| Brinell, min. | 201 | 223 | 248 | 269 | 293 | 311 | 341 | 363 | 444 |

| | Chemical Composition, % | | | | | Limiting Diameter in mm in which above mechanical properties are available from oil quenching | | | | | | | | | Oil Harden °C | Temper °C |
|---|---|---|---|---|---|---|---|---|---|---|---|---|---|---|---|---|
| BS970 No. | Mn | Ni | Cr | Mo | | R | S | T | U | V | W | X | Y | Z | | |
| 530M40 | 0·6/0·90 | | | | | 100 | | | | | | | | | 860/890† | 550/700 |
| 605M36 | 1·3/1·7 | | | 0·22/0·32 | | 150 | 63 | 30 | | | | | | | 840/870 | 550/680 |
| 640M40 | 0·6/0·9 | 1·1/1·5 | 0·5/0·8 | 0·15/0·25 | | 150 | 100 | 63 | 30 | 30 | | | | | 820/850† | 550/660§ |
| 945M38 | 1·2/1·6 | 0·6/0·9 | 0·4/0·6 | 0·15/0·25 | | 150 | 100 | 63 | 30 | 30 | | | | | 840/870† | 550/680 |
| 708M40 | 0·7/1·0 | | 0·9/1·2 | 0·4/0·55 | | 150 | 100 | 63 | 30 | 30 | | | | | 860/890† | 550/700 |
| 608M38 | 1·3/1·7 | | | | | 150 | 100 | 63 | 30 | 30 | | | | | 840/870† | 550/680 |
| 653M31 | 0·45/0·7 \|\| | 2·75/3·25 | 0·9/1·20 | | | | 150 | 100 | 63 | 30 | | | | | 820/850 | 550/660§ |
| 816M40 | 0·45/0·7 \|\| | 1·3/1·7 | 1·0/1·4 | 0·1/0·2 | | | 150 | 100 | 63 | 30 | | | | | 820/850† | 550/660 |
| 817M40 | 0·45/0·7 \|\| | 1·3/1·7 | 1·0/1·4 | 0·2/0·35 | | | | 150 | 100 | 63 | 30 | | | | 820/850 | 660 max |
| 830M31 | 0·45/0·7 \|\| | 2·75/3·25 | 0·9/1·2 | 0·25/0·35 | | | | 150 | 150 | 100 | 63 | | | | 820/850 | 550/660 |
| 722M24 | 0·45/0·7 \|\| | | 3·0/3·5 | 0·45/0·65 | | | | 150 | 150 | 100 | 100 | | | | 880/910 | 570/700 |
| 826M31 | 0·45/0·7 | 2·3/2·8 | 0·5/0·8 | 0·45/0·65 | | | | 150 | 150 | 150 | 100 | 63 | ‡ | 63 | 820/880 | 660 max |
| 826M40 | 0·45/0·7 | 2·3/2·8 | 0·5/0·8 | 0·45/0·65 | | | | | 150 | 150 | 150 | 150 | 150 | 100 | 820/850 | 660 max |
| 855M30 | 0·45/0·7 | 3·9/4·3 | 1·1/1·4 | 0·2/0·35 | | | | | | | | | | 150 | 810/840 | 200/280 |

*Max. range of TS is 150 N/mm² above minimum.
†Water or oil quench.
‡E = 5; Izod = 11.
§ = Cool rapidly.
\|\| = Suitable for nitriding.

tensile range. Used for connecting-rods, steering-levers, swivel arms, axle-shafts and gears. 608M38 with its higher molybdenum content should be used for the 100 mm sections but it has a tendency to air harden in thin sections, and when softening is necessary for clipping flashes from drop forgings or for machining or for cold processing bars or tubes, it is best done by tempering at 650 °C. Care should be taken to get an *efficient oil quench by separating bars* if the best properties are to be obtained since the effect is also reflected in the properties after tempering. A thin martensite layer is sometimes formed by grinding.

Both $1\frac{1}{2}$ Mn Ni Cr (945M38) and $1\frac{1}{4}$ Ni Cr (640M40) provide the steel-maker with a means for the effective use of nickel–chromium scrap and are of national importance since the importation of alloying elements is minimised.

*709M40.* 1% chromium–molybdenum steel is alternative to 608M38 for oil-hardened and tempered parts, and is a useful steel for *local* hardening. Used for automobile gears and propeller-shaft joints. A wide range of mechanical properties can be obtained, depending on the tempering temperature employed.

*1000–1850 N/mm² tensile steels*

Increased alloy content has to be used to counteract mass effect and this increases the risk of cracks in thin sections or surface defects arising in the various stages of manufacture, particularly where drop stampings are involved. The simplest way of providing reasonable machinability is to temper the steels at 650 °C. Air cooling after tempering reduces risk of distortion and is permissible with the steels containing molybdenum.

*653M31, 830M31.* 3% nickel–chromium (molybdenum) steels are very good for 927–1180 N/mm² range but are more highly alloyed than other recommended steels, and are somewhat more liable to 'temper-brittleness' in large sizes unless a high molybdenum content is present.

*817M40, 826M31, 826M40.* $1\frac{1}{2}$%, $2\frac{1}{2}$%, $2\frac{1}{2}$%(high C) nickel–chromium–molybdenum steels; for high-duty transmission parts and direct hardened gears, air-screw spiders.

Three common ranges of tempering are:

200–220 °C for maximum hardness (gears).
500–550 °C for TS of 1080–1390 N/mm² for stiffness and toughness (connecting-rods, axles).
550–650 °C for TS of 1000–1158 N/mm² with high ductility (E 18%; Izod
   68 J)—bolts, stub axles, swivel arms.

*835M30.* Nickel–chromium *air hardening* steel is suitable for parts requiring freedom from distortion in heat-treatment where the ruling section exceeds 63 mm; oil hardening is necessary to get the 1544 N/mm² tensile condition.

## Ultra-high tensile structural steels

Interest in (1544–2160 N/mm²) tensile steels for use in the aircraft industry is leading to the development of modified steels which can be used up to 400 °C in supersonic aircraft and which possess adequate ductility and notch impact strength. Vacuum arc or electro-slag melting of some of these steels reduces inclusions and impurities and gives a more suitable cast structure. As a result, the *transverse* properties of the resultant forgings tend to be superior to those from conventionally air cast ingots. A few typical compositions are given in Table 30. These are based on Ni–Cr–Mo or use silicon with nickel or copper. Silicon reduces the expansion in $\gamma \rightarrow$ martensite change and hence reduces risk of quench cracking; while copper is useful for producing secondary hardening at 450–550 °C.

TABLE 30   TYPICAL 0·37% C, 1850 N/mm² STEELS

|  | Si | Mn | Ni | Cr | Mo | V | Temper | El |
|---|---|---|---|---|---|---|---|---|
| 3% Cr–Mo–V | 0·2 | 0·5 | 0·2 | 3 | 0·9 | 0·2 | 300 | 8 |
| Rex 539 Si–Mn–Ni | 1·6 | 1·6 | 1·8 | 0·1 | 0·4 | 0·2 | 350 | 7 |
| H50 | 1·1 | 0·5 | — | 5·0 | 1·3 | 1·1 | 560 | 7 |
| SAE 4340 | 0·3 | 0·6 | 1·8 | 0·8 | 0·25 | — | 220 | 9 |
| Si–Cu steel | 2 | Cu = 2 | — | — | 0·7 | 0·2 | 400 | 9 |

With such low alloy contents protection against rust by cadmium plating or aluminium spraying is necessary. Unfortunately, cleaning and plating processes can introduce hydrogen into the metal and embrittlement at the high strength level can be serious. Such embrittlement becomes evident under sustained loading, for example a $2\frac{1}{2}$ Ni Cr Mo steel hardened to 2220 N/mm² tensile broke within 100 hours at 1235 N/mm² and in 12 hours at 1544 N/mm² sustained load in air and much shorter times in corrosive media. With such high tensile steels it is highly desirable to (*a*) avoid notches, (*b*) to reduce internal stresses by tempering at as high a temperature as possible, (*c*) minimise the introduction of hydrogen (abrasive blasting preferred). Cadmium coats should be avoided for service above 250 °C. For other high strength steels refer to precipitation hardening transformation and controlled transformation stainless steels.

## Maraging Steels

These use the martensitic reaction which end high hysteresis occurs in Fe–Ni alloys, and shown in Fig. 147. They are iron based alloys containing 18 Ni 8 Co 5 Mo with small amounts of Al and Ti and less than 0·03% C, which makes such a difference to fracture toughness and ease of welding. The strength is maintained with increase in section thickness and also up to 350 °C. Alloy cost is balanced by lower production cost, virtually no risk of decarburisation, distortion or cracking. These steels are used for air frame and engine components, injection moulds and dies. Typical composition and properties are given in Table 31.

TABLE 31  PROPERTIES OF MARAGING STEELS

| C | Ni | Co | Mo | Ti | 0·2 P N/mm² | TS N/mm² | E | RA | HV |
|---|---|---|---|---|---|---|---|---|---|
| *Solution annealed 800–900 AC*—all grades | | | | | | | | | |
| | | | | | 725 | 1035 | 15 | 72 | 300 |
| *820 °C + Maraged 3 hr 480 °C* | | | | | | | | | |
| 0·03 | 18 | 8·5 | 3 | 0·2 | 1344 | 1390 | 11 | 45 | 450 |
| 0·03 | 18 | 8 | 5 | 0·4 | 1620 | 1700 | 9 | 40 | 520 |
| 0·03 | 18 | 9 | 5 | 0·6 | 1807 | 1930 | 7 | 35 | 570 |
| *Stainless Maraging Steel* 12% Cr 870° WQ 4 hr—76 °C 6 hr 440 °C Vacuum remelted | | | | | | | | | |
| 0·03 | 4·4 | 14 | 4·6 | 0·5 | 1700 | 1850 | 10 | 40 | |

On cooling from the austenitic condition the alloy transforms to a fine lath type martensite, and precipitation hardening is induced by 'maraging' at 480 °C. Many types of precipitates have been reported (e.g. $Ni_3 TiAl$) but the main hardener is probably orthorhombic $Ni_3 Mo$, the solubility of which is probably reduced by Cobalt.

The steels have high fracture toughness, $K_{1c}$ due to a combination of fine grain size of the martensite and the high dislocation density, leading to fine precipitation. The steels can be nitrided.

The corrosion resistance is only slightly improved but the 12% Cr variety has been developed for corrosion resistance. (For marforming see p. 226.)

## Hair-line cracks

Many alloy-steel ingots and large forgings are susceptible to the formation of small silvery cracks or flakes in their interior. These cracks often form at room temperature after an incubation period and the cause of them is not completely known but is related to the cracking of welded high-tensile steels (p. 404) in that hydrogen has a large influence in promoting embrittlement which increases with the tensile strength of the steel. Less trouble is experienced with acid open-hearth steel (usually containing 4 cc per 100 gm) than with basic electric steel containing about 6–8 cc per 100 gm of hydrogen in the ladle. Slow cooling and also isothermal transformation at about 600 °C tends to reduce the incidence of hair-line cracks and this is materially assisted by the vacuum melting. To reduce the above hydrogen concentrations to about 1 cc per 100 gm requires heat-treatments of the following magnitude at 650 °C:

| Dia bar (metre) | 0·025 | 0·25 | 0·5 | 1 |
| --- | --- | --- | --- | --- |
| Hours | 1 | 100 | 400 | 1600 |

The problem is therefore more acute with large ingots.

*Alloy spring steels* of the chromium–vanadium (Cr, 1; V, 0·2; C, 0·6) (Ni, 0·5; Cr, 0·5; Mo, 0·2; C, 0·6) and silico–manganese (Si, 2; Mn, 1) types are also oil hardened from 80 °C and tempered at 480 °C to give a Vicker's hardness of 400, tensile strength of 1390 N/mm$^2$, with appreciable ductility.

Springs are made from steel treated as follows:

(a) Cold drawn patented steel wire (p. 200).
(b) Cold worked annealed steel.
(c) Quenched and tempered (0·5/1·0% C) steel (VPN 340–430). After having been formed, the springs, (a, b, c) are only given a low temperature temper (170–300 °C) to relieve forming stresses.
(d) Annealed steel. After having been formed, the spring is then quenched and tempered. All heavy springs are formed hot and frequently hardened immediately after forming. Aero-engine valve-spring steel must be free from any kind of surface defect.

Springs for watches and aircraft instruments, Bourden tubes, diaphragms etc. are often made from Ni-span containing 42 Ni, 5 Cr, 2·3 Ti, 0·02 C, 0·55 Al which has low mechanical hysteresis and can be heated to reduce the effect of changes in service temperatures.

Faults in springs are due to:

(1) Decarburisation due to annealing, etc., affecting the fatigue properties.
(2) Segregations forming lines of weakness in the material which may open up into splits in service.
(3) Internal cup and cone fractures due to overdrawing.
(4) Mechanical damage, such as rolling laps, deep grooves, scratches due to wire drawing, vice marks and scoring due to winding.
(5) Incorrect tempering, especially in chromium–vanadium steels.

*High and low thermal expansion steels*

There are cases in engine construction where steel has to work in conjunction with light alloys, such as cylinder-head bolts, valve seatings, or cylinder liners in aero engines. The comparatively high thermal expansivity of aluminium leads to looseness unless the steel has a similar coefficient of expansion.

The austenitic steel of the following composition

$$C, 0·59; Ni, 12; Mn, 5·1; Cr, 3·4$$

has a thermal expansion of 0·000021 per degree C up to 400 °C, which is only slightly lower than that of aluminium, and it combines good mechanical properties with resistance to abrasion.

Cold rolled austenitic stainless steel is another alternative.

Where an abnormally low coefficient of expansion is required, Invar, containing 36% nickel, is used (p. 219).

*Ball-race steel.* A typical composition is C, 1·0; Mn, 0·5; Cr, 1·36%. After quenching in oil from 810 °C the steel is usually tempered at 100–200 °C to (*a*) reduce hardening stresses, (*b*) reduce cracks in grinding. Tempering at 100 °C also increases the hardness slightly, e.g.:

| Tempering temperature | nil | 100 | 200 | 250 |
|---|---|---|---|---|
| VPN | 800 | 876 | 750 | 736 |

*Creep-resisting steels for use at steam temperatures* The use of higher temperatures and pressures in modern power stations has necessitated the use of special steels for the pipe-lines and other parts. The essential characteristic of these steels is higher resistance to creep at temperatures varying from 400° to 565 °C. A common steel used for this purpose is one containing approximately 0·5% molybdenum with a carbon content of approximately 0·15–0·2%. Failures of pipes have occurred in America, which have been traced to the formation of a network of graphite in the heat-affected zone of the pipe adjacent to the weld. Heating the steel to approximately 750 °C appears to accelerate the formation of this graphite, which is also largely affected by the process of making a steel, particularly as regards the use of aluminium as a deoxidiser, which is more commonly used in America than Great Britain. It has been found that a small addition of a carbide stabiliser, such as 1·0% chromium, is beneficial in minimising this trouble.

The addition of 25% V raises the creep resistance still further. The 1% Cr Mo V steels currently in use are chiefly of two types, those for steam-chest castings where due to welding considerations the carbon level is limited to 0·15%; and those used in HP/IP rotors in which the need for improved hardenability with large rotors has necessitated carbon levels of 0·25–0·30%.

Creep seems to be related to the *uniformity* of the $V_4C_3$ and its *interparticle spacing* which should be less that 1000Å ($10^{-5}$ cm). This distribution is affected by cementite which promotes regions denuded of $V_4C_3$ and to minimise this problem the cooling rate from the austenitising temperature must be such as to give an entirely *upper bainitic structure*.

Low-carbon 3 and 6% chromium steels containing 0·5% molybdenum have carbides which resist hydrogen attack and embrittlement at elevated temperatures and pressures and are useful in synthetic ammonia plants and oil refineries. Columbium or titanium is sometimes added to minimise the air-hardening tendencies in the steel by forming carbides less soluble at heat-treating temperatures. Air-hardening still occurs after welding but air-cooling from 800 °C is sufficient to soften the steel.

Resistance to corrosion and oxidation increases with the chromium content; the elongation at rupture is also increased (p. 260).

## Alloy tool and die steels

BS 4659:1971 groups tool steels into six types—high speed, hot work, cold work, shock resisting, special purpose and water hardening. The designations follow the AISI with the addition of B. Thus BT1 and BM1 designates high speed steel of tungsten and molybdenum grades respectively.

*Chisel steels*

TABLE 32 CHISEL STEELS

| C | Percentage Ni | Cr | W | Uses | Harden, °C | Forge, °C |
|---|---|---|---|---|---|---|
| 0·4 | — | 1 | — | Coal-cutter picks | 800 WQ | 1050 |
| 0·5 | — | 2 | — | Chisels and other tools | 900 OQ | 1000 |
| 0·4 | — | 1·8 | 2 | Pneumatic chisels and punches | 900 OQ | 1000–1100 |
| 0·4 | 3 | 0·6 | — | Hand tools resistant to shock | 900 OQ | 950–1000 |
| | | | | Chisels (edge), sates, drifts | No temper | |

*Non-shrinking Steels*

This term refers to steels which show little change in volume from the annealed state when hardened and tempered at low temperatures.

Usually the following volume changes occur.

$$\text{Pearlitic} \rightarrow \text{austenitic state, contraction}$$
$$\text{austenitic} \rightarrow \text{martensitic state, expansion}$$
$$\text{martensitic} \rightarrow \text{sorbitic state, contraction}$$

In non-shrinking steels the volume changes counterbalance each other, and such steels are required for master tools, gauges and dies which must not change size when hardened after machining in the annealed condition.

The cheapest non-shrinkage steel contains 0·9% carbon and about 1·7% manganese. A better steel is,

C, 1·0; Mn, 0·95; W, 0·5; Cr, 0·75; V, 0·2

Both steels are oil quenched from 780° to 800°C and tempered 224–245°C. High carbon 5% and 12% chromium steels are also used for non-distortion.

# ALLOY TOOL AND DIE STEELS

*Finishing tool steel*

While high-speed steels are very efficient with heavy cuts and high speeds they are incapable, at slow speeds and lighter cuts, of holding the keen edge necessary for obtaining a very smooth finish on certain articles. Special steels have been produced for this purpose, known as finishing steels, which are capable of retaining a keen cutting edge for much longer periods than carbon steel used under similar conditions.

The usual type has the approximate composition:

C, 1·1 to 1·4; W, 4; Cr, 0·7 to 1·5; V, 0·3

After preheating to 650 °C it is water hardened at 820–840 °C and immediately tempered at 150–180 °C. Anneal at 750 °C.

Tungsten steels containing 1 to 5% and 1 to 1·3% carbon are used for twist drills, taps, milling cutters, drawing dies and also tools for rifling gun barrels, boring cylinders and expanding tubes, which require long continuous cutting without interruption for regrinding. They are tempered at 200–230 °C.

*Cold die steels*

The standard oil hardening die steels contain 1 C, 1 Mn, 0·3–1·6 W, 0·5 Cr, hardened from 800 °C and immediately tempered at 170–250 °C. For cold obtrusion punches high-speed steels are satisfactory, e.g. 6W6 Mo.

*High carbon-chromium* (A)

| C | Cr | Mn | Si | Harden °C | Temper °C |
|---|----|----|----|-----------|-----------|
| 2 | 13 | 0·25 | 0·6 | OQ 950 or AC 1000 | 480—2 hrs |

This steel has good resistance to oxidation at elevated temperatures, high hardness and good wearing properties. It is suitable for intricate sections, dies for blanking, coining, roller threading and drop forging hard materials. The structure is martensitic on cooling in air but the carbides can be precipitated and the steel softened by very slow cooling from 840 °C.

*Hot-working die steels*

Many articles of iron, steel and non-ferrous metals are formed in the hot condition, and the tools used for shaping become hot from contact with the article. Carbon steels and pearlitic alloy steels are thus tempered and rendered too soft for service. To overcome this defect a class of alloy steel has been developed known as hot-working die steels; typical examples are given below:

*High Tungsten–chromium Steel*

| C | Mn | W | Cr | V | Mo | Harden °C | Temper °C | Anneal °C |
|---|---|---|---|---|---|---|---|---|
| 0·3 | 0·3 max | 10 | 3 | 0·3 | 0·3 | OQ 1150 | 570 | 850 |

This is the best type of steel for hot work except where resistance to scaling or oxidation is important. It is used for hot-drawing, hot-forging, extrusion dies and dies for die casting aluminium, brass and zinc alloys. Die-casting die steels often fail through surface cracking caused by cyclic expansion and contraction, aggravated by the erosive action of the molten metal. Increased die life necessitates regular maintenance and careful preheating before use. Typical examples of alloy steels for die casting are given in Table 33.

TABLE 33   DIE STEELS FOR DIE CASTING

| | C | Si | W | Cr | Mo | V | Pre-heat °C | Harden °C | Temper °C | VPN |
|---|---|---|---|---|---|---|---|---|---|---|
| B | 0·35 | 1·2 | 5 | 5 | 0·5 | — | 750 | AC 1010 | 570—4 hrs | 450 |
| C | 0·40 | 1·1 | — | 5 | 1·4 | 1·1 | 800 | AC 990 | 550—4 hrs * | 460 |
| D | 0·47 | — | — | 1·8 | — | 0·2 | 670 | OQ 850 | 570—2 hrs | 300 |
| E | 0·35 | 1·0 | 1·3 | 5 | 1·5 | 0·45 | 800 | AC 1020 | 580—2 hrs * | 580 |
| F | 1·0 | 0·3 | — | 3·5 | — | — | 830 | AC 1050 or OQ 870 | 450—650 | 500 |
| G | 0·30 | — | — | 1 | — | 0·2 | 670 | WQ 840 | 200—1¼ hrs | 500 |
| H | 0·28 | 0·45 | 6 | 1·3 | Ni 3 | 0·2 | 770 | OQ 1070 | 570 | 450 |

*Double temper advisable

Steels B and C are particularly used for aluminium pressure die casting dies and many other tools such as extrusion dies, piercing tools and press forging dies. Low alloy steel D is used for lower temperature work, e.g. zinc base alloy die-casting dies.

The physical properties of steel E compare favourably with some of the more expensive high tungsten hot-working steels and it is used for extrusion dies, piercing and heading tools and dies for forging or pressing brass. Steels F and G are suitable for tools subjected to a *succession* of rapid blows which normally cause fatigue failure e.g. rivit snaps, nut-piercers, mandril bars and hot dies.

Sensitivity of die steels to distortion during heat-treatment is largely affected by directionality and particle size of the carbides in the microstructure. Expansion is greatest in the direction of carbide stringers. Fine random distribution of carbides are therefore desirable. For die casting and extrusion dies molybdenum containing 0·5 Ti + 0·08 Zr is useful in

# ALLOY TOOL AND DIE STEELS

critical applications. Thermal conductivity, resistance to thermal shock and attack by molten metal is high and no heat treatment is required. Nimonic 80(*a*) and 90 have also been used satisfactorily for dies and inserts.

*Die block steels* for drop forging have been standardised into four type (BS 224, 1938). These are (*a*) 0·6 carbon steel, (*b*) 1% nickel, 0·6 C, (*c*) 1·5 Ni, 0·7 Cr, 0·6 C, (*d*) 1·5 Ni, 0·7 Cr, 0·6 C, 0·25 Mo. Hardness ranges from 425/455 for dies with shallow impressions to 298/355 for very large forgings.

*Shear blades*

Some examples of alloy steels used for shearing are given in Table 34.

*High-speed steels*

The evolution of high-speed cutting tools commenced with the production of Mushet's self-hardening tungsten–manganese steel in 1860. The possibilities of such steels for increased rates of machining were not fully appreciated until 1900, when Taylor and White developed the forerunner

TABLE 34  SHEAR BLADE STEEL

| Type of Work | C | Si | Cr | V | W |
|---|---|---|---|---|---|
| Cold shearing for heavy materials | 0·85 | — | — | 0·2 | — |
|  | 0·55 | 2 | Mn = 0·8 |  | — |
| Cold shearing for light materials | 1·0 | — | — | 0·2 | — |
|  | 0·7 | — | 0·9 | 0·2 | — |
|  | 0·6 | — | 4 | 1 | 18 |
|  | 2·2 | — | 12 | — | — |
| Shears for hot work | 0·5 | — | 1·2 | 0·2 | 2 |
|  | 0·4 | — | 3·5 | 0·4 | 10 |

of modern high-speed steels. In addition to tungsten, chromium was found to be essential and a high hardening temperature to be beneficial. The steel resisted tempering up to 600 °C. This allowed the tool to cut at speeds of 80–50 metres per minute with its nose at a dull red temperature and it was one of the astonishing exhibits at the Paris Exhibition of 1900.

The main constituents in high-speed steel are 14 or 18% tungsten, 3 to 5% chromium and 0·6% carbon. Other elements are frequently added to modern steels which vary considerably in composition and cost. 0·09–0·15% sulphur is sometimes added to give free machining for unground form tools, e.g. gear hobs in 6–5·2 $M_2S$. Typical compositions are given in Table 35.

Vanadium improves the cutting qualities of the tools and increases the tendency to air hardening. Cobalt, often added to the 'super high-speed' steel, raises the temperature of the solidus and enables a higher hardening temperature to be used, with consequent greater solution of carbon. 'Secondary hardness' is marked in such steels (p. 188), and this permits the use of deep cuts at fast speeds. The molybdenum steel is susceptible to decarburisation. The high vanadium steel is somewhat brittle, but is excellent for cutting very abrasive materials.

TABLE 35 TYPICAL HIGH-SPEED STEELS

| BS | C | Cr | V | W | Mo | Co | Harden °C* | Temper °C* | Hardness 20°C | HV 550°C |
|---|---|---|---|---|---|---|---|---|---|---|
| *Light Duty* | | | | | | | | | | |
| BT21 | 0·65 | 3·9 | 0·5 | 14 | | | 1280 | 2 × 560 | 775 | 520 |
| *Normal Duty* | | | | | | | | | | |
| BT1 | 0·75 | 4·1 | 1·1 | 18 | | | 1280 | 2 × 560 | 875 | 555 |
| BM1 | 0·80 | 4·1 | 1·1 | 1·5 | 8·5 | | 1210 | 2 × 540 | | |
| BM2 | 0·85 | 4·1 | 1·9 | 6·4 | 5·1 | | 1220 | 2 × 560 | 910 | 600 |
| *Fast Cutting* | | | | | | | | | | |
| BT4 | 0·75 | 4·1 | 1·1 | 18 | | 5·0 | 1290 | 3 × 560 | 880 | 595 |
| BT5 | 0·80 | 4·1 | 1·9 | 19 | | 9·5 | 1300 | 3 × 560 | | |
| BT6 | 0·80 | 4·1 | 1·5 | 20·5 | | 12 | 1300 | 3 × 560 | 915 | 610 |
| *Hard Cutting* | | | | | | | | | | |
| BT15 | 1·5 | 4·6 | 5·0 | 12·5 | | 5·0 | 1240 | 3 × 560 | | |
| BM15 | 1·53 | 4·7 | 5·0 | 6·6 | 3·0 | 5·0 | 1220 | 3 × 550 | 885 | 580 |
| BT42 | 1·33 | 4·1 | 3·0 | 9·0 | 3·0 | 9·5 | 1230 | 3 × 560 | 965 | 645 |
| BM42 | 1·05 | 3·9 | 1·2 | 1·5 | 9·5 | 8·0 | 1190 | 3 × 530 | 950 | 610 |

*average ± 10°C

The study of the structures of such highly alloyed steels is complex, but it can be simplified by converting the amounts of the various elements to an equivalent percentage of tungsten as regards the effect on the closed $\gamma$-loop:

| 1% of | Mo | V | Cr |
|---|---|---|---|
| Equivalent percentage of tungsten | 1·5 | 5·0 | 0·5 |

Hence 18 W, 4 Cr, 1 V is equivalent to 25% tungsten and the section of the Fe–W–C equilibrium diagram is shown in Fig. 150.

In the ingot the structure is similar to cast iron (Fig. 154), but the cementite consists of mixed carbides $(Fe, W\ Cr, V)_6C$ with the balance of the elements in solution in the ferrite. In this condition the steel is extremely brittle and the eutectic net-work has to be broken up into small globules, evenly distributed by careful annealing, followed by forging. 'Strings' or laminations of carbides should be avoided, otherwise cracks are liable to form during hardening.

*Annealing*

High-speed steel is softened by annealing at 850 °C for about four hours, followed by slow cooling. The steel must be protected against oxidation. After forging, tools should be heated to 680 °C for $\frac{1}{2}$ hour and air cooled before hardening in order to reduce risk of fracture. The annealed structure consists of carbide globules in a matrix of fine pearlite.

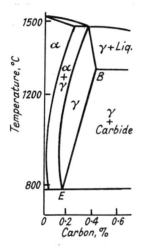

FIG. 150 Section of the Fe–W–C equilibrium diagram at 25% tungsten

*Hardening*

From Fig. 150 it will be seen that on heating, austenite forms at about 800 °C, but contains only 0·2% carbon (eutectoid E). Quenching produces martensite, which tempers readily and has no advantage over carbon tools. More carbide dissolves on heating, as indicated by line EB, and quenching produces structures of increasing red-hardness, due to the effect of the larger amounts of alloying elements in solution, which render the steel sluggish *to tempering*. Even at 1300 °C, when melting occurs, only 0·4% carbon (B) is dissolved and the remainder exists as complex carbides (Fig. 155). It will be seen, therefore, that to attain maximum cutting efficiency sufficient carbon and alloying elements must be dissolved in the austenite and this necessitates temperatures little short of fusion, usually 1150–1350°

C. Grain growth and oxidation occur rapidly at such temperatures. Hence the tools are carefully preheated up to 850 °C, then heated rapidly to the hardening temperature and quenched in oil or cooled in an air blast without soaking. To reduce the severe stresses set up by quenching, the following modifications can be used to reduce the temperature gradient from outside to centre prior to the austenite–martensite transformation:

(a) cool in salt bath at 600 °C until temperature is uniform; then quench in oil,
or
(b) oil quench to 425 °C, then air cool to room temperature.

*Tempering*

When quenched from high temperatures high-speed steels contain an appreciable amount of retained austenite which is softer than martensite. This is decomposed by tempering, or by sub-zero cooling to $-80$ °C. Multi-tempering is often more effective than a single temper of the same duration.

Tempering at 350–400 °C slightly reduces the hardness but increases toughness. Tempering at 400–600 °C increases the hardness, frequently to a value higher than that produced by quenching. This phenomenon is known as *secondary hardening*.

The structure of the hardened high-speed steel consists of isolated spherical carbides embedded in an austenite–martensite matrix. Dark etching grain boundaries are frequently evident (Fig. 155). Tempering produces a general darkening of the matrix (Fig. 156).

'*Stellite*' type alloys consist of a cobalt base with about Cr, 30; W, 15 with other additions, including carbon. The structure consists of a cobalt matrix with complex tungsten–chromium carbides. It has a high resistance to corrosion and to tempering and is used for tools, gauges, valve seatings and hard facing.

**Hard metal or cemented carbide**

Certain carbides of transition metals have very high hardness, high melting points and are difficult to fabricate by traditional methods, e.g.

|  |  | VPN | MPt °C |
|---|---|---|---|
| Tungsten carbide | WC | 2080 | 2600 |
| Titanium carbide | TiC | 3200 | 3200 |
| Tantalum carbide | TaC | 1790 | 3700 |

Cast products are brittle but by use of powder metallurgy techniques (see p. 135) a very useful group of tool materials can be made in which the main phase is one or more of the hard carbides bonded together with cobalt metal. As shown in Fig. 132 the carbides are present as particles between 0·8 and 10 μm diameter which constitute usually 80–90% of the volume of the alloy. The cobalt metal in which the carbides are slightly soluble not only enables the carbide to be consolidated at a relatively low temperature into useful shapes but also imparts such a useful measure of ductility.

151        152        153

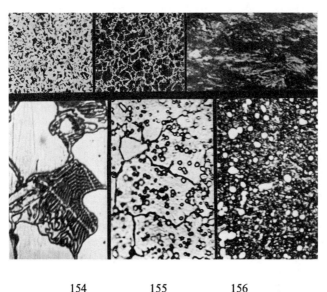

154        155        156

Fig. 151  0·3% carbon steel, normalised  (× 100)
Fig. 152  Similar to Fig. 151, but containing 3% of nickel. Greater volume of pearlite  (× 100)
Fig. 153  High tensile, drawn steel wire (C, 0·4) etched  (× 500)
Fig. 154  High-speed steel as cast: complex carbides  (Eutectic)  (× 800)
Fig. 155  High-speed steel, hardened by air-cooling from 1250°C: carbides in austenite–martensite matrix. Grain boundaries of original austenite. Etched—3% ammonia, 5 minutes  (× 1200)
Fig. 156  As Fig. 155, but after tempering at 600°C breakdown of austenite  (× 1200)

A useful range of properties is obtained by varying the proportions of the carbide and cobalt present, and by varying the grain size of the carbide, as shown in Table 36. The wear resistance of alloys used as tools for cutting steel can be greatly improved by substitution of other carbides, notably

TiC and TaC for part or all of the WC. The TiC has low solubility in steel and prevents cratering wear by the steel swarf. Similar improvement is obtained by vapour deposition of a thin layer of TiC or TiN on the tip surface.

The outstanding properties of these alloys are their exceptionally high hardness, wear resistance, higher compressive strength, higher modulus of elasticity (approx. 600 kN/mm²) than any other engineering material, and

TABLE 36   TYPES OF HARD METAL

| Co | WC | TaC | TiC | Grain Size | Hardness VPN | Use |
|----|----|-----|-----|------------|--------------|-----|
| 6  | 94 |     |     | 1 | 1750 | Cutting cast iron, non-ferrous metals and low impact wear resistance. |
| 6  | 94 |     |     | 3 | 1500 | |
| 9  | 91 |     |     | 4 | 1250 | High impact wear resistance. Rolls, rock drills, coal picks. |
| 20 | 80 |     |     | 2 | 1000 | Highly stressed impact tools and dies. |
| 10 | 70 | 10  | 10  | 2 | 1520 | Cutting steel at high speeds, moderate feed. |
| 10 | 79 | 5   | 3   | 2 | 1480 | Cutting steel at lower speeds with interruptions of cut. |
| 10 | (10% Mo) |  | 80 | 3 | 1650 | Cutting steel or cast iron at very high speeds 180/300 m/min low feed rate. |

the retention of hardness and strength to relatively high temperatures. They cannot be softened or hardened by heat treatment.

The many applications for these alloys, apart from those mentioned, include rolls for Sendzimir and other rolling mills, punches for cold forming, thread guides for textile machinery, burrs for dental drills and balls for ball-point pens.

With the development of throw-away tips, ceramic tools, mainly pure $Al_2O_3$, have been developed, and cast iron can be machined with them at very high speeds ($> 300$ m/min).

# 12 Stainless, Creep and Heat-resisting Steels

**Stainless steels**

*Composition*

The stainless steels owe their resistance to corrosion to the presence of chromium. Brearley discovered this fact more or less accidentally in 1913. Today, there is a range of steels from the plain chromium variety to those containing up to six alloying elements in addition to the usual impurities. A simple classification of the steels follow:

*Hardenable alloys*

(1) 12–14% chromium, iron and steels, whose mechanical properties are largely dependent on the carbon content. High strength is combined with considerable corrosion resistance:

   (a) *Stainless iron*, 12 Cr, 0·1C (410S21)* can be welded and easily fabricated and is useful for turbine blades.
   (b) *Stainless steel*, 12 Cr, is difficult to weld, resistant only when hardened and polished.
   (i) mild, (420S29) C, 0·17—used for steam valves, piston rods, but not in contact with non-ferrous metals or graphite packing owing to galvanic action (therefore use steel 4)
   (ii) medium, 0·32 C (420S45)—used for table cutlery and tools and also parts working at elevated temperatures
   (iii) hard 0·4–2·0 C—used as springs and ball bearings, which work under corrosive conditions.

(2) *Secondary hardening* 10–12% chromium iron, 0·12 C, with small additions Mo, V, Nb, Ni. A 927 N/mm² steel used for gas turbine blades, e.g. H46, Jethete, FV448 (0·12 C 1 Ni, 0·7 Mo, 0·15 V, 0·4 Nb, AC 1150 T650–700 °C).

(3) *High chromium steel*, 17 Cr, 0·15 C, 2½ Ni (431S29, S80). It has a higher resistance to corrosion than iron 1a, due to higher chromium content. It is used for pump shafts, valves and fittings subjected to high temperature and high-pressure steam, but is unsuitable for acid conditions.

*In revised BS 970 the first three digits indicate alloy type, usually corresponding to AISI No; last two digits indicate variations with a type and not the carbon content.

High carbon, 0·8 C, 16·5 Cr, 0·5 Mo steel OQ 1025 °C T 100 °C to give hardness of 700 is used for stainless ball bearings and instruments such as scalpels.

Cast, 25 Cr, 5 Ni + Mo and Cu has a corrosion resistance similar to 18/10/3 Mo but higher hardness, 325 VH, makes it suitable for seals and valve trim.

*Ferritic iron*

(4) (*a*) 16/18% chromium rustless iron with low carbon content (430 S15). It has high resistance to corrosion but low impact and cannot be refined by heat-treatment alone. Prolonged service at 480 °C can cause embrittlement. It is used for motor car trim.

(*b*) 25/30% chromium iron for furnace parts, resistant to sulphur compounds. Forms sigma phase (p. 258) additions of Nb and Mo prevent excessive grain growth.

*Austenitic steels*

(5) *Plain 18/8 Austenitic Steels.* 15–20 Cr, 11–6 Ni, 0·05–0·15 C (302). Commercial varieties are Anka H, Staybrite FST, Krupp V.2.A. (These may be subject to weld decay after welding.)

(6) *Soft Austenitic*, 12 Cr, 12 Ni. Used for table wear and ornamental goods; less difficult to cold work.

(7) *Decay-proof Steels.* These are of similar composition to (5), but specially designed for welding purposes with either low carbon or small additions of silicon, titanium, molybdenum and triobium; exemplified by Weldanka, Staybrite FDP, FCB 347 S17.

(8) *Special Purpose Austenitic Steels.* Similar to type (7) with addition of copper (2%), and molybdenum ($1\frac{1}{2}$–4%), (316, 317), nickel (10–18%), etc., to improve resistance to ammonium chloride, sulphuric and sulphurous acids. Selenium or sulphur is added to improve machining. Carbon is 0·03 max and 0·07 max. Titanium is also added in 320 S17, with 0·08 C.

(9) *High Manganese Steel* to conserve nickel 18 Cr, 5 Ni, 8 Mn, 0·15N (202), e.g. Staybrite FSM (17–4–6), can be cold rolled to 1080 N/mm$^2$ and used in rocket casings.

(10) *Heat-resisting Steels.* These are chromium steels with high nickel content (10–65%) together with tungsten, silicon and other elements designed specially for resisting oxidation at elevated temperatures, e.g. 25/20 (310 S24). 50/50 NiCr alloys resist fuel ash corrosion. Extra low carbon 0·02% 25 Cr 20 Ni and 18 Cr 24 Ni, 4·7 Mo steels are also used for resisting severe corrosion conditions.

(11) *Precipitation-hardening high tensile steels.* Three main groups based in their structure after solution anneal are:

(a) *martensitic*, 17 Cr, 4 Ni, with Cu, Al, Mo for ageing (480–590 °C); 17 Cr, 7 Ni + ageing additions.
(b) *Semi Austenitic* which require refrigeration or an austenite conditioning temper (700 °C) to form martensite prior to ageing at 450–560 °C.
(c) *Austenitic*, 17 Cr, 10 Ni, 0·25 P. Ageing at 680 °C causes precipitation of carbide and lattice strain induced by the phosphorus.

*Heat-treatment*

The hardening alloys possess critical ranges comparable with ordinary carbon steels, and can, therefore, be hardened, tempered and refined by heat-treatment which does not depend on recrystallisation after cold working.

The ferritic and normal austenitic steels, on the other hand, are not amenable to such treatment. Only cold work with subsequent heat-treatment involving recrystallisation can be employed to refine large grained material.

*Effects of chromium and nickel*

It will be readily appreciated that chromium is the chief alloying element in iron and steel for inhibiting corrosion. This resistance is not due to the inertness of the chromium, for it combines with oxygen with extreme rapidity, but the oxide so formed is very thin and stable, continuous and impervious to further attack. This property is, fortunately, conferred upon its solid solution in iron, becoming very marked as the amount exceeds 12% in low carbon steels. Thus in oxidising environments, such as nitric acid, the high chromium steel is initially attacked at the same rate as ordinary plain steel, but it rapidly builds up an oxide film, known as a self-healing *passive film*, which efficiently protects the underlying metal. This film has actually been isolated by U. R. Evans. The thickness of the film and its $Cr_2O_3$ content increases with the degree of polish.

In oxidising media any defect in the film which may arise through abrasion will be quickly repaired and such steel is quite satisfactory in the atmosphere, but the film does not offer sufficient permanent resistance to the less oxidising action of hydrochloric and sulphuric acids, except in very dilute solutions. Nickel has a low solubility in these acids and thus, with 8 to 10% of nickel in addition to chromium, the steel is immune from attack by nitric acid and the resistance to the other acids is markedly increased. Hence it is very evident why the 18/8 steels have such extensive uses. Their resistance to particular acids have been further improved by additions of elements such as molybdenum and copper.

## Mechanical properties of stainless steels

Some typical mechanical properties of various stainless steels are given in Table 37. A 12 Cr 6 Ni steel is useful for water turbines, but its properties depend on the unusual tempering above $Ac_1$ so that some austenite, enriched in carbon, is formed and about 25% is retained on cooling to room temperature. The presence of stable austenite increases fracture toughness and weldability by forming sinks for hydrogen. Alloy 420S29 shows the effect of tempering while 431S29 shows the beneficial influence of a small quantity of nickel. The low yield point is characteristic of the austenitic alloys. Steels with higher proof stress values are available by adding about 0·2% nitrogen to the standard alloys or by controlled warm working and low temperature softening. The pronounced hardening due to cold work is shown by alloy 301S21. This work-hardening is of much importance in press work and in machining, for which a sharp tool should always be used without any rubbing effect, otherwise the surface layers may harden to such an extent as to render subsequent machining difficult. The higher yield (up to 93 N/mm$^2$) produced by cold work is useful in stress carrying applications. The soft austenitic alloy does not work-harden so rapidly as the normal 18/8 steel and is used for the construction of many pressings for indoor use.

*Free-cutting stainless steels*

Machinability, especially of castings, is improved by adding about 0·2% selenium or 0·35% sulphur to form manganese molybdenum or zirconium sulphides. One would expect that these inclusions might weaken the passive film and promote pitting. Where absence of pitting is of prime importance a potential user is advised to test the steel under service conditions.

## Precipitation hardening stainless steels (see also p. 246)

The development of high speed aircraft and missiles has created a demand for high tensile stainless steels which can be formed and welded and which are able to retain their strength up to 400 °C. The cold-worked austenitic or ferritic types or the hardenable steels present fabrication difficulties, but a new variety of precipitation hardened stainless steels offer interesting possibilities.

The *martensitic class* possesses a delta ferrite-martensitic ($M_s$ is 130 °C) structure in solution treated condition (AC, 1050 °C), when they are readily machined (HV 350) but are not suitable for cold working (except AM 355). They are used in the form of bars, forgings and castings used for spindles, blades, valves and pumps. After machining optimum mechanical properties are obtained by a low temperature ageing at 450 °C, 1240 N/mm$^2$, HV 425. Typical examples are given in Table 38 together with the properties of FV 520B in three heat-treated conditions.

TABLE 37 TYPICAL MECHANICAL PROPERTIES OF STAINLESS STEELS

| BS 970 | Type | C | Cr | Ni | Treatment °C | YP N/mm² | TS N/mm² | El | RA | Izod ft lbf J | | HV |
|---|---|---|---|---|---|---|---|---|---|---|---|---|
| 403S17 | Ferritic | 0.06 | 13 | | Ann 700–780 | 280 | 416 | 20 | | | | 170 |
| | Ferritic | 0.06 | 13 | 6 | T 600 | 510 | 1158 | 16 | | 6 | 8 | 260 |
| 430S15 | Ferritic | 0.1 | 16.5 | | AC 750 | 340 | 540 | 28 | 50 | | | 175 |
| 410S21 | Martensitic | 0.12 | 12.5 | | OQ 1000, T 750 | 370 | 570 | 33 | 75 | 98 | 134 | 172 |
| | | | | | OQ 1000 | 1190 | 1850 | 2.5 | 6 | 5 | 7 | 371 |
| 420S29 | Martensitic | 0.16 | 12.5 | | OQ 960, T 400* | 1360 | 1498 | 18 | 50 | 12 | 16 | 480 |
| | | | | | OQ 960, T 700 | 630 | 757 | 26 | 64 | 69 | 94 | 223 |
| 420S45 | Cutlery | 0.32 | 13 | | OQ 980, T 180 | | 2780 | | | | | 600 |
| 431S29 | S80 | 0.16 | 16.5 | 2.5 | OQ 975, T 650 | 695 | 880 | 22 | 55 | 25 | 34 | 270 |
| 302S17 | Plain Austenitic | 0.08 | 18 | 9 | WQ 1050 | 250 | 618 | 50 | 50 | 80 | 108 | 170 |
| 301S21 | Cold rolled | 0.1 | 17.5 | 7.5 | Rolled full hard | 927 | 1235 | 5 | | | | 360 |
| | FMSI | 0.15 | 17 | 4 | Mu 7 Rolled full hard | 1000 | 1235 | 14 | | | | |
| | Soft Austenitic | 0.1 | 12 | 12 | WQ 1000 | 230 | 587 | 50 | 50 | | | 150 |
| 321S12 | Stabilised | 0.08 | 18 | 9 | Ti 0.5 AC 1050 | 260 | 649 | 45 | 50 | 80 | 108 | 175 |
| 316S16 | Molybdenum | 0.06 | 17.8 | 10 | Mo 2.8 AC 1050 | 278 | 618 | 50 | 50 | 80 | 108 | 180 |
| 317S12 | Molybdenum | 0.03 | 18 | 15 | Mo 3.5 AC 1050 | 216 | 525 | 40 | 50 | 80 | 108 | 170 |
| 310S24 | 25/20 Cr/Ni | 0.12 | 25 | 20 | AC 1050 | 216 | 540 | 40 | | | | 200 |

AC = air cool  OQ = oil quench  T = temper  *T 400 for Springs  T700 Gen. Eng. Parts  T 100 for knives

Table 38  Martensitic precipitation-hardening steels

|  | C | Cr | Ni | Cu | Mo | Nb |
|---|---|---|---|---|---|---|
| 17/4 PH | 0·04 | 17 | 4 | 4 | — | 0·3 |
| 17/4 PH.Mo | 0·04 | 15 | 4 | 4 | 2 | 0·3 |
| AM 355 | 0·13 | 15 | 4 | — | 2·7 | 0·1 |
| FV 520 (B) | 0·05 | 14 | 5·6 | 1·8 | 1·7 | 0·3 |

Heat-treatment and properties of FV520(B)

| Heat-treatment | TS N/mm² | E | HV | Izod ft/lbf | J |
|---|---|---|---|---|---|
| Softened 620 AC |  |  | 310 |  |  |
| Normalised |  |  |  |  |  |
| *950 AC + 3 hr 450°C* | 1344 | 10 | 420 | 18 | 24 |
| *950 AC + 2 hr 570°C* | 1035 | 12 | 330 | 45 | 61 |
| Primary hardened, aged |  |  |  |  |  |
| 1050 AC + 2 hr 850 AC |  |  | 330 |  |  |
| *1050 AC + 3 hr 450 AC* | 1205 | 12 | 400 | 35 | 47 |
| Doubled over-aged 750/550* |  |  |  |  |  |
| 1000 AC + 2 hr 750 AC + 2 hr 550 | 1004 | 15 | 310 | 75 | 102 |

User heat-treatment italic.
*Suitable for use—100 to 450°C without further treatment by user.

The *semi-austenitic steels* have been developed with a balanced alloy content which depresses the $M_s$ point to slightly above or just below room temperature with a $M_s$–$M_f$ range of about 130°C, so that the austenite is retained with up to 10% delta ferrite. In this condition they can be readily cold worked, although they tend to work-harden more rapidly than the 18/8 steel. Accurate control of the martensite transformation within the normal variations of commercial steel making, is by the solution annealing temperature which controls the amount of alloy and carbon taken into solution. The transformation to martensite, accompanied by expansion of 0·003 in per in, can be accomplished in three ways:

(1) By refrigeration to −78 °C, i.e. to near $M_f$.
(2) By a primary temper or 'austenite conditioning' at 700–800°C, thus precipitating an alloy carbide, which raises the $M_s$–$M_f$ range above room temperature for the impoverished matrix. Upon cooling to room temperature the steel then transforms to martensite, which is a somewhat softer than in (1) due to its lower carbon content.

(3) Cold working can also be used to form a martensite-like structure. Incomplete transformation to martensite adversely affects the mechanical properties.

## PRECIPITATION HARDENING STAINLESS STEELS

TABLE 39  SEMI-AUSTENITIC STEELS

|           | C    | Cr   | Ni  | Cu  | Mo  |        |
|-----------|------|------|-----|-----|-----|--------|
| 17/7 PH   | 0.07 | 17   | 7   | —   | —   | 1·2 Al |
| PH 15/7 Mo| 0.07 | 15   | 7   | —   | 2·5 | 1·2 Al |
| FV 520(S) | 0.06 | 15·7 | 5·5 | 1·8 | 1·7 | 0·1 Ti |
| SF 80     | 0.07 | 17   | 3·5 | 1·3 | 2   | 2 Co   |
| ESC D50   | 0·2  | 12   | 6   | —   | —   | 2 V    |

A secondary temper is then used to produce slight secondary hardening at 450–560 °C with such elements as Mo and age-hardening with elements such as Cu, Al, Cr, Ti and Nb. In the fully treated condition the steels are magnetic and have good corrosion resistance, low thermal expansion, high proof stress and tensile strength with reasonable ductility and weldability.

The essential requirements of these steels are, therefore, low carbon with Ti or Nb to prevent weld decay, Cr for corrosion resistance, Ni for austenitic structure, Mn for controlling transformation temperature, Mo for tempering resistance and Cu, Al, Co for ageing. Examples are given in Table 39.

TABLE 40

*Steel A:* 0·1 C, 17·3 Cr, 4·2 Ni, 2 Mo

|                                           | 0·1 PS N/mm² | TS N/mm² | E  |
|-------------------------------------------|--------------|----------|----|
| 1050°, 1 hr AC                            | 216          | 444      | 18 |
| 1050°, 1 hr AC + 2 hr 700 °C              | 556          | 973      | 17 |
| 1050°, 1 hr AC + 2 hr 700 °C + 1 hr 450 °C| 834          | 1081     | 19 |
| 1050° 1 hr + 2 hr −78 °C                  | 664          | 1405     | 14 |
| 1050° 1 hr + 2 hr −78 °C + 1 hr 450 °C    | 926          | 1282     | 18 |

*Steel: FV 520(S)*

|                          | PS N/mm² | TS N/mm² | E  | HV  |
|--------------------------|----------|----------|----|-----|
| Softened 1050 AC         | 310      | 830      | 30 | 200 |
| *Overaged 540 °C*        |          |          |    |     |
| 750°–2 hr AC             |          |          |    |     |
| 0°–2 hr                  |          |          |    |     |
| 540°–1 hr AC             | 1020     | 1080     | 12 | 320 |
| *Aged 450 °C*            |          |          |    |     |
| 750°–2 hr AC             |          |          |    |     |
| 0°–2 hr                  | 1080     | 1270     | 10 | 400 |
| 450°–2 hr AC             |          |          |    |     |

The high carbon D 50, in the fully heat-treated condition will give a tensile strength of 1544 and a 0·1% PS of 1190 N/mm², but work-hardens too rapidly for sheet production.

Table 40 shows the trend of mechanical properties in these steels, especially the effect on proof stress. The Izod is somewhat low. The microstructures of the solution treated and fully hardened 17/7/1·1 Al steel are shown in Figs. 162, 163.

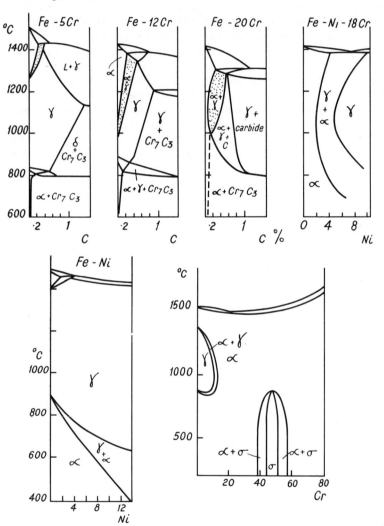

FIG. 157 Equilibrium diagrams showing effect of Fe, Cr, C, Ni

## Structures of individual alloys

Some of the general structural characteristics of stainless steels can be best understood by considering the binary Fe–Cr and Fe–Ni binary diagrams and the modifying effects of additional elements, Figs. 157 and 158 (see also Fig. 280). The chromium, as a ferrite stabiliser, tends to form a closed loop diagram with ferrite a stable phase at about 12%, and it also forms a sigma phase from about 38 to 57% (p. 258). Carbon and nickel tend to form austenite. The combined effects of nickel and chromium are shown in the ternary diagram at 700°C, see also Fig. 280.

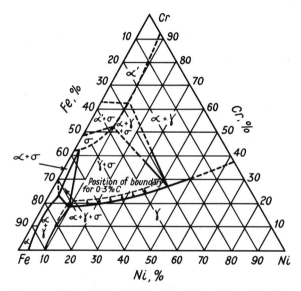

FIG. 158 Equilibrium diagram showing effect of Fe, Cr, C, Ni

*Stainless iron*, 0·1% carbon, 11–14% chromium. When annealed at high temperatures (1000°C) the structure consists of ferrite and pearlite with tendency to form isolated carbides as shown in Fig. 159, which exhibits equal areas of the two constituents, eutectoid being 0·3% carbon. Coalescence of the carbides frequently produces strings of particles round the grain boundaries.

On heating to 925–1000°C the alloy transforms to austenite and when air cooled or quenched from this temperature consists entirely of martensite (like Fig. 160). After tempering at 650–751°C, the structure consists of ferrite with fine carbides dispersed through it and the iron can be machined readily.

When heated above a minimum temperature depending on the chromium

Fig. 159 0·13% carbon, 13% chromium iron. Slowly cooled — pearlite + chromium ferrite (× 1000)

Fig. 160 13% chromium stainless steel (C, 0·28) quenched from 950°C martensite (× 500)

content decarburisation of stainless iron leads to the production of a sharply defined layer of coarse columnar crystals. In oil refineries a steel containing Cr, 12, Al, 0·1–0·4% is favoured instead of the 17% chromium iron, since embrittlement in service at 480°C is largely avoided.

*Cutlery stainless steel*, 0·25–0·30% carbon, 11–13% chromium. This alloy of approximately eutectoid composition and transforms completely to austenite on heating to about 1000°C. The effect of quenching or air cooling from this temperature is shown in Fig. 160. The annealed structure consists of large particles of carbide in ferrite.

The structures found in this class of steel are summarised in Table 41. Locally heating the heat-treated steel to about 820°C such as occurs in brazing a handle to a knife-blade produces a band which is more readily corroded.

TABLE 41  STRUCTURES OF CUTLERY STAINLESS STEEL

| Soaking Temperature °C | Rate of Cooling | Brinell Hardness | Structure |
| --- | --- | --- | --- |
| 1000 | Very slow, $1\frac{1}{2}$ hr | 200 | Pearlitic like ordinary steel |
| 1000 | Slow, $\frac{3}{4}$ hr | 200–400 | Mixture of martensite and fine pearlite |
| 1000 | Slow, $\frac{1}{4}$ hr | 400 | Martensite with rapidly etching areas or troostite |
| 850 | Quenched | 450 | Carbides in martensite |
| 850 | Very slow | 190 | Granular: carbides in ferrite |

## 16–18% chromium iron

Carbon-free material would be entirely without critical ranges, consequently the ferrite undergoes grain growth at the high temperature (low Izod) and cannot be refined by heat-treatment. Commercial alloys, however, contain 0·05–0·15% carbon, and consequently small carbon-rich regions transform to austenite at elevated temperatures. In the annealed condition, in which this alloy is generally used, the structure consists of polyhedral ferrite grains in which are embedded fine carbide particles. The laminated structure due to rolling is very persistent in these irons (Fig. 161).

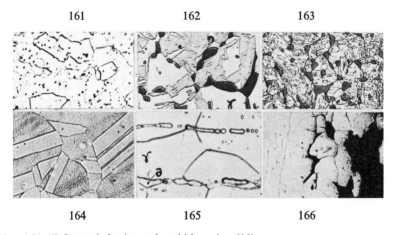

FIG. 161 17 Cr steel, ferrite and carbides (× 600)
FIG. 162 17/7PH. Softened 1050°C AC. White = austenite (VPN = 163); grey = martensite and austenite (VPN = 213); black = delta ferrite (× 500), etched 10% HCl in alcohol
FIG. 163 17/7PH hardened 750° AC, 560° AC. Grey area = aged martensite, dark areas = delta ferrite (× 500), VPN % 435, Izod = 3 ft lb
FIG. 164 18/8 steel, softened 1100°C AC note twins (× 500)
FIG. 165 18/8 + 2·8 Mo, 1050°C AC. Shows iron and carbides in austenite (× 500)
FIG. 166 Intergranular corrosion of 18/8 steel (× 200) unetched

(*Figs. 161–165 by courtesy of Brown Firth Research Laboratory*)

The addition of 2% nickel enables the 17 and 25 Cr steels to respond to heat-treatment and thus allows the combination of high corrosion-resistance of the high chromium iron with the mechanical properties of the lower chromium product.

*Austenitic chromium nickel steels* These steels cannot be hardened by any form of heat-treatment; in fact quenching from 1000–1100°C merely softens them and this treatment is used to obtain these steels in the best condition to resist corrosion, i.e. with all carbon and chromium in solid solution. The microstructure produced by this 'softening' treatment consists of polyhedral grains of austenite with thin grain boundaries, similar to all solid solutions. A distinctive characteristic of these alloys is the frequent twinning of the crystals (i.e. parallel markings), shown in Fig. 164.

Although austenitic steels are not amenable to hardening by heat-treatment, an inordinate increase in hardness (from 90 to 400 Brinell) occurs when this alloy is subjected to cold deformation. The higher the proportion of austenite to ferrite stabilising alloys (p. 215) the greater the stability of the austenite and the lower the rate of work-hardening. The microstructure of cold worked austenite reveals the usual distortion of the polyhedral grains, in which are many markings resembling slip bands.

On heating in the range 500–950 °C carbides, rich in chromium, are precipitated as particles usually in the grain boundaries in unworked material, but also in the region of the slip planes when the steel has received cold work.

*Duplex steels*

When silicon or molybdenum or other ferrite-forming element (p. 215) is added to the steel in sufficient quantity it is found that ferrite is staple even after fairly slow cooling. This ferrite exists as little islands, as shown in Fig. 165 (see p. 255).

*Weld decay*

Besides the instability under cold work, these steels are also unstable under certain forms of heat-treatment, namely, when reheated in the range 500–800 °C, they undergo a structural change which is detrimental to their corrosion resistant properties, although their mechanical properties are not affected to an appreciable extent. The conditions necessary to produce this change are easily realised during welding, when portions of the metal at some distance from the weld are heated in this dangerous zone (Figs. 166, 278). Hence the term 'weld decay.'

If such a welded sample is immersed in certain corrosive media a narrow band of the plate, 1–2 mm from the welded edge is found to be attacked, but not in a general manner. The centre of the grains are unaffected but the boundary material is removed. Consequently, loss in weight measurements give little indication of the extent of the damage done and the material may be likened to a brick wall, the mortar of which has disintegrated. A slight mechanical deformation serves to produce a series of

# STRUCTURES OF INDIVIDUAL ALLOYS

intercrystalline cracks in the material, which in bad cases will actually crumble to a powder.

The cause of the defect is due to the precipitation at the grain boundaries of very thin films of chromium rich carbides containing as much as 90% chromium taken from the layer of metal immediately adjacent to the grain boundary, whose corrosion resistance is therefore seriously impaired by this impoverishment of its chromium content and by the electrochemical cell action.

*Tests*

The liability of decay can be tested by boiling in the Hatfield solution containing 98 gm of concentrated sulphuric acid and 111 gm of copper sulphate in one litre of water. To prevent loss by evaporation a reflux condenser should be used. The usual time is 72 hours.

Initially this defect was a serious handicap to the application of these steels in welded chemical plant and the methods of overcoming the defect are as follows.

*Prevention of weld decay*

(1) *Heat-treatment.* Complete immunity from decay can be obtained by raising the whole article to a temperature of 1000–1150 °C and rapidly cooling by quenching, but this treatment causes distortion of finished parts and is impossible in the case of large welded vessels.
(2) *Reduction of carbon content to 0·03%.*
(3) *The addition of ferrite-forming elements.* Certain elements—silicon, molybdenum, tungsten and vanadium—when added to the steel tend to form ferrite islands in which the carbon is precipitated without forming continuous films.
(4) *The production of other carbides.* An alternative method of preventing impoverishment of chromium is to add to the alloy an element which has a greater affinity for carbon than has chromium. The carbides which then form contain little chromium and the defect does not arise. Such carbide forming elements are titanium and niobium. A sufficient amount of each must be present, and for titanium it is six times the carbon content in excess of 0·02%. It must be pointed out that these alloys offer far greater resistance to intergranular corrosion if the heat-treatment prior to final fabrication is such as to allow the alloying element to accomplish its purpose of combining with the carbon. This can be obtained by using a lower heat-treatment than that employed for straight 18/8 steels or by a subsequent anneal at 900 °C. Titanium is readily oxidised during welding and the use of niobium is preferable.

## Stress corrosion

Stress-corrosion failures have been reported in stainless steels in solutions containing chlorides, but the occasions of such failures have been relatively few, and the residual stresses have undoubtedly been high. Solutions of magnesium, zinc or lithium chlorides and moist ethyl chloride in combination with a high stress cause stress-corrosion cracking. Of these, a boiling concentrated solution of magnesium chloride is the most corrosive and will cause the austenitic steels to fail in short periods of time when the applied stress is near the yield strength of the steel. Immunity is obtained with 16–17 Cr ferrite steels, often with addition of 2 Ni 1 Mo 0·5 Nb.

## Sigma phase

In addition to ferrite, austenite and carbides, another micro-constituent is found in many stainless steels—called sigma phase. This phase is hard (800 VPN), *brittle*, non-magnetic and is formed from *ferrite* with a considerable diminution in volume which causes fine cracks in it. Sigma appears to be the intermetallic compound FeCr, with a complicated body-centred cubic lattice structure, which can dissolve small amounts of other elements.*

Sigma phase forms when a steel is heated in the range 500–1000 °C, very slowly in pure iron–chromium alloys, somewhat quicker in Fe–Cr–Ni alloys and much faster in Fe–Cr–Si. In 18/8 alloys the addition of ferrite-forming elements—silicon, niobium, molybdenum, tungsten—make the steel susceptible to sigma formation unless the austenite is stabilised by raising the nickel content. The effect of nickel content on the Izod value of a steel containing Cr 18, Mo 2, Nb 1 % after reheating to 800 °C for 4 hours is as follows:

| Nickel, % | 6 | 8 | 12 | 14 | 20 |
|---|---|---|---|---|---|
| Izod, J | 28 | 28 | 87 | 95 | 115 |

The change in properties of an 18/8/3/1 Cr–Ni–Mo–Ti steel is as follows:

TABLE 42  EFFECT OF REHEATING ON 18/8/3/1 STEEL (Smith and Bowen)

| Reheating Temp. °C | Nil | 550 | 650 | 750 | 850 | 950 | 1100 |
|---|---|---|---|---|---|---|---|
| Ferrite, per cent | 43 | 43 | 35 | 11 | 3 | 25 | 40 |
| VPN | 200 | 199 | 200 | 229 | 237 | 210 | 184 |
| Izod Joules | 163 | — | 106 | 11 | 13 | 48 | 161 |

Fig. 167 shows ferrite plus austenite while Figs. 168–170 show the presence of sigma in various stages in a Cr–Ni–Mo–Ti steel (18/8/3/1).

*10% oxalic acid in water used electrolytically outlines and then actively dissolves the sigma grains, but does not attack ferrite.

(The fineness and form depend on the temperature of formation; fine troostitic type at 550°C, acicular type at higher temperatures.) With this steel a few minutes heating at 850°C produces some sigma phase and after a few hours practically all the ferrite is converted into sigma phase and

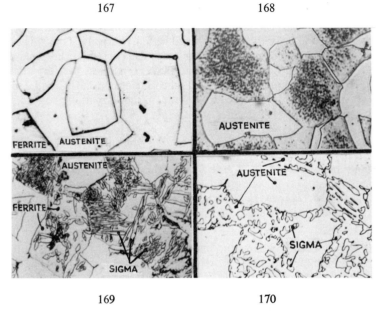

FIG. 167  18/8/3/1—Cr/Ni/Mo/Ti steel (× 400) (Smith and Bowen); etched 10% oxalic acid as quenched from 1200°C; austenite and ferrite (66%)
FIG. 168  As Fig. 167. Reheated 600°C, 4 hours, ferrite 50%. Ferrite darkened by fine dispersion of sigma particles, mainly in centre
FIG. 169  Reheated 850°C, 5 min, ferrite 43%, larger particles of sigma
FIG. 170  Reheated 850°C, 16 hours, ferrite 3%. Austenite and larger particles of sigma

austenite, but this 'new' austenite may be of different composition from the original austenite as shown by its etching characteristics.

It would appear that as sigma forms, local alloy *depletions* occur as a result of the sigma phase becoming enriched in ferrite forming alloys, particularly chromium (Fig. 168) and depleted in nickel. At 850°C longer reheating enables chromium and other elements to diffuse into the impoverished areas, with consequent improvement in corrosion resistance and uniformity of etching (Fig. 170).

Heating at 650°C may render the steel more susceptible to corrosion because diffusion of the alloys is slow. It is important not to over-emphasise the effect of sigma phase on the corrosion resistance of this steel since much welded equipment has given excellent service. A greater significance should be given to the possible effects of the low Izod value—as low as 7 Joules after prolonged heating—in the design of parts in heat resisting steels, especially as the embrittling effect of the sigma phase appears to be as potent while the steel is at the elevated temperature.

*'475' Embrittlement.* When ferritic chromium steels are heated in the range 400–540 °C, embrittlement occurs within relatively short times and hardness increases continuously over extended periods. The maximum lowering of impact ductility occurs at 475 °C regardless of alloy content but embrittlement can be removed by reheating at 600–850 °C and cooling quickly. The cause of the brittlement is due to the formation of chromium-rich zones along the {100} planes of the alloy.

## Heat-resisting steels

Steels are now used for a wide variety of conditions entailing heat and corrosion under both static and dynamic stresses, such as aero engine valves, furnace conveyers, retorts, oil cracking units and gas turbines. Three important properties are necessary in material used at elevated temperatures:

(1) Resistance to oxidation and to scaling (Chap. 8).
(2) Retention of strength at the working temperature (see creep, Chap. 1).
(3) Structural stability as regards carbide precipitation, spheroidisation, sigma formation and temper embrittlement.

Other properties may also be important in particular applications, such as specific resistance and temperature coefficient for electrical purposes, coefficient of expansion for constructional units and resistance to penetration by products of combustion in many furnace applications. In the case of gas turbine steels additional characteristics have to be considered; internal damping capacity and fatigue strength, notch sensitivity and impact strength (hot and cold), machining and welding characteristics, especially as large rotors may have to be built of small sections welded together.

The scale which forms on iron is porous and loosely adherent, but it is rendered adherent and protective by the addition of certain elements to the steel. These are usually chromium, silicon and aluminium, and they are characterised by their great affinity for combining with oxygen, but the reaction is rapidly stifled by the formation of the inert oxide films.

The resistance of mild steel to oxidation is vastly improved by forming an aluminium–iron alloy on the surface. This is done by heating at 1000 °C in contact with powdered aluminium (calorising) or by metal-spraying the steel surface with aluminium, coating with bitumastic paint to prevent oxidation and heating to 780 °C (aluminising).

*Improved creep strength* may be attained by

(a) raising the softening temperature by the solution of alloying elements;
(b) judicious use of precipitation hardening at the working temperature without readily over-ageing. Second phase hardening is critically dependent on the degree and uniformity of the dispersion achieved and creep rate is related to a critical range of particle spacing.
(c) controlled degree of work-hardening in the appropriate temperature ranges which often reduces the extent of the primary creep stage;
(d) variations in the process of manufacture; deoxidisers and particles in the crystal boundary can have a marked effect on creep properties.
(e) vacuum melting permits the use of advantageous compositions which cannot be melted by conventional methods. It also improves ductility in the transverse direction.

*Mechanical properties* are improved by the addition of various elements.

Cobalt, tungsten and molybdenum cause the steels to withstand the action of tempering. High alloy austenitic steels have no change points and therefore do not harden by air cooling, but their resistance to wear resistance is not great. A sufficiency of elements such as silicon and chromium raise the $Ac_1$ point to temperatures above those reached in service and prevent the steel from air hardening on cooling.

Steels with high nickel content should not be used at high temperatures in contact with gases containing sulphur dioxide or other sulphur compounds, since intercrystalline films of nickel sulphide are formed.

In high chromium steels the carbides coalesce into large particles which have less obstructive action on grain growth of the ferrite at temperatures above 700 °C. The excessive grain growth reduces still further the toughness which these steels possess. Grain growth also occurs in the austenitic steels above 1000 °C, but no trouble arises since they remain tough and ductile even in the coarse grained condition. When heated in the range 500–900 °C austenitic steels precipitate carbides along the boundaries of the austenite and as a result intercrystalline cracks are liable to develop if the steel is stressed continuously in tension in this temperature range (p. 256). With certain compositions both ferritic and austenitic steels are embrittled by the formation of sigma phase (p. 258).

*Scale resisting steel*

Typical compositions of heat-resisting steels are given in Table 43, together with the heat-treatment, micro-structure and creep index. The choice of a steel will depend on:

(1) maximum working temperature;
(2) form of the material and operations to be performed on it;
(3) strength at elevated temperature.

Steels Nos. 1 to 4 are used essentially for exhaust valves in automobile, aero and diesel engines. The silchrome steel (2) has a satisfactory resistance to scaling and wear, but the impact figure at room temperature is only 4–8 Joules, and it is weak, although tough at service temperature. Steel 3 (silchrome XB) gives improved resistance to scaling and lead attack and is stable up to about 900 °C. Alloy 3A (known as XCR) is austenitic and can

TABLE 43  HEAT-RESISTING STEELS

| No. | Composition | | | | | Condition °C | Structure at 15°C | Time Yield at 800°C N/mm² |
|---|---|---|---|---|---|---|---|---|
| | C | Si | Ni | Cr | W | | | |
| 1 | 0·6 | — | — | 10 | — | OQ 950, T 700 | F, C | — |
| 2 | 0·4 | 3·9 | — | 8 | — | OQ 950, T 750 | F, C | 1·24 |
| 3 | 0·8 | 2 | 1·4 | 20 | — | OQ 1050, T 725 | F, C | — |
| 3A | 0·4 | 0·8 | 4 | 22 | Mo 2 | WQ 1000, T 780 | A, sigma | — |
| 4 | 0·5 | 1 | 14 | 13 | 3 | AC 950 | A, C | 5·5 |
| 5 | 0·24 | — | — | 15 | — | OQ 950, T 750 | F, C | 0·86 |
| 6 | 0·2 | — | — | 33 | — | AC 1000 | F, C | 2·4 |
| 7 | 0·25 | 1·7 | 12 | 23 | 3 | AC 1050 | F, A, C | 8·25 |
| 8 | 0·35 | 1·7 | 7 | 20 | 3 | AC 1050 | F, A, C | 8·25 |
| 9 | 0·12 | 1·5 | 21 | 24 | — | AC 1050 | A, C | 6·2 |
| 10 | 0·3 | 2 | 63 | 12 | 2 | AC 1050 | — | 9·0 |
| 11 | 0·07 | — | 65 | 15 | — | AC 1050 | A | — |
| 12 | — | — | 80 | 20 | — | AC 1050 | — | 75·7 |

A = austenite.  F = ferrite.  C = carbide. AC = air cool.  OQ = oil quench. Time yield is approximately equivalent to a creep rate of 1 part in a million.

be precipitation hardened with sigma phase to 420 VPN. Its room temperature impact strength is poor. The nickel–chromium–tungsten steel No. 4 is tough at all temperatures, resists lead bromide from leaded fuels and the low hardness of the stems can be improved by nitriding.

The chromium alloy No. 5 resists oxidation up to 750 °C, but possesses no special strength at high temperatures. The high chromium steel No. 6 is suitable up to 1100 °C. Nitrogen (0·2%) is often added to refine the grain structure and to inhibit grain growth at elevated temperatures.

The alloy has been used for thermocouple sheath retorts, baffles in furnaces, and is often preferred, especially in the presence of sulphur-rich hot gases.

The use of the alloy has been limited because of its relatively poor strength at elevated temperatures, its tendency towards embrittlement on slow cooling through or on heating in the vicinity of 475 °C; its notch sensitivity at room temperature, which, however, is reduced at 150 °C. Designs should, therefore, be simple; keyways, re-entrant angles, sharp radii and surface scratches should be avoided.

Complex steels Nos. 7, 8 and 9 possess excellent ductility at ordinary temperatures and appreciable mechanical strength at elevated temperatures and can be used up to 1100 °C.

Steels 7 and 8 can be used where resistance to stress as well as resistance to oxidation is required. They can be welded and are obtained as castings, forgings, sheet and wire, but not solid-drawn tubes. No. 7 steel has been used for conveyer bars and skid bars in furnaces, and pyrometer sheaths. No. 9 is easier to manipulate and seamless tubes can be made.

No. 10 can be used up to 1150 °C, but is mostly supplied in the form of castings for heat-treatment and case-hardening boxes.

Nickel–chromium alloys 11 and 12 exhibit good resistance to oxidation up to about 850 °C and 1150 °C respectively, combined with good mechanical properties at elevated temperatures. They are mainly used for electrical resistance units in domestic fires, stoves and industrial furnaces.

An 80/14/6 Ni–Cr–Fe alloy—Inconel—is often used for aero-engine exhaust manifolds.

*Steels for gas turbines*

The development of creep-resistant steels has been given a great incentive by the advent of the jet and gas turbine engine. Vastly improved alloys have been developed, each with a fairly well defined field of application and an alloy suitable for use at 700–800 °C may offer no advantage over older steels at 400–500 °C. Most of the new alloys are subject to solid solubility changes at high temperatures, and the associated precipitation hardening effects are of immense use. In general, the strengthening effect of iron carbide, although marked at low temperatures, is less useful at high temperatures. The strengthening effects of (closed $\gamma$-loop) carbide-forming element such as molybdenum, tungsten, titanium and niobium, are extremely powerful through their influence on the precipitation phases (see G18B example).

In addition to iron, nickel and cobalt have also been used extensively as basis metals, since they enable a ductile austenitic structure to be retained which is capable of producing precipitation hardening systems. They also increase the temperature at which softening occurs after strain hardening.

The melting, mechanical and thermal treatments of most of the alloys are extremely important in the production of optimum properties. For demanding applications vacuum melting for ESR are used increasingly.

Stain hardening increases the strength below the recrystallisation temperature and it also has an accelerating effect on the precipitation of phases from solid solution. Prior heat-treatment of stain-hardened material greatly affects its strength and ductility. Above 730° C the strength depends largely upon precipitation effects (p. 97) and is affected by

(a) amount of alloy going into solution and its homogeneity, i.e. solution temperature and time;
(b) distribution of precipitate. Rate of cooling should be rapid to prevent excessive precipitate forming at grain boundaries;
(c) particle size and coherency are influenced by temperature of formation of precipitate.

Typical compositions of steels for gas turbines are given in Table 44.

*Iron base alloys*

Although austenitic alloys are most favoured for turbine discs there is also an interest in ferritic steels for temperatures up to 550 °C. The C–Cr–Mo–V steel has a creep stress of 325 N/mm² for 0·1 % strain in 300 hours at 550 °C. The 10 % Cr steel (FV 535) has been improved by the addition of 6 Co for service at higher stresses in the 500–600 °C range.

The austenitic Cr–Ni–Nb steel has sufficient nickel and carbon to minimise the formation of sigma phase, but experience has revealed some shortcomings and the addition of $1\frac{1}{2}$% Mo in lieu of Cr has produced FV548 with some advantages as regards creep, rupture strength and ductility. With higher steam temperature $2\frac{1}{4}$ Mo 17/10 austenitic steel is used for piping (FV 555).

Steel 7 in Table 43 has also been used successfully for nozzle blades.

After G18B steel is 'solution' heat-treated at 1330 °C (air cooled or oil quenched), it is 'warm-worked' by subjecting it to plastic deformation in the range 650–800 °C. The combination of work-hardening and precipitation-hardening increases the 0·1 % proof stress at room temperature to about 380–460 N/mm² with some increase in creep strength.

The American Timkin 16–25–6 steel is similarly reduced 22 % at 650 °C by rolling and stress relieved at 650 °C.

Rex 326F is similar to G18B but contains no tungsten. G68 is precipitation hardening alloy for large discs calling for good properties in the range 600–750 °C.

*Nickel base*

'Nimonic'* alloy 75, which is the 78/20 Ni–Cr alloy with a small titanium addition, is very suitable for combustion tubes. It is annealed at 900–1050 °C.

'Nimonic' alloys 80(A), 90, 105 and 115 are nickel–chromium alloys

*'Nimonic' is a registered trade mark.

TABLE 44  TYPICAL GAS TURBINE STEELS

| Name | C | Cr | Ni | Co | Mo | Ti | W | Nb | Al | Others | Uses |
|---|---|---|---|---|---|---|---|---|---|---|---|
| C–Cr–Mo–V | 0·25 | 3·0 | — | — | 0·5 | — | 0·5 | — | — | V 0·75 | Discs and blades |
| FV 535 | 0·07 | 10·5 | 0·3 | 6·0 | 0·75 | — | — | 0·45 | — | V 0·20 | |
| FV 467 | 0·2 | 12·0 | 9·5 | — | 2·0 | 0·8 | — | — | — | Cu 2·5 | * |
| FV 548 | 0·08 | 16·2 | 11·5 | — | 1·5 | — | — | 1·0 | — | — | |
| Timkin 16–25–6 | 0·15 | 16·0 | 25·0 | — | 6·0 | — | — | — | — | N 0·18 | Turbine discs |
| G18B | 0·4 | 13·5 | 13·5 | 10·0 | 2·0 | — | 2·5 | 2·8 | — | — | Turbine discs |
| G68 | 0·07 | 15·0 | 25·0 | — | 1·3 | 2·1 | — | 0·25 | — | V 0·3 | Tubes, discs, blades |
| Nimonic 80A | 0·1 | 20·0 | Balance | 20·0 | — | 2·4 | — | — | 1·4 | Si 1·0, Mg 1·0 | Blades |
| Nimonic 90 | 0·11 | 20·0 | Balance | 18·0 | — | 2·7 | — | — | 1·5 | Si 1·5, Mg 1·0 | |
| Nimonic 105 | 0·15 | 18·0 | Balance | 20·0 | 5·0 | 1·1 | — | — | 4·7 | Si 1·0, Mg 1·0 | |
| Nimonic 115 | 0·15 | 15·0 | Balance | 15·0 | 4·0 | 4·0 | — | — | 5·0 | Si 1·0 | |
| 713C | 0·12 | 15·0 | Balance | — | 4·5 | 0·8 | — | 2·0 | 6·0 | B 0·01, Zr 0·1 | |
| M21 | 0·12 | 15·0 | Balance | — | 2·0 | — | 11·0 | 1·5 | 5·9 | B 0·02, Zr 0·12 | |
| G32 | 0·3 | 19·0 | 12·0 | 45·0 | 2·0 | — | — | — | — | — | Blades |
| Vitallium | 0·24 | 28·7 | — | Balance | 5·6 | — | — | — | — | Fe 1·0 | Blades |

*1230°C, OQ, age 16 hr. 700/750°C

hardened by precipitation of $Ni_3(TiAl)$ and have become standard material for rotor blades of aero gas turbines, in the temperature range 700–1000 °C. The alloys of this series have been progressively improved by the use of optimum levels of aluminium and titanium, the addition of cobalt and molybdenum, and improved methods of production, including extrusion and vacuum melting. At the same time it has been necessary to reduce the chromium content from 20% in 'Nimonic' alloy 80(A) to 15% in 'Nimonic' alloy 115 to maintain structural stability. Chromium contents of the alloys are important and low chromium contents favour increased creep rupture strength at high temperatures, while high chromium contents increase the resistance to hot corrosion by sulphur-containing fuels. Cast alloys such as 713C, IN–100 and M21 are generally similar materials designed for

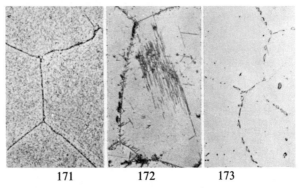

FIG. 171  $Ni_3$ (TiAl) phase in Nimonic alloy  ($\times$ 500)
FIG. 172  Acicular precipitate of carbide in Nimonic 80A  ($\times$ 500)
FIG. 173  Grain boundary $Cr_7C_3$ in Nimonic 80A  ($\times$ 500)

(*By courtesy of the International Nickel Co.*)

precision castings and usually have lower chromium contents to obtain the highest possible strength levels. However, these alloys are normally given a protective coating to ensure adequate corrosion resistance.

The micro-structures of these alloys contain precipitates of $Ni_3(TiAl)$ (Fig. 171), chromium carbide, TiC, TiN and titanium cyanonitride particles. The chromium carbide can form an acicular precipitate within the grains (at 800 °C—Fig. 172) or masses at the grain boundaries (Fig. 173) associated with chromium impoverishment of matrix (at 1080 °C). Such boundaries are weak and permit relaxation of creep stresses and hence increase creep extension before rupture. For maximum creep resistance the $Ni_3$ (Ti Al) should form submicrosopically at low ageing temperatures (700 °C). A triple heat-treatment, therefore, is often necessary to produce both of these effects. The solution temperature of $Ni_3$ (Ti Al) varies with the different alloys as shown in Table 45, which also indicates the solidus

TABLE 45

| 'Nimonic' alloy | Ni₃(TiAl) solution temp. °C | Solidus temp. °C | Heat Treatment | TS 15°C N/mm² | Mod.* kN/mm² | °C | Stress for 0.1% creep N/mm² | Rupture in 1000 hr stress N/mm² | Ext. % |
|---|---|---|---|---|---|---|---|---|---|
| 80A | 885 | 1360 | 8 hr 1080 AC + 16 hr 700 AC | 1065 | 186 | 650 | 324 | 417 | 2.5 |
|  |  |  |  |  |  | 815 | 59 | 86 | 7 |
| 90 | 960 | 1360 | As for 80A | 1235 | 193 | 650 | 345 | 459 | 1.5 |
|  |  |  |  |  |  | 815 | 57 | 117 | 10 |
| 105 | 1025 | 1290 | 4 hr 1150 AC + 16 hr 1060 AC + 16 hr 850 AC | 1158 | 220 | 750 | 216 | 368 | 7 |
|  |  |  |  |  |  | 815 | 72.5 | 216 | 11 |
|  |  |  |  |  |  | 870 | 37 | 131 | 10 |
| 115 | 1125 | 1260 | 1½ hr 1190 AC + 6 hr 1100 AC | 1235 | 214 | 750 | — | 448 | — |
|  |  |  |  |  |  | 850 | — | 216 | 8 |
|  |  |  |  |  |  | 950 | — | 93 | 10 |

*Young's Modulus × 10⁶ kN/mm²

temperature. Hot working is normally carried out between these two temperatures. The solution treatment serves (a) to anneal the matrix and the higher the temperature the better the creep properties, (b) to dissolve the $Ni_3$ (Ti Al). When relatively high solution temperatures are used it is desirable to cool or reheat in the range 900–1100 °C to form boundary carbide followed by ageing at 700 °C. Such treatment increases time to fracture and creep extension. Turbine blades can suffer severe corrosion from a combination of sulphur (0·6%) in the fuel and sea salt. Aluminised coatings protect nickel alloys but much more work is needed on suitable coatings.

*Cobalt base.* Many alloys have been tried in America; G32, S 816 types and precision cast-alloys of the Vitallium type are examples.

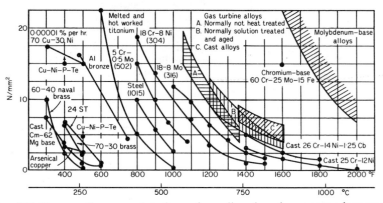

FIG. 174 Comparative creep-data for various alloys based on a secondary creep rate of 0·0001% hr  (*After Deuble*)

*Chromium base* alloys are also being tried in America, e.g. 60 Cr, 15 Fe, 25 Mo. Some chromium-based alloys have good creep properties at 900–1000 °C, coupled with high resistance to oxidation, but they have extremely low ductility at room temperature.

*Molybdenum base* alloy with 0·5 Ti has very good creep properties greatly in excess of the Nimonics (Fig. 174), but its use is prevented by the volatility of the MoO. Various forms of surface coatings (e.g. sprayed Al–Cr–Si and Ni–Cr–Si–Fe–B) are being tried in order to overcome this trouble. Molybdenum alloys have poor low temperature toughness but considerable rigidity.

*Cermets.* Much attention is being devoted to the study of cermets, consisting of a refractory and a binding metal to improve thermal conductivity and toughness. Typical examples are 70 $Al_2 O_3$ + 30 Cr or TiC + Ni–Cr binder. $Si_3N_4$ is promising up to to 1200 °C and has low expansion coefficient and a relatively high resistance to thermal shock.

Silicon powder can also be flame sprayed to give free standing hollow bodies which can be nitrided. Silicon is also isostatically pressed prenitrided, machined and fired at 1450 °C to produce furnace parts, hooks and stoppers. As regards future developments, better steels will undoubtedly become available and greater attention will be paid to the study of steels over extended periods since the design life of a jet engine is about 300 hours as compared with, say, 100 000 hours for a gas turbine.

*Glass—ceramics*

These have found commercial applications in thermal shock resistant cook-ware, space vehicles and under-sea research. They are made by melting a suitable glass batch containing an appropriate nucleating agent, forming the article by normal glass making techniques, and allowing to cool to a temperature at which nucleation occurs. When sufficient nuclei have formed, the temperature is raised to complete the recrystallisation process. $TiO_2$ is one nucleating agent and lithium oxide is often an important constituent in promoting internal recrystallisation.

# 13 Cast Irons

Flake graphite iron finds use due to:
  (1) its cheapness and ease of machining;
  (2) low-melting temperature (1140–1200 °C);
  (3) ability to take good casting impressions;
  (4) wear resistance;
  (5) high damping capacity;
  (6) a reasonable tensile strength of 108·340 N/mm² associated with a very high compressive strength, making it very suitable for applications requiring rigidity and resistance to wear.

The different types vary from grey iron which is machinable to either mottled or white iron which is not easily machinable. The white irons of suitable composition can be annealed to give malleable cast iron. During the last thirty years much development work has taken place and it has been found worth while to add even expensive elements to the cheap metal because vastly improved properties result. The new irons formed by alloying or by special melting and casting methods are becoming competitors to steel.

The various irons can be classified as shown in Fig. 175, based on the form of graphite and the type of matrix structure in which it is embedded. The metallurgical structure, composition and section of the casting largely govern the engineering properties.

One of the differences between cast iron and steel is the presence of a large quantity of carbon, generally 2–4%, and frequently high silicon contents. While carbon in ordinary steel exists as cementite ($Fe_3C$), in cast iron it occurs in two forms:
  (1) stable form—graphite;
  (2) unstable form—cementite, analysed as combined carbon.

Graphite is grey, soft, and occupies a large bulk, hence counteracting shrinkage; while cementite is intensely hard, with a density of the same order as iron. On the relative amounts, shape and the distribution of these two forms of carbon largely depend the general properties of the iron.

The factors mainly influencing the character of the carbon are:
  (1) The rate of cooling.
  (2) The chemical composition.
  (3) The presence of nuclei of graphite and other substances.

# CAST IRONS

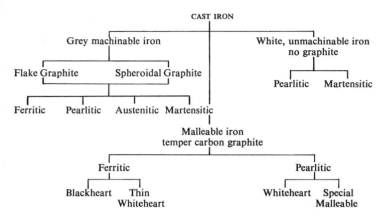

FIG. 175.—Classification of cast iron (Pearce)

(1) *Rate of cooling.* A high rate of cooling tends to prevent the formation of graphite, hence maintains the iron in a hard, unmachinable condition. If the casting consists of varying sections then the thin ones will cool at a much greater rate than the thick. Consequently, the slowly cooled sections will be grey and the rapidly cooled material will be chilled. These points are illustrated in Fig. 176, which shows the variation in hardness of a step casting.

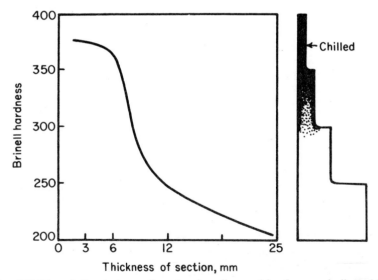

FIG. 176 The relation between the rate of cooling and hardness as indicated by sections of varying thickness

(2) *The effect of chemical composition.* (a) *Carbon* lowers the melting-point of the metal and produces more graphite. Hence it favours a soft, weak iron.

(b) *Silicon* slightly strengthens the ferrite but raises the brittle transition temperature (p. 14). Indirectly, however, it acts as a softener by increasing the tendency of the cementite to slip up into graphite and ferrite. Fig. 177 shows the relation between the carbon and silicon contents in producing the different irons for one rate of cooling. It will be noted that either a high carbon and low silicon or low carbon and high silicon content give grey

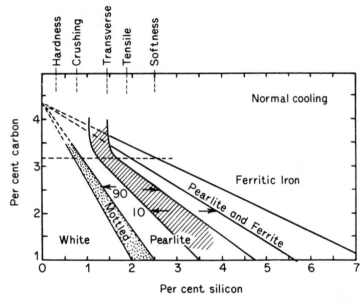

Fig. 177 Diagram indicating the structures of iron resulting from variation of silicon and carbon contents (*Maurer*)

Shaded area shows compositions common to 10 mm and 90 mm sections

iron; the fracture can, therefore, be misleading as to analysis, especially if the rate of cooling is not considered. The amounts of silicon, giving the maximum values for various properties, are also shown in Fig. 177. The percentage of silicon is varied according to the thickness of the casting.

(c) *Sulphur and manganese.* Sulphur can exist in iron, as either iron sulphide, FeS, or manganese sulphide, MnS. Sulphur as FeS tends to promote cementite producing a harder iron.

When manganese is added, MnS is formed which rapidly coalesces and rises to the top of the melt. The first effect of the manganese is, therefore, to cause the formation of graphite due to its effect on the sulphur. The direct effect of manganese is to harden the iron, and this it will do when it exists in amounts greater than that required to combine with the sulphur—1 part sulphur to 1·72 part manganese.

(d) *Phosphorus* has a little effect on the graphite–cementite ratio; but renders the metal very fluid indirectly through the production of a low-melting constituent, which is readily recognised in the micro-structure (Fig. 183). In the production of sound castings of heavy section, phosphorus should be reduced to about 0·3% in order to avoid shrinkage porosity.

(e) *Trace elements* not normally considered in routine analyses can exert a profound influence upon the characteristics of cast iron. Examples are 0·1% of aluminium graphitises, antimony embrittles, lead, tellurium promotes carbide but reduces strength of iron; 0·003% of hydrogen can greatly affect soundness of castings and tends to coarsen graphite. Nitrogen behaves as a carbide stabiliser; oxygen has no specific effect.

*The carbon equivalent value.* From Fig. 101 it will be seen that the eutectic E is at 4·3% carbon and irons with a greater carbon content will (under suitable conditions) start freezing by throwing out kish graphite of large size. With carbon contents progressively less than 4·3% normal graphite is formed in diminishing quantities until a mottled or white iron range is reached. Naturally other elements, especially silicon and phosphorus, affect the composition of the eutectic point in a complex alloy and a carbon equivalent value is suggested as an index which converts the amount of these elements into carbon replacement values.

$$\text{Carbon equivalent value (CE)} = \text{Total C\%} + \tfrac{1}{3}(\text{Si\%} + \text{P\%})$$

For a given cooling rate the carbon equivalent value, therefore, determines how close a given composition of iron is to the eutectic (CE 4·3) and therefore how much free graphite is likely to be present, and consequently the probable strength in a given section: the carbon equivalent value is also a useful guide to chilling tendency of a given section, although it must be borne in mind that pouring temperature, cooling rate and alloying elements have a marked influence.

## Relation between CE structure and mechanical properties

While Fig. 177 was a useful first attempt to relate composition and structure it had limited use in the foundry, and Fig. 178 shows a more useful relationship between CE value, structure, tensile strength in 30 mm dia bars and section size. A cylindrical test bar of given dia cools more rapidly than a flat plate of equivalent thickness, hence the section is expressed as bar diameter or section thickness. Line H is the boundary of unmachinable irons while line P is the boundary between soft and pearlitic irons. Thus an iron of carbon equivalent 4·35 (grade 12) should not be made thicker than 20 mm as a bar or 10 mm as a plate to attain a pearlitic iron. To avoid

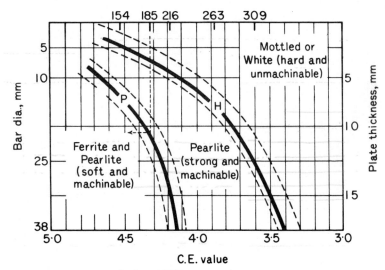

FIG. 178 Diagram relating section size, CE value, tensile strength and structure
(*After BCIRA*)

an unmachinable chilled casting the bar should not be less than 8 mm dia or plate less than 4 mm thick.

A melting furnace usually produces iron of a constant CE value and silicon is the element normally used to control chill. Alloying elements are added to cast iron to confer special properties and also to control the chill. A suggested allowance is given in Table 46 as silicon replacement for 1% of each element.

TABLE 46   EQUIVALENT SILICON CONTENT FOR 1% ELEMENT

| Graphitisers | | | | | Carbon Promoter | | | |
|---|---|---|---|---|---|---|---|---|
| C | P | Ni | Cu | Al | Mn | Mo | Cr | V |
| +3 | +1 | +0.3 | +0.3 | +0.5 | −0.25 | −0.35 | −1.2 | −1 to −3 |

*Formation of graphite*

*Flake.* Neglecting the effect phosphorus, and the presence of primary austenite dendrites, the successive stages in the growth from the liquid of flake graphite is shown in Fig. 179(*a*). The eutectic begins to solidify at nuclei from each of which is formed a roughly spherical lump, referred to as a *eutectic cell*. In this cell there has been simultaneous growth of austenite and graphite, the latter being in continuous contact with the liquid. The normal appearance of graphite in a micrograph suggests that the structure is made up of a number of separate flakes, but now it is considered that within each eutectic cell there is a continuous branched skeleton of graphite, like a cabbage. The skeleton is branched more frequently with a rapid

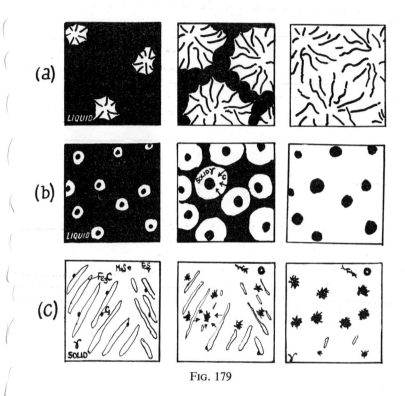

Fig. 179

radial growth of the cell such as occurs when increasing the rate of cooling of an iron which produces undercooling, and therefore finer graphite in the micrograph (Fig. 181).

The diameter of a eutectic cell, therefore, has a major effect on mechanical properties, e.g. the greater the number of cells per unit volume the higher the tensile strength, but soundness is affected adversely.

Superheating or holding time of the molten iron reduce the number of nuclei, while inoculants such as ferro-silicon and also sulphur increase nuclei.

*Spheroidal.* Figs. 179(b) and 192 shows the growth of spherulitic graphite in a magnesium-treated iron. In this case the spherulitic graphite is quickly surrounded by a layer of austenite and growth of the spheroid occurs by diffusion of carbon from the liquid through the austenite envelope. If

FIG. 180 Coarse graphite flakes. Matrix unetched ( × 60)
FIG. 181 Medium size graphite outlining dendrites ( × 60)
FIG. 182 Temper carbon in a malleable iron; ferrite crystals etched ( × 100)
FIG. 183 Common grey iron showing ferrite (F), pearlite (P) and phosphide eutectic (PH) ( × 250). Ferrite is associated with the graphite. Note banded structure in the phosphide eutectic
FIG. 184 Hyper-eutectic white cast iron ( × 100). White primary crystals of cementite in eutectic (cementite and pearlite)

*(Courtesy of the BCIRA)*

diffusion distances become large there will be a tendency for the remaining liquid to solidify as white iron eutectic, hence inoculation in this iron is highly desirable in order to increase the number of graphite centres.

*Temper carbon nodules.* At the malleabilising temperature (800–950 °C) the solid white iron consists of eutectic matrix of cementite, austenite and sulphide inclusions. Nucleation of graphite then occurs at austenite cementite interfaces and at sulphide inclusions. The cementite gradually dissolves in the austenite and the carbon diffuses to the graphite nuclei. The MnS tends to form a flake aggregate and the FeS a spherulitic nodule (Fig. 179(c)).

*Micro-structure of cast iron*

In preparing the specimens care is required, otherwise, erroneous results might arise. The graphite is readily removed during polishing and in this case the cavities can be either burnished over or enlarged.

The various types of micro-structure can be classified into groups without considering the presence of phosphorus.

The graphite can vary in size and form as illustrated in Figs. 180–182. The coarse flaky graphite is found in common iron, while the fine curly type, frequently outlining the dendrites, is found in high-class iron, especially when superheated before casting. Spheroidal graphite is found in magnesium treated irons (Fig. 192). The nodular form is found in annealed irons in which the cementite has decomposed at 800–950 °C. Thus we have:

| | |
|---|---|
| Pearlite + cementite (i.e. eutectic cementite in hypoeutectic irons (Fig. 185) and primary and eutectic cementite in hypereutectic irons (Fig. 184) | — white, hard, unmachinable. |
| Cementite + graphite + pearlite | — mottled, difficult to machine. |
| Graphite + pearlite (Fig. 186) | — grey, machinable, high strength. |
| Graphite + pearlite + ferrite (Fig. 183) | — grey, soft, weaker, |
| Graphite + ferrite | — grey, very soft, easily machined. |

The ferrite is of course much less pure than that in carbon steels.

*Phosphide eutectic*

Most cast irons contain phosphorus in amounts varying from 0·03 to 1·5%, consequently another micro-constituent is frequently present in the structure, in addition to those phases mentioned above. It occurs in white irons as a laminated constituent (ternary eutectic), consisting of:

| | |
|---|---|
| Iron, 91·19% | Ferrite (with a little phosphorus). |
| Carbon, 1·92% | Cementite, $Fe_3C$. |
| Phosphorus, 6·89% | Iron phosphide, $Fe_3P$. |

The melting-point is in the region of 960 °C, consequently it is the last constituent to solidify and forms islands in the interstices of the dendrites.

Although this constituent is very brittle it does not unduly weaken the iron when in small amounts (up to 1%) due to the fact that continuous cells are not formed round the grains. The structure is illustrated in Fig. 183 which shows the structure of the phosphide eutectic, together with graphite, ferrite and pearlite. Phosphorus will thus form this additional constituent in any of the 'grouped' structures already discussed.

### High-strength irons

It has been shown that the structures of grey cast irons are similar to those of ordinary steels but with the addition of graphite flakes which break up the continuity of the iron. Thus with a totally pearlitic structure cast iron should approach in tensile strength and toughness the properties of a 0·9% carbon normalised steel; the limiting factor being the shape and distribution of the graphite and fineness of the pearlite (Fig. 186).

Such irons have tensile strengths of up to 370 N/mm².

Modification of the micro-structure and properties of cast iron can be brought about by:

(1) The use of special melting and casting technique.
(2) The addition of alloying elements.
(3) Heat-treatment, particularly of white iron.

### *1. High-duty irons due to casting technique*

The gradual introduction of so-called *semi-steel* during 1914–18 marked the real commencement in improved properties. It is made by adding to the cupola steel scrap which slightly reduces the carbon content and in particular the amount of free graphite together with the production of a pearlitic matrix.

Other methods consist of superheating the molten metals in a separate furnace, whereby the graphite is greatly refined. Alternatively, an iron which would normally cast white can be graphitised by inoculation with ferro silicon (75% Si), sometimes with addition strontium in the ladle to give strength of 370 N/mm².

### *2. Addition of alloying elements*

The most common of the special elements added to cast iron are nickel, chromium, copper and molybdenum. Nickel tends to produce grey iron, in which respect it is less powerful than silicon. Consequently in castings of widely varying section the silicon can be reduced slightly and nickel added to prevent chilling in the thin sections, but still retaining a close structure in the thick ones.

On the other hand, chromium, by forming carbides, acts in the opposite way to nickel, but at the same time it exerts a grain refining action. These

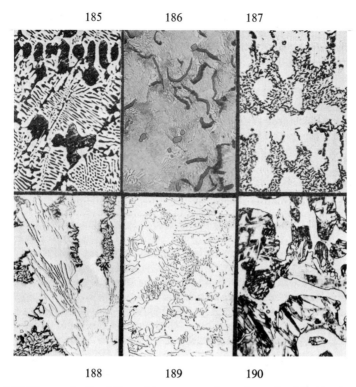

Fig. 185 Hypo-eutectic white cast iron, cementite and pearlite (black) ($\times$ 100) BH = 500

Fig. 186 Grey iron. High duty; pearlite and graphite ($\times$ 200)

Fig. 187 Silal (Si, 6), ferrite and fine graphite (black) ($\times$ 200)

Fig. 188 Austenitic iron. Graphite, chromium carbide in austenite matrix ($\times$ 200) (Nicro-silal)

Fig. 189 High chromium iron. Ferrite and cementite ($\times$ 200)

Fig. 190 Martensitic iron (Ni-hard). Cementite (white masses) in martensite–austenite matrix ($\times$ 200) BH = 700

*Figs. 186–190 by courtesy of the BCIRA*

elements, singly or together, are commonly found in motor cylinder irons.

Molybdenum strengthens the matrix by promoting a fine pearlite, but it is used preferably with other elements such as nickel to produce acicular structures.

A rough classification of the types of alloy iron is:

(1) *Pearlitic Irons.* $\frac{1}{2}$–2% nickel (chromium up to 0·8% and molybdenum up to 0·6%). Used for many general castings. The addition of tin in amounts up to 0·1% promotes a fully pearlitic matrix.

High carbon Ni–Cr–Mo cast iron is useful for resisting thermal shock in applications such as die-casting moulds and brake-drums. The nickel and chromium give the desired closeness of grain and molybdenum helps to strengthen the matrix. The considerable graphite reduces the tendency to 'crazy crack'.

Chromium (0·6)-molybdenum (0·6) irons are useful for engine liners, press sleeves, dies, etc., where wear resistance in relatively heavy sections is important. Cast iron with 1% each of chromium and molybdenum is used for piston-ring pots which are heat-treated to give a high transverse breaking strength coupled with a high elasticity value (DTD 485).

(2) *Acicular Irons.* Carbon 2·9–3·2, nickel 1·5–2·0, molybdenum 0·3–0·6%. Copper can replace nickel up to 1·5%. This rigid, high-strength, shock-resisting material is used for diesel crankshafts, gears and machine columns.

With the correct amounts of nickel and molybdenum *correlated with the cooling rate of a particular casting* the pearlitic change point can be suppressed and an acicular intermediate constituent (ferrite needles in austenitic matrix) can be produced with high mechanical properties. Acicular cast iron is very much tougher than any of the pearlitic cast irons of lower strength. The tensile strength of acicular cast iron with a carbon content of about 3·0% will vary from 380 to 540 N/mm² but these figures can be maintained in quite large sections. Phosphorus should not exceed about 0·15% in the presence of molybdenum, otherwise a compound is formed which impoverishes the matrix of molybdenum. Quite large variations in silicon content can be tolerated, but chromium in excess of 0·4% is harmful.

The structure changes rapidly at 600–750 °C and these irons should not be used at temperatures greater than 300 °C.

(3) *Martensitic Irons.* 5–7% nickel with other elements. Very hard irons used for resisting abrasion (Fig. 190), e.g. metal working rolls.

(4) *Austenitic Irons.* Non-magnetic, with 11–33% nickel but below 20% it is necessary to add about 6% copper or 6% manganese to maintain fully austenitic structures e.g. Nomag irons contain 11% Ni with 6% Mn.

These have a good resistance to corrosion and heat, e.g. Ni-Resist.

The outstanding characteristics of the austenitic cast irons, as compared with ordinary cast iron, are:

(*a*) resistance to corrosion;
(*b*) marked resistance to heat;
(*c*) non-magnetic, with suitable compositions;
(*d*) a high electrical resistance coupled with a low temperature coefficient of resistance;
(*e*) a high coefficient of thermal expansion;
(*f*) no change points.

(5) *Spheroidal graphite cast iron.* The production of spheroidal graphite

# HIGH-STRENGTH IRONS

as in Fig. 191 in the as-cast state is an outstanding development of a new iron, initially due to the use of cerium by Morrogh (BCIRA, 1946 BP 645862) and later, magnesium by the International Nickel Co. (1947 BP 630.070). The use of magnesium, to give 0·04–0·06% residual content proved to be the more adaptable and economic of the two processes. The production of spheroidal structure is prevented, however, by certain trace elements, e.g. 0·1 Ti, 0·009 Pb, 0·003 Bi, 0·004% Sb, but their effect can be eliminated by 0·005–0·01% cerium. For most raw materials the combined use of cerium and magnesium followed by ferro-silicon as an inoculent is used to produce spheroidal graphite iron. Remelting causes a reversion to flake graphite due to loss of magnesium. Magnesium treatment desulphurises the iron to below 0·02% before alloying with the iron, and for economic reasons the sulphur content should be as low as possible. A typical composition of SG iron is given in Table 47, but SG irons can be made in a wide range of compositions depending on the section size of the casting and the properties required. The SG iron can be used with a pearlite matrix or ferrite after a short annealing or with an acicular or austenitic matrix when suitably alloyed.

The stress strain curve is similar to that of steel, with measurable elongation. The ferrite grade of SG iron has a strength of 370 N/mm² with 17% El whereas a normalised pearlitic SG iron has a strength of 700 N/mm² with a minimum of 2% El. The strength can be increased to 925 N/mm² by special heat treatment or by the addition of alloying elements. Damping capacity is lower but shock, heat and growth resistance and weldability are higher than for flake graphite iron. SG iron can, therefore, compete successfully with malleable iron for thick sections, cast steel and alloy flake-graphite cast iron. SG cast irons are not so section sensitive as normal iron, e.g. a variation of 25–150 mm section causes grey iron to change from 278 to 154 N/mm² whereas a SG iron would change from 664 to 587 N/mm².

A new iron contains fine vermicular graphite similar but finer than undercooled graphite. It has a worm-like form which enables high strengths to be obtained with 2–3% El. Very precise production control is necessary and this limits commercial production at the moment. The sulphur content must be below 0·002% and casting must be cooled rapidly.

Some typical compositions for a few of these special irons are given in Table 47.

The first two cast irons illustrate the way in which the silicon content is altered with size of section and the third iron shows the use of high silicon and high phosphorus contents for these thin castings which are not highly stressed in service.

The Ni–Hard iron shown in Table 47 and Fig. 190 indicates the attempts made to increase the hardness of chill-irons for rolls and wear-resisting castings. The excess cementite is not affected by the nickel addition, but the matrix is changed from pearlite to hard martensite.

TABLE 47  COMPOSITIONS OF SOME CAST IRONS

| Type | TC | Si | Mn | S | P | Ni | Cr | |
|---|---|---|---|---|---|---|---|---|
| Light section machinery | 3·2 | 2·2 | 0·5 | 0·1 | 0·35 | — | — | |
| 20 tonne hydraulic cylinder | 3·2 | 0·9 | 1·0 | 0·1 | 0·35 | — | — | |
| Switch-boxes | 3·0 | 2·8 | 0·8 | 0·1 | 1·2 | — | — | |
| Lorry cylinder | 3·3 | 2·1 | 0·9 | 0·09 | 0·15 | — | 0·3 | |
| High pressure pump | 3·2 | 1·4 | 0·5 | — | — | 1 | 0·4 | Mo, 0·73 |
| 100 mm liner | 3·1 | 2·2 | 0·9 | 0·1 | 0·1 | — | 0·6 | Mo, 0·6 |
| Alloy chill roll | 3·1 | 0·5 | 0·3 | 0·07 | 0·15 | 4 | 1·2 | Mo, 0·4 |
| Acid resisting iron | 1 | 12 | 0·7 | — | — | — | — | |
| Ni-Hard | 2·8 | 0·5 | 1·2 | 0·1 | 0·3 | 4·5 | 0·75 | Mo = 0·25 |
| Silal | 3·0 | 6 | 0·7 | 0·1 | 0·3 | — | — | |
| Nicrosilal | 1·9 | 6 | 1·0 | 0·1 | 0·05 | 18 | 1·4 | |
| Ni-Resist | 3·0 | 1·5 | 1·0 | 0·1 | 0·2 | 14 | 2 | Cu = 7 |
| Whiteheart malleable | 3·3 | 0·6 | 0·3 | 0·2 | 0·1 | — | — | } As Cast |
| Blackheart malleable | 2·8 | 0·9 | 0·3 | 0·05 | 0·05 | — | — | |
| SG iron | 3·6 | <2·6 | <0·3 | 0·015 | 0·07 | — | — | Mg, 0·04–0·06 |
| Cryogenic SG iron | 2·4 | 2·2 | 4 | 0·01 | 0·07 | 22 | — | Mg, 0·1 |

TC = Total Carbon

## Stress relief of grey cast iron

Stress is completely removed at 650 °C, but grain growth commences at 550 °C and is serious at 600 °C. Current practice is to heat slowly to 475–500 °C, hold at temperature for 1 hour per 25 mm section and cool in furnace to 300 °C.

## Heat-treatment

Various matrix structures can be produced by heat-treatment, especially of SG irons, using the same basic principles outlined for steel. Usually it is desirable to heat slowly to 690 °C to avoid cracking, then raise to 850–900 ° to dissolve carbides ($\frac{1}{4}$–6 hr) followed by:

(a) air cool to form *pearlite*;
(b) or oil quench to produce *martensite* (in thin sections) which can be tempered to desired properties;
(c) or cool to and soak at 690–700 °C to effect complete graphitisation associated with a *ferrite matrix* (a sub-critical anneal *only* at 690–720 °C gives poor impact values).

Heat-treatment of a ferritic iron at 900 °C dissolves some graphite in the austenite and air cooling produces some pearlite.

The aims of annealing are (a) to break down accidental chill, (b) to stabilise an iron to be used at 540–550 °C in service, (c) to increase the *machinability* several times with a ferritic structure.

## Malleablising cast iron by annealing

The castings are initially made as white iron of a suitable composition which varies with the process of malleablising employed. Two such methods are available known as Whiteheart—common in England for thin sections, and Blackheart. These names refer to the respective fractures of the irons after annealing, the former being white and crystalline due to the presence of pearlite and the latter grey or black due to graphite.

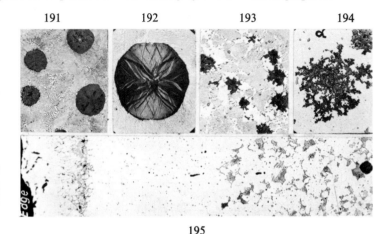

FIG. 191 Spheroidal cast iron. Spheroidal graphite in pearlite matrix ( × 200)
FIG. 192 Enlarged view of graphite spheroid. Polarised light ( × 600)
FIG. 193 Blackheart malleable iron. Graphite aggregate nodules in ferrite and pearlite ( × 200)
FIG. 194 Enlarged view of aggregate nodules in balanced sulphur malleable iron ( × 600)
FIG. 195 Whiteheart malleable iron. Ferrite and sulphides at the edge, ferrite and pearlite (and little spheroidal graphite) at centre ( × 80) (*BCIRA*)

In the whiteheart process, the white iron castings are packed in boxes with a mixture of used and new hematite ore (2–16 to 1) and heated to 900–950 °C for 5–6 days, and finally cooled very slowly. Modern gas annealing reduces the anneal to 50 hrs. As a result, a portion of the carbon is oxidised leaving in thin sections, a ferrite structure, while in thick sections the edge of the casting is ferritic, and gradually changes to a steel-like structure of ferrite and pearlite, with interspersed nodules of graphite (Fig. 195). The sulphur content is not critical. Properties are given in Table 48.

In the blackheart process the castings are annealed at 850–950 °C in a neutral packing or atmosphere, consequently little oxidation of the carbon occurs. But due to the careful balance of the silicon and sulphur contents the cementite breaks down to carbon aggregates in ferrite (Fig. 193).

TABLE 48  MECHANICAL PROPERTIES

*Grey Cast Iron Castings* (BS 1452–1961) have grades 10, 12, 14, 17, 20, 23, 26 corresponding to the minimum tensile strength tonf/in$^2$ for a 30 mm dia bar. i.e. 154, 185, 216, 262, 309, 555, 402 N/mm$^2$. Elastic modulus ranges from 100–145 kN/mm$^2$ fatigue limit = 0·45–0·38 TS

*Malleable Iron (min. prop.)*

|    | BS 309 Whiteheart Grades | | BS 310 Blackheart Grades | | | ASTM 45 010 to 80 002 Pearlitic |
|----|------|------|------|------|------|------|
|    | W22/4 | W24/8 | B18/6 | B20/10 | B22/14 |  |
| TS | 309–340 | 340–370 | 278 | 309 | 340 | 448–690 |
| YP | 185–200 | 200–216 | 170 | 185 | 200 | 309–553 |
| El | 4–6 | 8–10 | 6 | 10 | 14 | 10–2 |

Elastic mod = 160–187 kN/mm$^2$   Fatigue limit = 0·4–0·6 TS

*Spheroidal or nodular graphite castings (min. prop.) BS 2789, 1961*

| Grade | TS | 0·5 Perm Set | El | Matrix |
|-------|-----|------|----|--------|
| SNG 24/17* | 370 | 232 | 17 | Ferrite, impact resistance |
| SNG 27/12 | 417 | 278 | 12 | Mainly ferrite, strength and toughness |
| SNG 32/7 | 494 | 340 | 7 | Ferrite–pearlite, reasonable ductility |
| SNG 37/2 | 571 | 386 | 2 | Pearlite, high strength but ductility and resistance to impact less important. Can be hardened and tempered |
| SNG 42/2 | 649 | 432 | 2 | |
| SNG 47/2 | 726 | 463 | 2 | |

*Impact value 13 Joules min. Elastic mod = 170–173 kN/mm$^2$. Fatigue limit 0·5–0·4 TS. Coding is based on letters indicating type of iron and figures min TS tsi and El%.

The annealing cycle involves three important steps. The first is a very slow heating (15 °C per hr) to 900 °C to produce graphite nuclei. The second consists of soaking (e.g. 40 hr) to graphitise the free cementite to a point on Acem Fig. 101, i.e. first stage graphitisation (Fig. 179(c)). The third, known as the second stage graphitisation, consists of a slow cool to allow austenite to precipitate the carbon (along line FP) to prevent a cementite network; followed by a very slow cool ($2\frac{1}{2}$–4 °C hr) through the eutectoid transformation, 720 to 670 °C to avoid the formation of pearlite. This second stage graphitisation is sometimes purposely avoided by more rapid cooling or by addition of 1 % manganese to form *pearlitic malleable iron*, used for wear resistance, e.g. gears, links. Properties are shown in Table 48.

*Strength*

In British Standard specifications no composition is specified for grey cast iron, but the mechanical properties are stated for various thicknesses of castings (Table 48).

*Growth of cast iron*

When cast iron is heated and cooled through the range 700–800 °C constitutional changes, accompanied by changes in volume, occur in a similar manner as described for steels in Chap. 9. The pearlite, however, tends to break down into ferrite and graphite, with a resultant increase in volume, which may be termed the growth of the iron for that cycle. Subsequently, hot gases penetrate into the graphite cavities, oxidise the iron to iron oxide with a decided volume increase, setting up a stress which sooner or later relieves itself in a minute crack. This produces the characteristic crazy cracking on the surface. Finally, the growth may reach the stage where the article, such as a firebar, is completely buckled and rendered useless. Coarse graphite assists growth.

Several irons have been developed to resist growth, for example Silal, Nicrosilal. The former contains low carbon, high silicon, and has a structure of ferrite with fine graphite, which tends to prevent ingress of oxidising gases (Fig. 187). The silicon increases the resistance to oxidation and raises the change points to over 900 °C, so preventing the occurrence of the constitutional volume changes in service.

The most serious disadvantage of Silal, which is relatively cheap, is its brittleness. The addition of nickel and other elements removes this brittleness and produces Nicrosilal, an austenitic iron with no critical change points; but the cost of this material is high (Fig. 188). Low carbon, high chromium (12–30 %) irons offer good resistance to heat, and especially to sulphur, hydrogen sulphide and sulphur dioxide (Fig. 189). For good service ordinary irons should not be used above 450 °C, Silal above 750 °C and austenitic irons above 950 °C.

# 14 Electrical and Magnetic Alloys

*Electrical conduction*

Electrical resistivity of materials varies over an exceptionally wide range from $10^{-6}$ to $10^7$ ohms/cm$^3$ for copper to polystyrene respectively, and three groups can be considered as conductors (metals), semi-conductors and insulators.

The first requirement for conduction is a supply of carriers that are free to wander through the material. In metals these carriers are valence electrons (Fig. 196(*a*)), i.e. free electrons in the outer electronic shell, (compare diamond and graphite p. 73).

As atoms are brought close together and begin to influence one another, the electrons are forced to have slightly different energies because, by the Pauli exclusion principle, a given quantum state cannot be occupied by more than two electrons with opposite spins (Fig. 196(*b*)). Within a metal containing some $10^{23}$ (e.g. 1 cm$^3$Cu) the energy levels become continuous bands (Fig. 196(*c*)) and the available electrons fill up these energy levels from the lowest up to a highest level, viz. the Fermi level (Fig. 196(*d*)).

Each band contains as many discrete levels as there are atoms in the crystal. With sodium or copper with one valency electron, the valence band is only half filled, and only a minute amount of energy is required to raise an electron to the next highest level, a requirement for conductivity, (Fig. 196(*d*) and (*e*)). Adjacent energy bands in materials do not always overlap so that an energy gap may be present. This does not affect the conductivity of the material if there are still vacant levels within a band, for example, aluminium (Fig. 196 (*f*)) has a gap between 2nd and 3rd bands, but the 1st and 2nd bands overlap and there are vacant levels in the second band.

It is thus quite easy for a lone electron to jump into the vacancy in a neighbouring orbit and so to drift inside the crystal.

In the absence of an external electric field the movement of one electron is offset by the movement of another in the opposite direction. When an electric charge is applied to a conductor there is a preferred drift of a cloud of electrons towards the positive plate.

The movement of electrons is affected by collisions with lattice atoms and other displaced electrons,* and this increases as the lattice becomes distorted by solution alloying and other crystal defects (is independent of

---
*This really requires a wave-mechanical picture of wave scattering.

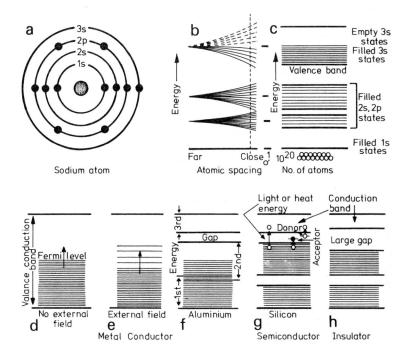

FIG. 196 (a) Diagram of sodium. Three inner shells are filled. Valence shell 3s could hold another atom
(b) Specific energy levels of electrons in each shell when far apart and when close together, at which point bands form
(c) Energy band levels at the vertical broken line in (b)
(d) Energy levels available to the valence electrons in a conductor (e.g. sodium, copper), with no external field. Can act as a conductor band
(e) Effect of external field in raising electrons to an empty state where it is free to move inside crystal and to conduct electricity
(f) Overlapping zones in aluminium with vacant sites in 2nd zone which allow electrons to move
(g) Semi conductor. Small gap between filled valence band and conductor band. Light or heat energy can raise electron into conduction band. Donor impurities contribute electrons to conduction band. Acceptors contribute electron 'holes' to valence band
(h) Energy band in an insulator. Note large forbidden gap

temperatures) and also as the temperature is raised to give atoms more thermal agitation which causes temperature dependence of conductivity. The jostling of the electrons causes a transfer of some of their energy to the vibrating atoms whose vibration increases, and this corresponds to an increase in temperature of the solid. With suitable resistance alloys a wire can be made red hot for an electric fire or incandescent for a lamp. The resistance of a metal increases with temperature and tends towards zero at a temperature of absolute zero (see superconductivity). Electrical and thermal conductivity are related and their ratio varies with absolute temperature.

When a material has four valence electrons, the first band is completely filled and the next band is separated by a forbidden gap, requiring energy to jump it.

If the energy gap is large, extremely high electric fields would be required to raise electrons across the forbidden gap to the next band, and the material is an insulator (Fig. 196($h$)).

The materials used for electrical insulation are mainly ceramics, such as porcelains and polymers. In polymers the main molecular backbone is formed from covalently bonded carbon, the valence electrons of which completely fill the valence band. In addition, however, the configuration of the molecules, like cooked spaghetti, does not allow a molecule to extend from one electrode to the other, and the chance of a valence electron passing from the end of one molecule to the next is small. Thus, there is only a limited number of circuits between electrodes.

Electrical conductivities for various materials relative to copper are given in Table 49.

TABLE 49  RELATION OF ELECTRICAL CONDUCTIVITY TO ENERGY BAND GAP

| Material | Energy gap eV | Conductivity Cu=1 |
|---|---|---|
| Copper | unfilled zone | 1·00 |
| Aluminium | overlapping zones | 0·67 |
| Iron | overlapping zones | 0·186 |
| Germanium | 0·7 | $4·55 \times 10^{-8}$ |
| Silicon | 1·10 | $1·14 \times 10^{-10}$ |
| Porcelain | >6 | $2·28 \times 10^{-20}$ |
| PVC | >6 | $2·28 \times 10^{-28}$ |

If the energy gap is small, e.g. 1–2 eV for Si, thermal or light energy may be sufficient to excite some electrons into the next higher band (called conduction band) and render conduction possible, Fig. 196($g$). i.e. Semiconductor.

## Semiconductors

In recent years increasing interest has been taken in semiconductor materials which are almost insulators but which become conductors if free conducting electrons are made available, for example by exposing the material to light or heat or by adding impurities, Fig. 196. Hence transistors, which are based on these materials, have to be sealed from light, not overheated, and are made from highly purified material. *Intrinsic semiconductors* which conduct when very *pure* are germanium, silicon, boron and tellurium. Most semiconductors, however, owe their conductivity to traces of impurities.

In addition to the ways atoms can associate, described on p. 72, germanium and silicon have *four* valency electrons, and each is shared with one from a neighbouring atom to form a covalent bond shown in Fig. 197(*a*). This stable electron arrangement can be altered by adding a

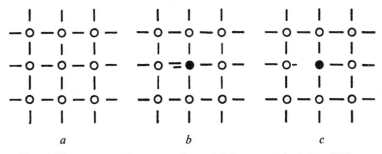

FIG. 197   *a*—pure   *b*—pentavalent addition   *c*—trivalent addition

minute quantity (e.g. 2 parts in $10^8$) of pentavalent or trivalent impurity. With the pentavalent impurity (arsenic or bismuth) shown in Fig. 197(*b*), one valency electron is surplus, and thus conduction is possible by *negatively* charged particles, and the doped semiconductor is called *n*-type. If, however, germanium is doped with a trivalent element (aluminium) one electron is missing (Fig. 197(*c*)), thus leaving a 'positive hole'. This hole can move (Fig. 46) by accepting electrons from surrounding atoms in the crystal (i.e. acceptor addition). Since the charge of the hole consists of a missing electron, this migration of holes behaves as if it was a current of *positive* charges, and the semiconductor is called a *p*-type.

It is also possible to have electrons in *p*-type semiconductors and holes in *n*-type, and these electrons or holes are called minority carriers. In any semiconductor the product of the numbers, of electrons and holes is constant, but this equilibrium can be affected by external means, e.g. light increases conductivity. When the external source is removed the time to reach equilibrium condition is the *life time* of the minority carrier. Certain impurity atoms, e.g. Cu, Ni, Fe in germanium affect the life-time of minority carriers, and have to be controlled (1 in $10^9$ maximum content) if suitable properties are to be achieved, hence the process of zone refining (p. 86). Further, imperfections, such as dislocations, boundaries and strain affect the properties of semiconductors, hence *single* crystals are used.

Other promising compounds appear to be: (*a*) gallium arsenide; (*b*) indium antimonide, which has interesting magnetic electrical properties; (*c*) silicon carbide, which retains its semiconducting properties up to about 500 °C; and (*d*) bismuth telluride, which has good thermoelectrical properties (p. 416).

*Transistors*

A transistor consists of dual p–n junctions formed in the same single crystal of base material. The junctions are formed by an alloying process, or by diffusion, and by selection and location of the impurities it is possible to fabricate transistors in either the 'p–n–p' or 'n–p–n' forms. The three active layers are known as the emitter, the base and the collector.

The operation of a transistor is by injection from the emitter of a disturbing electric field which causes the carriers—electrons or 'holes'—to diffuse across the base to within the field of the collector which is provided with an appropriate voltage from an external source to sweep the carriers into the collector circuit.

Transistors are basically current operated devices and work at low impedances, unlike electronic valves which are higher-impedance devices and perhaps more easily thought of as voltage-operated.

*Conductors*

Pure copper and aluminium are used for many conductors, but in order to increase the strength or to facilitate machining while still retaining high conductivity, a number of alloys are available to the engineer. See aluminium alloy. For increasing machinability of copper, tellurium or sulphur is added.

*Machinable copper*

Although the electrical properties of pure copper are attractive, it has never been considered an easy metal to machine. This disadvantage is greatly reduced by the addition of 0·5 % tellurium to copper. The presence of this element only decreases the electrical conductivity to about 98 % that of copper, and at the same time elevates the softening temperatures by 100 °C. The machinability of this alloy approaches that of free turning brass, and it is finding increasing application for the manufacture of special electrical parts where high electrical and high thermal conductivity are required, coupled with the necessity for mass production. This alloy can be quite readily machined at a speed of 100 metres per minute, giving a very smooth finish.

Recently, due to the rising cost of tellurium, an alloy containing about 0·5% sulphur has been produced. This has the same virtues as the tellurium copper, but is cheaper and is becoming widely accepted as a substitute.

*Strengthened copper*

Small quantities of silver, cadmium, oxygen, lead and zinc produce only a slight lowering of the electrical conductivity of copper; while phosphorus, silicon, arsenic and iron have most deleterious effects. Cold working also lowers the conductivity some 3%. For trolley wires an alloy containing 0·7% cadmium is frequently used, cold worked to give a tensile strength of 695 N/mm², with 94% conductivity. The addition of 0·08% silver raises the softening point of copper some 100–150°C and increases the elastic limit. This alloy is useful for applications which require hardness and high conductivity at somewhat elevated temperatures, where work-hardened high conductivity copper would become softened (switchgear contacts, commutators, radiator grilles).

*Precipitation hardened copper*

The conductivity of copper, in common with other metals, is particularly lowered as an alloying element is dissolved in it to form a solid solution. Conductivity is raised if the element is thrown out of solution by precipitation. Several precipitation hardened alloys are available, and development is proceeding on dispersion hardened material, for example Cu + 2% $Al_2O_3$; Cu + 0·8% Beu to give high strength and resistance to softening.

*Copper–chromium (0·5%)*

The solubility of chromium in copper is about 0·02% at 400°C and 0·5% at 1050°C. To retain the chromium in solution the alloy is rapidly cooled from about 1000°C, although a separate treatment may be unnecessary with thin castings. Tempering at 430–500°C for 2–3 hours increases the hardness from 70 to 140 VPN.

In the fully hardened condition the tensile strength is approximately 463 N/mm², proof strength 340 N/mm², and the elongation approximately 15%. The great advantages of this alloy are that the electrical conductivity is approximately 85% that of copper, and that the mechanical strength is maintained even after operating for prolonged periods at temperatures of 350–400°C. Because of these characteristics the alloy is finding extensive application for spot-welding electrodes, seam-welding electrode discs, electrical switchgear, and for various types of heat interchangers.

*Copper—0·15% Zirconium* is similar to Cu Cr alloy, and both retain a useful proportion of their strength after brazing at 650°C for a short time. It is solution treated at 900°C and aged at 450°C produces a strength of 460 N/mm² and 10% El.

*Copper–cobalt–beryllium—97/2·6/0·4.* This alloy is 'solution' heat-treated for 1 hour at 900°C, quenched in water and hardened by heating to 500°C for 2 to 4 hours.

The hardening appears to be due to the precipitation of a CoBe compound. In the hardened condition the alloy has a strength of 617 N/mm², with 10–20% elongation, elastic modulus of 117 kN/mm², and a conductivity of about 45% of that of copper; and can withstand a service temperature in excess of 400°C.

*Copper–nickel–silicon.* A typical heat-treatable alloy contains Ni, 2·5; Si, 0·5, with a tensile strength of 300–700 N/mm², after quenching in water from 700°C and ageing at 450°C for $\frac{1}{2}$ hr.

*Copper–nickel–phosphorus—An alloy 1Ni, 0·2P,* is also susceptible to temper hardening and has good electrical conductivity after water quenching from 900°C and ageing at 450°C.

For resistance alloys see pp. 313, 325.

*Superconductivity*

Conduction by the drift of electrons in the direction of the applied field will be opposed by irregularities within the lattice, i.e. by alloying (see p. 212) deformation, nuclear irradiation and also by the thermally vibrating ions of the metal, the degree of which decreases as temperature is lowered approaching zero resistance at 0K.

There is a class of materials, called superconductors, whose resistance abruptly disappears at a temperature, characteristic of the particular material, termed the critical temperature, Tc, which may be as high as about 20K.

In addition to losing its resistance below the critical temperature, the superconductor has another basic property; its critical current density (that is the highest current per unit area that can be supported by the superconductor before it reverts to the normal state) decreases rapidly with increase in the strength of the magnetic field to which it is subjected.

In 1961 certain alloys and compounds were discovered which were able to support high critical current densities ($J \sim 10^5$ amps/cm²) in the presence of magnetic fields approaching 10 tesla; these materials have been called high field superconductors.

Soon after the discovery of high field superconductors, materials like the compound $Nb_3Sn$ and the alloy Nb 25%Zr were processed into tape and wire configurations suitable for winding electromagnets. The development of superconducting magnets, however, was delayed by the fact that magnet coils wound with superconducting materials reverted to the normal state when supporting currents well below those that could be carried by short samples of the particular superconductor.

It has since been shown that this 'degradation' can be caused for a variety of reasons. The chief reason for this effect arises through changes in magnetic flux distribution in the superconducting windings when the coil is being charged. The discontinuous motion of flux in a superconductor, which has been termed a flux jump, results in local heating of the superconducting wire, which may raise the temperature of regions of the superconductor and bring about reversion to the normal conductive state. This effect has been countered by including a good normal conductor in close contact with the superconductor. Copper is the most popular normal component of a composite because of its excellent thermal and electrical properties at a low temperature and its metallurgical compatibility with modern superconducting alloys. The normal component of the composite provides a shunt for current impeded by local normal regions and in addition, absorbs (or conducts away) heat generated locally by flux jumps. This is the basis of 'stabilisation' of the superconductor and such composite materials have been used with good reliability, to build devices which generate high magnetic fields.

The superconducting alloys in the NbTi system have superceded NbZr alloys because of their superior critical fields (the maximum magnetic field at which the material will still be superconducting) and its relative ease of fabrication particularly with composites containing wrought high purity copper. $Nb_3Sn$ (the first high field superconductor to be discovered) is more difficult and more expensive to produce and is used chiefly for constructing devices which have to operate in very high magnetic fields (greater than 10 tesla).

The first generation of superconducting composites, comprising copper and NbTi included sufficient copper to fully stabilise the conductor against any thermal or electrical effects. Fig. 198(a) shows some typical composite Cu/NbTi superconductors.

Typical composites comprise filaments of NbTi alloy of the order of 0·025 mm diameter enclosed in high conductivity copper matrix. The properties of the two components, which are processed together from a very early stage in the manufacturing process in order to produce a metallurgical bond between them, are 'optimised' by a terminal heat treatment. On the one hand the terminal heat treatment greatly increases the current carrying capacity of the superconductor and on the other hand decreases the resistivity of the copper. The number and size of superconducting filaments, and the quantity of copper in a composite shape may be varied in order to satisfy design parameters of a specific requirement such as

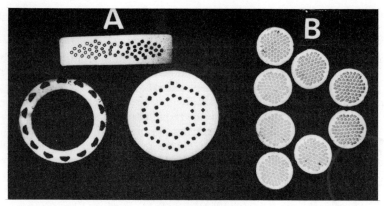

Fig. 198 (A) Early Cu/NbTi composites (× 2–10)
(B) Intrinsically stabilised super conductors (× 45)
(*Courtesy IMI Ltd*)

critical current, section, size and shape, degree of stabilisation and degree of mechanical strength.

The second generation of superconducting composite still contain NbTi alloys and copper but advances in theoretical physics have led to theories on intrinsic stability and the development of NbTi copper composites in which the NbTi filaments are stable in their own right; a minimal amount of the copper component being included for protection purposes. Fig. 198(*b*) shows typical cross-sections of intrinsically stabilised superconductors (0·25 mm diameter) which have been wound into cables.

The next generation of superconducting composites is likely to include conductors which can operate in alternating current conditions without large power losses.

*Magnetic alloys*

The strong attraction by a magnetic field observed in iron, cobalt and nickel is closely related to the presence of electrons with unpaired spin in the inner 3d shell, but the atoms must also be relatively far apart (e.g. Cr has atoms too near, and adjacent electrons with unpaired spin oppose each other). This is shown diagrammatically in Fig. 199(A). The alignment of electron spins in a ferromagnetic crystal tend to be along certain preferred directions of *easy magnetisation*, e.g.

| Iron | b.c.c. lattice | [100] | 1043 K Curie point |
|---|---|---|---|
| Nickel | f.c.c. lattice | [111] | 631 K |
| Cobalt | h.c.p. | [0001] | 1393 K |

The alignment of the magnetic moment is affected by temperature, and as it increases the thermal agitation increases to a point where random orientation occurs—known as the Curie temperature (Fig. 199(B)) and material becomes non-magnetic.

This alignment of electron spins might be expected to cause every piece of iron to be a permanent magnet. It does not do so because the iron

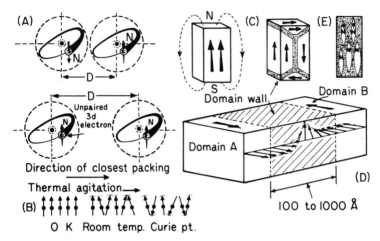

FIG. 199 (A) Effect of atom spacing on the electron spin alignment of the transition metals. When closely spaced (top-D) the spins of unpaired electrons in adjacent atoms are opposed; when widely spaced (bottom-D) spins are aligned
(B) Effect of temperature on the alignment of the atomic magnets
(C) Shows the elimination of all external lines of force by the division of a uniformly magnetised material into four domains
(D) The zone between domains. Some north poles which are turning through 180° come to the surface of the material and can be revealed by powdered iron oxide
(E) Shows ferro magnetic powder too small to split into multiple domains. Forms very stable magnet

crystal is spontaneously subdivided into many smaller units known as domains. Fig. 199(C) shows how the external lines of force (dotted) are eliminated by four domains and energy reduced. The boundary zone (approx. 100Å) between domains, known as Block wall, is shown in Fig. 199(D) in which some north poles are turning through 180° and the line where they outcrop the surface can be shown by magnetic dust, thus revealing the pattern of the domains.

When iron is placed in a magnetic field those domains which are favourably orientated grow at the expense of those not orientated with the field until almost all are aligned in the same direction (i.e. saturation, in Fig. 200). Further magnetisation only pulls the domains more exactly into the field direction.

Thus magnetic properties are greatly influenced by the ease of domain boundary movement which is affected by similar variables which control movement of dislocations. Small needle-like particles unable to contain a domain wall are very difficult to demagnetise and thus form the basis of a permanent magnet as a precipitate or bonded in a non-magnetic medium (Fig. 199(E)).

The Curie temperature and the saturation magnetism are independent of micro-structure whereas the shape and area of the hysteresis loop, which determine the practical application of the material, are almost completely determined by the micro-structure.

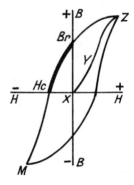

Fig. 200 Magnetisation curve for steel

Magnetic alloys can be divided into two groups:

(*a*) magnetically hard—for permanent magnets; (*b*) magnetically soft—for transformer cores. There is a general connection between mechanical hardness and magnetic hardness. Pure, well-annealed metals are magnetically soft; while increase in internal strain, e.g. by cold working, alloying, quench or precipitation hardening increases magnetic hardness.

When a ferromagnetic material is placed in a magnetic field of intensity H lines of force per square centimetre it becomes magnetised and a greater density of lines of force pass through it; this flux density is termed B and is the measure of the degree of magnetism. The relation between B and H is shown by curve XYZ (Fig. 200) and at Z the curve tends to become horizontal since the metal becomes magnetically saturated. If the magnetising field H is removed the demagnetisation of the iron lags behind, giving rise to curve ZBr, and the value XBr is known as the remanence of the magnet, that is, the residual flux density in the material when the magnetising force is removed. A negative magnetising force is now required to completely

# MAGNETIC ALLOYS

demagnetise the material, and this is represented by XHc, the coercive force. The values of XBr and XHc and the curve BrHc are very important in connection with permanent magnets. The products of values of B and H corresponding to points on curve BrHc varies, and the maxium value, termed $B–H_{max}$, is an important criterion of the properties of a permanent magnet and is useful in determining the volume of steel required for a magnet for a specific purpose.

For *permanent magnets* high values of both Br and Hc are required and these are obtained in quenched high-carbon steels, but for powerful magnets the addition of chromium, tungsten and cobalt is necessary, as shown by the compositions in Table 50. In 1930 the aluminium–nickel steel was discovered by Japanese workers. This alloy contains about 13% aluminium and 25% nickel with a maximum of 0·06% carbon. Up to 12% of cobalt (Alnico) is added in some cases. Above 100°C 'Alni' alloy consists

TABLE 50  MAGNET ALLOYS

| Type | C | W | Cr | Co | | Remanence Gauss | Coercive Force Oersteds | BH max. cgs units $\times 10^6$ |
|---|---|---|---|---|---|---|---|---|
| C | 0·6 | — | — | — | | 9000 | 50 | 0·18 |
| 6% W | 0·6 | — | — | — | | 11 000 | 65 | 0·32 |
| Cr | 1·0 | — | 6 | — | | 9500 | 60 | 0·27 |
| *3% Co* | 1·0 | — | 9 | 3 | | 7200 | 130 | 0·35 |
| *15% Co* | 1·0 | — | 9 | 15 | | 8600 | 180 | 0·65 |
| *35% Co* | 0·9 | 4 | 6 | 35 | | 9000 | 250 | 1·0 |
| | Ni | Al | Cu | Co | Nb | | | |
| Alni | 25 | 13 | 4 | — | — | 6300 | 500 | 1·2 |
| Alnico I | 18 | 10 | 6 | 12 | — | 7300 | 500 | 1·7 |
| Alcomax II | 11 | 8 | 6 | 25 | — | *13 000* | *550* | *5·4* |
| Alcomax IV | 13 | 8 | 3 | 24 | 2·5 | 11 200 | 750 | 4·3 |
| Columax | 13 | 8 | 3 | 24 | 0·7 | *13 500* | *740* | *7·5* |
| *Hycomax IV* | *14·5* | *8* | *2·7* | *36* | *7·87* | *7800* | *1900* | *5·8* |
| *Hynico II* | *20* | *8* | *4* | *20* | *4 T₁* | *6000* | *900* | *1·8* |
| Co-Pt | | | Pt 77, Co 23 | | | 4500 | 2700 | 4·0 |
| *Feroba I* | | | BaO 6 Fe₂O₂ | | | *2300* | *1700* | *1·0* |
| *Feroba III*** | | | | | | *3400* | *2500* | *2·7* |

*Cobalt steels frequently receive a triple treatment: (1) AC 1150°C; (2) 780°C AC; 930°C oil quench or 1000°C AC.
Tungsten steels should not be soaked at 750–1250°C.
Alnico alloys are AC 1100–1300 with an optional temper at 550–650°C.
Crystal oriented.
To convert to SI or MKSA units: gauss $\times 10^{-4}$ = tesla or Weber/in²
oersted $\times 10\frac{3}{4}$ = amp Tarn/m.

of $Fe_3NiAl$, which tends to split up on cooling into two body-centred structures, one of which is practically pure iron, the other having approximately equal atomic proportions of iron, nickel and aluminium. The prime cause of high energy and coercivity arises from internal strain and the small size and elongated shape of the iron rich islands that causes them to be spontaneously magnetised in the direction of their length and the additions of the other elements mainly influence the degree and conditions of maximum development of hysteresis.

In 1939 it was found that cooling a generally similar alloy with 25% cobalt content from 1200°C in strong magnetic field greatly increased the subsequent magnetic properties along the field axis at the expense of the other two axes which are not used. This anisotropic alloy is known as Alcomax and is capable of over twice the energy of Alnico. These alloys have proved exceedingly stable in service under the influence of temperature and shock. Their coercive force is very much greater than cobalt steels, consequently these magnets are shorter but of greater cross-section. These magnets find use in the non-electric chucks, and especially in magnetos and small dynamos. The designs must be suitable for casting or in the case of small magnets by sintering powder.

In 1947 it was discovered that the coercivity of Alcomax could be increased by the addition of a few per cent of niobium. A further advance was produced by orienting the crystal axes by uni-direction solidification from a chill. The cube edge is oriented in the direction of magnetisation, hence 'Columax' with a large increase in energy. Several of the alloys can be made by sintering. Hycomax alloys were developed to improve coercivity with reasonable BH.

Other magnetic alloys include cobalt-platinum, pressed fine powder of iron and cobalt, manganese bismuthide and also ferrites consisting of mixed oxides of iron and other metals pressed and fired or bonded with rubber or PVC to form flexible magnets. Feroba is a typical hexagonal barium ferrite permanent magnet with high coercivity but low remanence and low density (5). Feroba is hard, brittle and sensitive to thermal shock but looses negligible flux in assembly but does so when subjected to low temperatures. Their high electrical resistivity minimises losses from eddy currents. A magnesium-manganese ferrite has a square hysteresis loop and is used for computer storage elements.

In the soft magnetic field a manganese–zinc–ferrous cubic ferrite (Ferroxcube A) is an example of the type $M.Fe_2O_4$ and is used as a low eddy current loss ferrite in applications from low frequencies to micro-waves. This material is made by powder metallurgy techniques.

Magnetically soft materials are necessary for such parts as transformer cores, and dynamo pole pieces which have to become magnetised and demagnetised at will. The ratio of $\frac{B}{H}$ is called permeability and its value changes with the magnetising field, rising from a low value to a maximum and then falling. A high value of permeability is required in a transformer core. A transformer works with alternating current and consequently the iron core is first magnetised, then demagnetised and remagnetised with reversed polarity, and this produces a curve ZBrHcMZ (Fig. 200), which is known as a hysteresis loop, the area of which represents the energy converted into heat every time the current is reversed, and this occurs many times in a second. It is therefore necessary to use materials for the core which give narrow loops (i.e. small hysteresis losses). The electrical resistance of the material should also be high to reduce the eddy current loss, and laminated cores assist in this way. Pure iron and silicon (4–5%) iron sheets are used for this purpose. Most ferromagnetic materials have different magnetic properties in different crystallographic directions. In iron the permeability is highest in the direction of the cube edge (100). If the cube edge of the crystals (100) can be aligned in the field direction in a transformer, greater induction can be obtained without excessive field and energy losses by induced currents.

A degree of orientation of the crystals is possible by heavy rolling followed by controlled annealing. Crystalloy is a typical oriented silicon–iron alloy and HCR an oriented nickel–iron (50/50) alloy characterised by a rectangular hysteresis loop. Improvements are obtained with some alloys by cooling them in a magnetic field through the magnetic charge point.

Another class of material is important because of its high permeability in small magnetic fields. Its use round submarine cables has resulted in sending six times the number of signals per minute, and it is also of considerable use in transformers working with very high frequency current. The binary alloy, known as Permalloy, contains 78·5% nickel and 21·5% iron. This alloy requires annealing at 900°C followed by air cooling from 600°C. Additions of copper, chromium or molybdenum are made to this alloy to render it less sensitive to heat treatment and for increasing the electrical resistance. Cobalt additions are beneficial in producing constant permeability over a range of magnetising force. The alloys are very sensitive to impurities, internal stresses and heat treatment.

The reduction in impurities such as sulphides in silicon iron by the use of double slags and the vacuum melting of nickel alloys improves the quality of the materials.

# 15 Copper and its Alloys

Copper is used extensively for electrical purposes, radiators, refrigerators, heat exchangers, expansion pieces, condenser plates and tubes owing to its high conductivity for electricity and heat; plumbing services, chemical and brewing plant on account of its corrosion resistance and it is also the basis of many industrial alloys.

The main grades of raw copper used for cast and wrought copper base alloys are 'Cathode', 'Electrolytic', 'Fire-refined' 'deoxidised' and 'oxygen free' covered by BS 1035–40. The highest grades only are used for billets for subsequent working and for high-grade cast alloys.

Typical compositions of fabricated copper are:

TABLE 51 TYPICAL COMPOSITIONS OF COPPER (%)

|  | As | O | P | Pb | Fe | Ni | Ag | Sb | Bi |
|---|---|---|---|---|---|---|---|---|---|
| High-conductivity copper ≮ 99·9 Cu | nil | 0·04 | nil | <0·005 | <0·005 | nil | 0·002 | nil | <0·001 |
| Arsenical tough pitch | 0·40 | 0·065 | nil | 0·02 | <0·020 | 0·15 | 0·006 | 0·01 | <0·005 |
| Deoxidised Cu, ≮ 99·85 | <0·05 | nil | 0·05 | 0·010 | <0·030 | 0·10 | 0·003 | 0·005 | <0·003 |
| Arsenical deoxidised | 0·4 | nil | 0·04 |  |  |  |  |  |  |
| Oxygen free copper | nil | nil | nil | 0·005 | <0·0005 | 0·001 | 0·001 | nil | <0·001 |

*High conductivity copper* (HC) is electrolytically or fire refined to a high degree of purity, subsequently melted under controlled conditions (sometimes deoxidised with calcium boride and lithium) and is used for electrical purposes.

*Arsenical copper* contains up to 0·5 % arsenic which slightly increases its tensile strength, especially at temperatures of about 400°C as well as reducing the tendency to scaling when heated.

*Tough Pitch Copper* This term is used to denote that some oxygen is present in the copper. For welding and tube manufacture, copper free from oxygen is most suitable. Hence we have:

*Deoxidised copper*, in which the oxygen has been removed from the liquid metal by treatment with phosphorus or another deoxidiser, is used for tube production and where welding is necessary. The amount of phosphorus required is small, but a residual quantity is essential to prevent absorption of oxygen as the metal is cast. The solution of this residual deoxidiser in the copper adversely affects the electrical conductivity; 0·04 % phosphorus reduces the conductivity to about 75 % of pure copper.

*Deoxidised arsenical copper* is frequently used for welded vessels, tanks, etc.

*Oxygen-free copper* is a very pure copper which is melted and cast in a special non-oxidising atmosphere, so that deoxidation is not subsequently required.

*Free-cutting copper* containing $\sim$ 0·5% tellurium or sulphur is useful where machining to fine tolerances together with high conductivity ($>$95% Cu) is essential, e.g. magnetron valves.

*Production of tough pitch copper*

In fire-refining copper the impurities are removed by oxidising the metal until about 4% copper oxide ($Cu_2O$) is absorbed. During this stage the impurities form oxides more readily than the copper and are removed as a slag or evolved as gas. The last impurity so removed is sulphur which is not completely driven off as sulphur dioxide by mere oxidation. To remove the last traces of sulphur and to reduce the oxygen to the correct level the metal has to be violently agitated by poling, i.e. introducing an unseasoned piece of wood under the surface. Small test castings or button castings are taken to indicate the state of the metal. With sulphur present the ingot spurts just as it goes solid due to the evolution of gas ($SO_2$), but as the sulphur is reduced in amount the surface of the ingot sinks in the manner normal to most metals. If a micro-examination is made of this metal it will be found to contain globules of copper oxide in the form of a eutectic (Cu–$Cu_2O$). A layer of crushed coal is then placed on the molten copper, and as poling continues the copper oxide is reduced and when a content of about 0·04 to 0·08% oxygen is reached the surface of the button remains level and the properties of the metal are good, in other words 'tough'. The lower the oxygen, the higher the so-called 'pitch' and vice versa, hence the name 'Tough Pitch'. As poling continues past this point the copper absorbs hydrogen from the furnace gases and when cast the metal rises on solidification.

These changes in behaviour, micro-structure and mechanical properties are due to the influence of hydrogen and oxygen on the copper.

*Relation of hydrogen and oxygen*

Some chemical reactions proceed in one way only, whereas others may proceed either way according to the conditions. Such actions are said to be reversible. Copper takes up hydrogen from the furnace gases and this reacts with the copper oxide thus:

$$Cu_2O + H_2 \rightarrow H_2O + 2\,Cu \quad\quad\quad (1)$$

but steam also reacts with molten copper

$$Cu_2O + H_2 \leftarrow H_2O + 2\,Cu \quad\quad\quad (2)$$

Reaction No. 1 tends to proceed with a certain speed towards the right while reaction No. 2 proceeds with a different speed towards the left. After a time these reactions become balanced at a stage where we have quantities of copper oxide, hydrogen and steam dissolved in the copper. The solubility of steam in copper is very slight and can be neglected and so we say that the reaction stops or is in equilibrium when there are certain quantities of $Cu_2O$ and hydrogen, expressed thus:

$$(\text{Concentration of } Cu_2O) \times (\text{Conc } H_2) = \text{Constant}$$

When the amount of copper oxide is large the quantity of hydrogen is small, no blowholes are formed but the presence of copper oxide lowers the toughness of the material.

When the concentration of copper oxide is small, the amount of dissolved hydrogen is relatively large and at the moment of solidification the hydrogen reacts with the oxygen to form steam which is trapped within the frozen crust of metal to form blowholes. Such copper is weak due to lack of continuity. Tough pitch copper represents the best compromise between the two defects. The amount of hydrogen in the melt is about 0·000018% and the oxygen is 0·04%. The hydrogen unites with some of the oxygen to form steam and the remainder forms oxide particles. The holes produced are sufficiently small to close up during hot working.

*Effects of impurities*

The main disadvantage of deoxidised copper is that it:

(1) Is harder to work by hand.
(2) Demands a higher standard of purity.

Bismuth, antimony and lead even in small quantities render copper brittle, particularly hot short, and in the absence of oxygen not more than 0·003% bismuth, 0·02% lead, 0·01% antimony are permissible.

If *oxygen is present* the embrittling effect is not nearly so marked, particularly if arsenic be present and complicated oxides are formed, e.g.

$$2Cu_2O.PbO; \quad Pb_3As_2O_8; \quad Bi_2O_3; \quad Cu_2O.Bi_2O_3; \quad Cu_2O.Sb_2O_5$$

In deoxidised copper bismuth contents as low as 0·001% cause some embrittlement at 540–600 °C, due to the segregation of bismuth at the grain boundaries. The hot shortness disappears at temperatures above 700 °C as the bismuth passes into solid solution in the copper, about 0·01% is soluble at 800 °C. A high residual phosphorus content (e.g. 0·4%) reduces embrittlement at room temperature but not at 450–600 °C. A lithium addition neutralises the embrittlement at room and at elevated temperatures probably due to the formation of $BiLi_3$ with a high melting-point of 1145 °C.

## Gassing

Tough pitch copper is liable to a defect known as 'gassing' and this is closely allied to the cause of unsoundness. This trouble occurs when the copper is subjected to a reducing atmosphere such as in annealing or during welding with a reducing flame. Hydrogen penetrates the metal and reacts with the copper oxide to form steam which, being unable to escape, forces the crystal grains apart. 'Gassed copper' is readily recognised by the thick boundaries and cracks, which develop on bending.

Tough pitch copper is therefore unsuitable for welding. Deoxidised and oxygen-free copper are not so liable to this defect (as shown in the following figures), and are recommended for articles to be welded.

TABLE 52   EFFECT OF ANNEALING IN REDUCING ATMOSPHERE

|  | TS N/mm$^2$ | El | RA |
|---|---|---|---|
| TP copper before annealing | 227 | 65 | 67 |
| TP copper after annealing | 139 | 11 | 6 |
| Deoxidised copper before annealing | 244 | 67 | 83 |
| Deoxidised copper after annealing | 235 | 63 | 75 |

*Mechanical properties of copper* vary widely according to the physical condition. As cast the tensile strength is about 154–170 N/mm$^2$, which is increased to 216 N/mm$^2$ with 50% elongation by working and annealing. The material is frequently sold in tempers. 'Soft' temper refers to annealed material, VP Hardness below 60; 'half hard' material is slightly cold worked to 70–80 hardness, and 'hard' temper is more heavily cold worked to yield a hardness of 100 or more.

## Copper alloys

Copper is alloyed with many elements, the most common of which are zinc, aluminium, tin and nickel. The α-solid solutions of all these alloys have the same type of micro-structure and are relatively tough and ductile.

When the quantity of zinc, tin or aluminium added exceeds that required to saturate the α-solid solution a β-intermediate phase appears in the structure and the alloy is stiffened considerably. The micro-structure in each alloy is similar and is associated with increased strength at the expense of ductility.

Further additions of the second metal give rise to hard and brittle constituents, which impair the toughness of the material.

Both bismuth and antimony when present in small quantities form brittle films in copper and brasses and aluminium bronzes which have a deleterious effect on the mechanical properties.

For strong high conductivity alloys see Chap. 14.

## Brasses

The copper–zinc alloys, with zinc varying from 0 to 50%, are particularly important to the engineer in view of their wide range of mechanical properties, their ease of working, their colour and resistance to atmospheric and marine corrosion.

The portion of the equilibrium diagram which covers the commercial

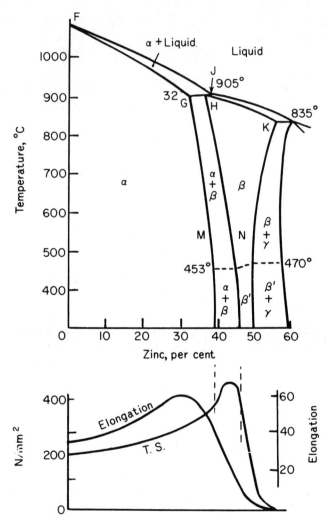

FIG. 201 Equilibrium diagram and mechanical properties of brasses

brasses is shown in Fig. 201, although it must be remembered that changes rarely reach completion during normal rates of cooling. The main features of the system are:

Compositions between F and G solidify as α-solid solution, usually cored. Compositions G to H commence by forming α- and finally form some β-solid solution (Fig. 202) of composition H due to the peritectic reaction. On cooling to room temperature the amount of the β constituent decreases according to the solubility lines GM and HN. Compositions between H and J form α-crystals which, at 905 °C, react with liquid of composition J to form β-crystals. On cooling, α-crystals are precipitated from the β when solubility line HN is passed. A Widmanstätten (p. 103) type of structure is formed (Fig. 203). Compositions between J and K solidify as β (Fig. 206); alloys containing less than 46·6% zinc precipitate some α-crystals; while alloys with more than 50% zinc precipitate γ-crystals on cooling to room temperature (Fig. 207). The range of composition for the three phases is:

TABLE 53

| Temperature | Zinc % | | |
|---|---|---|---|
| | α | α + β | β |
| At room | 0–35 | 35–46·6 | 46·6–50·6 |
| At solidus | 0–32·5 | 32·5–36·8 | 36·8–56·5 |

It is interesting to note that the solubility of zinc in α-solid solution increases (39%) as the temperature falls to 453 °C which is contrary to general behaviour; then decreases down to room temperature.

Between 453 °C and 470 °C the β-phase transforms to a low temperature modification known as β′. This transformation is due to the zinc atoms changing from a random to an ordered arrangement on the lattice.

The *properties* of a brass will depend on the volumes of the phases present and Fig. 201 serves to show the general connection between the structure and the properties.

The tensile strength and the elongation increase in the α region, the ductility reaching a maximum at 30% zinc. The presence of β-phase causes a considerable drop in elongation, but rapidly increases the tensile strength up to a maximum, when the alloy contains all β. The strengths falls rapidly at the appearance of the weak and brittle γ constituent. The resistance to shock decreases while the hardness increases by the presence of β-phase, and the alloy becomes hard and extremely brittle when γ-phase appears. Consequently the presence of the γ constituent is avoided. The only commercial alloy likely to contain γ is one containing 50% zinc and used as a brazing solder because of its low melting point. In brazing, however, some of the zinc is lost by volatilisation and some diffuses into the metals being united.

The β-phase is much harder than α-phase at room temperature and will withstand only a small amount of cold deformation. It begins to soften suddenly at 470 °C (i.e. disordered change) and at 800 °C is very much easier to work than α-phase. Thus it is advantageous to hot work β-phase, and it is possible to classify brasses into two groups:

(1) *α-brasses*, specially suited for cold-rolling into sheets, wire-drawing, tube manufacture and pressing; (2) *α-β-brasses* for extrusion and hot-stamping.

*Alpha brasses for cold working*

Annealed α-brasses will withstand a remarkable degree of deformation by cold work without the slightest sign of fracture. The impact resistance of α-brass at 350 to 650 °C is extremely poor, but the resistance to creep at elevated temperatures is superior to that of α-β-brasses. Consequently, there is no appreciable advantage in hot working α-brass, except that when breaking down large ingots into strip, the handling costs are reduced and repeated annealing is unnecessary. If hot working is used low melting point impurities such as lead and bismuth must be kept to low levels (e.g. 0·02, 0·002%) but zirconium and uranium additions offset the embrittlement.

*Cap Copper* contains about 2–5% zinc and is used as a deoxidiser. It is very ductile and is used for ammunition priming caps.

*85Cu/15Zn brass or gilding metal*  This is used for bullet envelopes, drawn containers, compact boxes and dress jewellery—because its colour resembles that of gold.

*70/30 Brass* is frequently termed cartridge metal because of its use for cartridge cases which require great ductility allied with strength. The highest purity is necessary, especially in regard to iron, lead, bismuth, arsenic and antimony, hence the use of the purest copper and zinc obtainable together with clean scrap.

Typical properties are:

TABLE 54

| Condition | 0·1% PS | TS | El | BH |
|---|---|---|---|---|
| Chill cast | 6<br>93 | 16<br>247 | 65 | 60 |
| Hard-rolled sheet | over 25<br>386 | 30–40<br>463–618 | 10–15 | 150–200 |
| Annealed sheet | 6<br>93 | 20–23<br>309–355 | 65–75 | 60 |

The ratio of endurance limit to tensile strength is 0·2.

As cast the structure is dendritic (as Fig. 65), but the coring can be removed by annealing (Fig. 67). The effect of cold working and annealing is the same as described in Chap. 7. The crystals after annealing must be uniform and of the right size to suit subsequent working processes. Since there are no critical ranges as in steels, the grain size depends on the laws of recrystallisation and grain growth discussed in Chap. 7. In practice it is

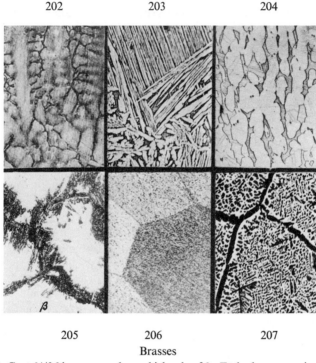

202     203     204

205     206     207
Brasses

Fig. 202 Cast 64/36 brass: cored α and islands of β. Etched—ammonia ( × 80)
Fig. 203 Cast 60/40 brass. α light, β dark ( × 80)
Fig. 204 Extruded and slightly cold drawn 60/40; elongated β ( × 80)
Fig. 205 60/40 brass, heated to 840 °C, and quenched in water: 'feather' shaped α plates precipitated at boundaries of β grains ( × 50)
Fig. 206 Brass containing 48% zinc. Uniform β. Etched—acid ferric chloride ( × 50)
Fig. 207 Brass containing 57% zinc. γ Network and particles in β matrix: etched—ferric chloride ( × 80)

convenient to anneal at a constant temperature of 600–650 °C and the initial degree of cold work can be varied, although the effect of this variation on the properties seems to be a minimum at 600–650 °C.

When annealed above 750 °C the grains coarsen and on subsequent pressing, 'orange peel' surfaces become evident. Such over-annealed brass can be rectified by cold working and annealing at 600 °C. Burnt brass, caused by annealing close to the solidus is friable and useless except for remelting (Fig. 217).

For condenser tubes either 1 % tin (Admiralty brass) or 2 % aluminium is added to improve the resistance to corrosion, in the latter case the zinc content is lowered to 22 % to avoid the formation of β and a small amount of arsenic is added to minimise dezincification (p. 145). For heavily polluted water ($H_2S$) Cu, 8Sn, 1 Al, 0·1 Si has been found some use.

*63/37 Brass* (Basis quality) is a ductile alloy used for cold press work, but is cheaper than 70/30 brass. As cast this alloy contains a small quantity of β constituent, owing to the lack of equilibrium (Fig. 202). After cold working and annealing the structure is similar to 70/30 brass (as Fig. 71).

Brasses are liable to season cracking (see p. 120).

(2) *Hot-working brasses* comprise the second group and are especially suitable for casting, hot-working by rolling, extruding or hot-stamping, although they are frequently cold worked as well.

*60/40 Brass* This alloy has the best combination of properties of the α + β alloys and is frequently called Muntz metal, after the inventor. Typical properties of the brass as cast are:

|  | TS $N/mm^2$ | El | BH |
|---|---|---|---|
| Cast | 340–386 | 40–45 | 90–100 |
| Hard | 494 | 17 | 131 |

When just solid at about 875 °C the structure consists of β-crystals. Between about 770° and 500 °C the α-constituent is deposited at the boundaries and along the cleavage planes of the β-crystals (Fig. 203).

The relative proportions of α and β can be varied by the rate of cooling. Quenching the alloy while in the wholly β condition (800 °C) may suppress the precipitation of α-crystals, but usually a little is deposited as 'feathers' along the β boundaries (Fig. 205). Such treatment increases the strength at the expense of the ductility.

The form of the α-constituent can be influenced also by hot-working. The best properties are usually obtained if the alloy is deformed while the α-constituent is being deposited (700–600 °C). This results in a fine grain and at the lower temperature yields a fibrous type of structure (Fig. 204). Below about 600 °C the brass becomes cold worked.

The grain size of Muntz metal can be refined by heating first into the wholly β region (just above line HN in Fig. 201) followed by slow cooling. The α-constituent then occurs as small rounded masses in the β matrix. The temperature of annealing is very critical since, once the α-phase is dissolved, the uniform β-crystals grow at an extremely rapid rate.

## Leaded brass

The presence of lead has a tremendous effect in improving the machining properties of brasses; common alloys used for hot stampings and extruded rods which have to be machined usually contain 56–61 % copper, 1·5–3·5 % lead, balance zinc. Segregation limits the lead content, but the introduction of continuous casting reduces the problem and allows higher lead contents to be used with great benefit. The tensile strength does not differ greatly from that of Muntz metal, but the elongation is slightly less and the impact properties are reduced. The lead is almost insoluble in both the liquid and solid states and appears as globules in the micro-structure (Fig. 218). The effect of these globules is to cause the turnings to break up into small pieces.

Successive cold drawing and annealing at 750 °C can increase the diameter of the lead particles three times and reduce machinability 25 %.

## Complex or high tensile brasses

Aluminium, iron, manganese, tin and nickel are frequently added to the 60/40 copper–zinc alloy to increase its tensile strength without seriously affecting its ductility. Such an alloy, containing one or more of these elements, is called 'high tensile brass'. There are two main classes of high tensile brasses, one owes its strength to the essential addition of tin (Admiralty propeller brass) and the other to aluminium which is used for strengths above 540 N/mm$^2$. The amount of tin in the presence of aluminium is limited. High tensile brasses are used in the extruded, forged, stamped and rolled condition, but their special application is in the form of castings which may weigh up to 50 tonnes in the case of marine propellers. This is due to the facility with which they may be cast into complex shapes with almost complete freedom from porosity coupled with the satisfactory mechanical properties which are not much inferior to those obtained on forged material.

The large variety of compositions made for the same mechanical properties, frustrated pre-war efforts to standardise limits of composition which, however, was eventually done in the BS schedule, 1400, in which a new feature is the specifying of alpha content minima for alloys of the α-β type. Compositions are given in Table 55.

TABLE 55  CAST HIGH TENSILE BRASSES

| BS 1400 | Cu % min | Mn % max | Al % max | Fe % max | Sn % max | Ni % max | Area α-phase | TS min N/mm$^2$ | El min |
|---|---|---|---|---|---|---|---|---|---|
| HTB 1 | 55 | 3 | 0·5–2·5 | 0·7–2 | 1·0 | 1 | 15 | 460 | 20 |
| HTB 3 | 55 | 4 | 3–6 | 1·5–3·25 | 0·2 | 1 | nil | 740 | 12 |

It should be noted that the beta-structure alloy is liable to stress corrosion cracking.

Typical uses of castings are: marine propellers, rudders, stern tubes, autoclaves, gun mountings and sights, while wrought sections are employed for pump rods, stampings and pressings for automobile fittings and switch gear.

Many of the elements added to high tensile brass affect the micro-structure in the same way as increasing the zinc content, except nickel which has the opposite effect. The following equivalents, due to Guillet, express the approximate zinc replacement capacity of the various elements when added to brass:

| Element | Si | Al | Sn | Mg | Pb | Fe | Mn | Ni |
|---|---|---|---|---|---|---|---|---|
| Equivalent Zn | 10 | 6–4 | 2 | 2 | 1 | 0·9 | 0·5 | −1·2 |

Thus, 1% of silicon has a similar effect to 10% of zinc, while 1% of nickel is similar to 1·2% of copper.

For example a brass containing:

|  | Copper | Zinc | Aluminium | Manganese |
|---|---|---|---|---|
| Percentages | 62 | 34 | 2 | 2 |

would have a zinc equivalent of

$$34 + (2 \times 6) + (2 \times 0.5) = 47$$

The total (copper and zinc) amounts to $62 + 47 = 109$ and the equivalent percentage of zinc in the alloy is $\frac{47}{109} \times 100 = 43 \cdot 1\%$. This brings the alloy well into the $\alpha + \beta$ region, although from the actual zinc content the alloy would appear to be an $\alpha$-brass. A convenient practical rule is that 1% zinc below 46·6% is roughly equivalent to 10% alpha, i.e. 35% alpha, in above example.

*Tin* produces a slight increase in tensile strength, but the ductility drops when more than about 1% is present owing to the appearance of a hard brittle constituent. Tin improves the corrosion resistance of brasses, and is present in Naval brass (Cu 61; Sn 1) for this reason.

*Iron* has a limited solubility (0·2%) in both $\alpha$ and $\beta$ brasses, and when the amount exceeds about 0·35% a bluish micro-constituent separates, which results in a marked refinement of the Widmanstätten structure. With 53% copper grain refining commences with 0·88% iron and is a maximum at 1·35%, then little effect up to 3%. The grain size of the wrought product is fine with about 0·25% Fe but higher quantities enable more lead to be tolerated without hot cracking. Up to 2% of iron has no appreciable effect on the mechanical properties, but the iron-rich phase is liable to gravitational segregation, above 1% Fe there is a risk of hard spots being formed if Si is picked up in the furnace.

*Manganese*, up to 2%, is often added (as ferro–manganese) to the 60/40 brasses, which are then termed manganese bronzes. Manganese is soluble in the solid brass, and its presence increases the amount of iron which can be dissolved and at the same time it acts as a good scavenger. Up to 2·5% it causes slight grain refinement.

*Aluminium* has a great influence on the tensile strength of brass (1% giving an increase of 88 N/mm$^2$), and the high strength of the modern high tensile brasses is largely due to it, the function of the iron and manganese being to increase the range of the strong $\beta$ alloys, to improve the stability and to prevent grain growth. A characteristic feature of brasses containing aluminium is the presence of a superficial film of aluminium oxide, which renders soldering and casting more difficult.

*Copper–zinc–nickel alloys—nickel silvers*

Nickel is added to copper–zinc alloys to form two classes of material:
(1) *The $\alpha$ alloys* contain 7–30% nickel with about 63% copper and the remainder zinc and are known as German silvers or nickel silvers, although they contain no silver.

The nickel has a remarkable decolorising effect on the brass, which with about 20% nickel appears silvery white. The alloys take a brilliant polish and have a good resistance to corrosion.

Typical mechanical properties for one alloy in the annealed condition, a second for a hard spring quality telecommunications and a third as sintered at 980 °C are:

TABLE 56

| No. | Ni | Cu | Zn | PS | TS N/mm$^2$ | El | BH |
|---|---|---|---|---|---|---|---|
| 1 | 18 | 62 | 20 | 123 | 370 | 50 | 75 |
| 2 | 18 | 55 | 27 | 695 | 850 | 3 | 240 |
| 3 | 18 | 64 | 18 | — | 220 | 15 | 50 |

The alloys can be cold rolled to give strengths of 617–927 N/mm$^2$ and annealing is usually carried out at 650–800 °C according to the nickel content. The nickel silvers are used as the basis of silver and chromium electroplated goods such as spoons (e.g. EPNS standing for electroplated nickel silver); extensively used in telephone and wireless industries for contactors and springs; in the form of tape and wire for electrical resistance; also for fittings for ships, lavatories and shops; spectacle parts, zip fasteners, and latch keys. A 45% Ni alloy is used for parachute harnesses because of its high shear strength.

(2) *The α-β alloy*, having a composition of about 45% copper, 45% zinc and 10% nickel, is available in the form of extruded sections for architectural and ornamental purposes because of its silvery colour. The addition of 2% of lead improves machining. These alloys are referred to as nickel brass and sometimes 'silver bronzes'.

## Copper–nickel alloys

Nickel and copper form a complete series of solid solutions as shown by Fig. 37, and in the annealed condition all the alloys have similar structures. The cupronickels are sensitive to oxygen, sulphur, lead and carbon, which cause embrittlement. The effect of lead on hot cracking can be neutralised by addition of mischmetall coupled with homogenisation at 670°C. Manganese is usually employed as a deoxidant. The alloys are softened at 550 to 690°C and should not be heated above 800°C, otherwise any carbon present is liable to be precipitated.

The addition of 1–2% iron to the alloys increases the resistance to corrosion, especially to impingement attack in fast-moving sea-water, while still allowing the alloy to be manipulated by ordinary smithy methods. Quenching from 900°C is desirable to disperse the iron.

Typical alloys and properties of the annealed material are:

TABLE 57

| Nickel % | TS N/mm$^2$ | El | Uses |
|---|---|---|---|
| 10 (Fe 1–2) | 385 | 40 | Condenser tubes, seawater piping |
| 20 | 355 | 41 | Condenser tubes, bullet envelopes |
| 25 | 385 | 39 | Coinage |
| 30 (Fe, 1) | 415 | 40 | High quality condenser tubes |
| 40–45 | 430 | 51 | Electrical resistances and thermocouples (Constantan, Eureka) |
| 68 (Fe, 2) | 570 | 40 | Monel; turbine blades |

The 80/20 and 70/30 cupronickel can be hot or cold worked, and are noted for their extreme malleability in the cold. Corrosion troubles in condenser tubes can be largely eliminated by the use of these alloys, preferably the one containing 30% nickel with about 1% iron, which are also free from season cracking. High strength alloys are being developed based on 70/30 Cu Ni with addition of about 0·4 Si and 1·3 Nb and on 64/36 Cu Ni containing 0·2 zirconium.

Monel varies somewhat in analysis. It has a good resistance to sea and estuary waters, alkalies, reducing acids such as sulphuric acid, oils containing brines and sodium sulphide, and alkaline solutions. A poor resistance is offered to nitric acid, ferritic iron solutions, chromic acid and cyanide solution.

Resistance to attack by superheated steam and the retention of strength at high temperatures enable the alloy to be used for turbine blading, valve parts, pump rod liners and impellers.

Monel K-500, containing about 3% Al and 0·5% Ti, retains the characteristic corrosion resistance of Monel, but which can be temper-hardened at 600°C (see p. 96). It is thus suitable for applications requiring high strength with good corrosion resistance.

Typical properties are:

TABLE 58

| Condition | PS N/mm$^2$ | TS N/mm$^2$ | El | BH |
|---|---|---|---|---|
| Softened, Q 800°C | 290 | 600 | 35 | 140 |
| Temper hardened, Q 800°C, T 590°C 4 hours | 665 | 925 | 30 | 270 |
| Cold worked and temper hardened | 925 | 1110 | 15 | 320 |

'S' Monel with 3–4% silicon is useful for resisting wear at high temperatures and pressures in steam plant and on oil tankers.

For marine engineering two high tensile alloys with good corrosion resistance are:

| | Ni | Al | Fe | Mn | TS | El | Izod J |
|---|---|---|---|---|---|---|---|
| Hiduron 191, wrought | 16 | 2 | 1 | 5 | 740 | 25 | 47 |
| Hiduron 501, cast | 14 | 2 | 5 | 10 | 500 | 25 | 47 |

The high manganese in the cast alloy is necessary for reasonable weldability and the Ni/Al ratio is important in both alloys. Heat treatment at 550°C increases strength by precipitation. The properties are fairly independent of section size and the low magnetic permeability is important in electrical application.

## Bronzes

The portion of the copper–tin equilibrium diagram which covers industrial alloys is shown in Fig. 208. The solubility of tin in the α-solid solution is 13·5% at 798°C and about 1% at room temperature. A peritectic reaction occurs at 798°C in alloys containing 13·5 to 25·5% tin, resulting in the formation of β intermediate solid solution (see Chap. 5). On cooling to about 586°C the β phase changes to γ. At 520°C the γ decomposes into a eutectoid consisting of α and δ, in the same way, as in steels, austenite changes to ferrite and cementite at 700°C. The δ-constituent consists substantially of the compound $Cu_{31}Sn_8$. It is pale blue in colour, hard and brittle (Fig. 211). The ε-phase appears to be $Cu_3Sn$.

Equilibrium is not easily reached even at 600°C and the duplex α plus ε structure is only obtained after prolonged low-temperature heat-treatment

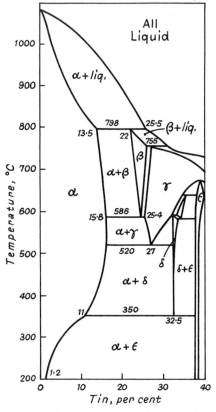

FIG. 208 Equilibrium diagram of copper–tin alloys

in alloys containing up to 15% tin. *Cast alloys* are far from equilibrium and *have a structure of α plus (α + δ) eutectoid*.

Owing to the wide separation of the liquidus and solidus curves pronounced coring occurs and diffusion takes place so slowly in the bronzes, that the δ-constituent begins to appear in commercial castings containing 7% of tin, and alloys with a greater amount of tin can only be rolled with difficulty unless they are annealed for prolonged periods in order to form a uniform solid solution.

# BRONZES

*Coinage bronze* is an α alloy, containing 3·5% tin and 1·5% zinc. The zinc acts as a deoxidiser and its effect on the structure is equivalent to about half that of tin at a much lower cost. Since 1959 British bronze coins contain tin 0·5 and zinc 2·5%. The structures of α-bronzes are shown in Figs. 65 to 72.

*Gun-metal* is a 10% tin alloy, with the addition of zinc which is the essential difference between gun-metal and a tin bronze. Admiralty gun-metal sand cast is:

| Cu | Sn | Zn | TS N/mm² | El |
|----|----|----|----------|----|
| 88 | 10 | 2  | 309      | 20 |

Gun-metal has been replaced by steel for ordnance and it is now used chiefly for castings requiring considerable strength coupled with corrosion resistance. For the best properties both very high and very low casting temperatures should be avoided; about 1200 °C is recommended for thin sections, 1130 °C for 37 mm sections. The structure is heavily cored and contains α + δ eutectoid islands (Fig. 210), the amount of which increases with the increased rate of solidification. The presence of the eutectoid renders the material unfit for cold work, but hot working is possible above 590 °C.

Although Admiralty gun-metal was formerly the standard composition for marine purposes, other alloys (BS 1400-LG4, LG2) are now frequently used as substitutes to economise in tin and to produce pressure-tight castings.

The *86/7/3/3* (Cu/Sn/Zn/Pb) alloy is used for general purpose castings for medium steam pressures above 0·7 N/mm² at temperatures below 260 °C; also for hydraulic pressure up to 14 N/mm² and lightly loaded, unlined bearings. It is an easy casting alloy. The *85/5/5/5* (Cu/Sn/Zn/Pb) alloy has a tensile strength of about 216 N/mm² and 15% elongation. It is used for general purpose castings for steam pressures up to 0·7 N/mm² at temperatures below 205 °C and for well-supported backings of certain lined bearings.

The addition of about 3% nickel is beneficial in sand castings, suitable for gears, valves and pumps.

## Phosphor-bronze

In making bronze some copper oxide may be formed as the copper is melted, but on adding the tin the soluble copper oxide is replaced by tin oxide which is heavy and insoluble in the molten bronze. The tin oxide therefore remains entangled in the melt. Such bronze is mechanically weak, porous and shows dirty patches on machining (Fig. 209). When zinc is present zinc oxide is formed which is fairly light and tends to float to the surface of the melt. Phosphorus is a more powerful deoxidiser than zinc and is therefore used for producing sound, clean castings. A deoxidiser

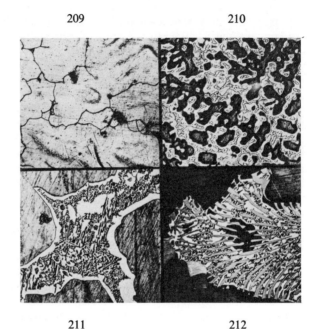

Photographs by courtesy of Messrs. D. Brown and Sons (Huddersfield) Ltd.

Bronze

FIG. 209 Porosity and intercrystalline oxide films ($\times$ 200)
FIG. 210 Centrifugally cast (10%) bronze: cores $\alpha$ and ($\alpha + \delta$) eutectoid ($\times$ 200)
FIG. 211 The $\alpha + \delta$ eutectoid: fringe of $\delta$ (light) due to absorption of $\alpha$ by primary dendrites ($\times$ 400)
FIG. 212 Triple eutectoid of $\alpha + \delta + Cu_3P$ (copper phosphide half-tone plates). Tin, 10; phosphorus, 0·7% ($\times$ 800)

reacts very slowly with insoluble oxides, consequently phosphor–copper should be added to the copper prior to the tin, so that no oxygen is present to form the insoluble tin oxides. Unfortunately, this is not often possible in practice because a certain amount of scrap bronze is used in the charge, and a slightly oxidising atmosphere is preferred to a *reducing one* which *causes gas absorption*.

Gas absorption increases with (*a*) phosphorus content, (*b*) pouring temperature, (c) moisture from the sand mould. Reaction with the latter is minimised by a coating of aluminium paint on the mould.

Several methods of removing gas are now well established; of these the most convenient appears to be the introduction of a small quantity of manganese ore into the bottom of the pot, where it sticks and evolves oxygen or carbon monoxide by reaction with the graphite of the crucible. This relatively inert gas, rising through the melt, carries with it the unwanted hydrogen. Alternatively, marble chips may be plunged beneath the surface of the molten alloy where they evolve carbon dioxide, which has an equally scavenging action. Scavenging with nitrogen is also employed. Good mechanical properties are also obtained with an oxidising flux consisting of equal parts of copper oxide, borax and silica sand thrown on the surface of the melt. Where pressure tightness is required, it is often preferable to retain an appropriate amount of dissolved gas in the alloy, e.g. melt under charcoal and deoxidise with phosphorus.

While it is impossible to lay down hard and fast rules applicable to all castings, a pouring temperature of from 1080 to 1180°C should be used for phosphorus contents between 0·01 and 0·01%, while for 0·25% phosphorus the pouring temperature may be from 1050 to 1150°C and for 0·5% phosphorus from 1020° to 1120°C.

Where phosphorus has been used solely as a deoxidiser only traces of phosphorus remain in the alloy which is misnamed phosphor–bronze. True phosphor–bronzes, however, contain more than 0·1% phosphorus, which acts as a definite alloying element; the alloys can be considered in two groups.

*Wrought phosphor–bronze* contains 2·5 to 8·5% tin and 0·1 to 0·35% phosphorus, which improves the tensile properties. In the form of rod, strip and wire the alloy is used for springs and high strength, corrosion resistant fasteners.

A common analysis for springs is:

| Copper | Tin | Phosphorus | TS | El |
|---|---|---|---|---|
| 93·7 | 6 | 0·2 | Cast 232 | 18 |
|  |  |  | Drawn 849 | 5 |
|  |  |  | Annealed 370 | 65 |

*Cast phosphor–bronzes* contain 5 to 13% of tin and 0·3 to 1% of phosphorus and are mainly used for machine parts which need a bearing surface capable of carrying heavy loads coupled with a low coefficient of friction, such as gear wheels and slide valves. A popular bronze for centrifugally cast gear blades is:

When the phosphorus exceeds about 0·3% in a 10% tin bronze hard $Cu_3P$ is formed which is usually associated with the δ-constituent (Fig. 212).

| Tin | Phosphorus | PS | TS | El | BH |
|---|---|---|---|---|---|
| 12 | 0·3 | 154 | 263 | 5 | 100 |

*Effect of other elements on bronze*

Small quantities of either aluminium or bismuth are objectionable in bronze. The aluminium forms alumina films which tend to retain gases and cause unsound castings, even 0·005% is sufficient to give localised patches of shrinkage porosity. Bismuth induces brittleness due to the formation of intercrystalline films. Hot rolling necessitates close control of impurities, e.g. Pb 0·0004, Bi 0·0004% and of rolling temperature 780 ± 20°C. More than 0·3% of either antimony or arsenic causes some deterioration of the bronze. When the iron exceeds 0·2% the bronze is hardened and rendered somewhat brittle. Sulphur up to 0·3% has no important deleterious effect on porosity, fluidity, strength or hot tearing of bronze.

Up to 2% of lead is added to the bronzes to improve machining; the turnings then chip away in small pieces. Larger quantities of lead reduce the strength and ductility. From 8 to 30% lead is added to bronzes for bearings. These alloys are known as plastic bronzes and they frequently contain 1% nickel which tends to give a more uniform distribution of the lead globules in the matrix (Fig. 219).

*Bearing bronzes* (see also Chap. 18). The alloy has to be selected according to conditions of service, taking into consideration such factors as shaft speed and material, load and impact, lubrication and temperature. In increasing order of hardness bearing bronzes are given in Table 59.

TABLE 59 BEARING BRONZE

| | Sn | Pb | P | El | BH | |
|---|---|---|---|---|---|---|
| Plastic bronze | 5 | 20 | | 2–15 | 50 | Useful where some measure of plasticity is desirable, where lubrication is imperfect and with 'soft' steel shafts |
| | 8 | 15 | | 4–10 | 60 | |
| Medium tin + lead | 10 | 10 | | 4–10 | 65 | |
| Low tin, Zn, 1 | 7·5 | 3·5 | 0·5 | 2–15 | 90–130 | Substitute for 10% tin alloy |
| Medium tin | 10 | | 1 | 2–15 | 90–130 | Standard bronze 2B8 |
| High tin | 14–20 | | | 1–3 | 100–140 | Rigidity, turntable bearing |

In tin and phosphor bronzes the increase of tin and the introduction of phosphorus serves to produce hard delta particles and phosphides in a softer matrix containing a fairly high content of tin. This leads to a limited ductility.

## Aluminium bronze

Copper-rich aluminium alloys are known as aluminium bronzes, although they contain no tin. The characteristic properties of these alloys are (*a*) high strength, (*b*) good working properties, (*c*) resistance to corrosion and wear, (*d*) high resistance to fatigue, (*e*) fine golden colour and (*f*) possibility of heat-treatment in manner similar to steels. One difficulty arises during casting, due to the formation of tenacious films of alumina, with the result that discontinuities occur in the metal. For casting ingots this trouble had been overcome by the Durville method of casting, which reduces turbulence to a minimum. The formation of oxide in castings is avoided by bottom pouring, using a form of dross trap and avoiding turbulence. The considerable shrinkage is often better controlled by the use of chills than by excessive use of large feeder heads. Another difficulty is 'self-annealing', discussed later. The welding difficulties have now been largely surmounted (p. 412) and fabrication by metal arc welding has extended the engineering applications of these alloys.

The portion of the equilibrium diagram covering the commercial alloys is shown in Fig. 213. The $\alpha$-solid solution contains 7·5% of aluminium at 1037 °C and 9·4% at room temperature, although in commercial alloys the $\beta$-phase appears at about 7% aluminium owing to the lack of equilibrium.

The $\beta$-constituent is harder and more brittle than the $\alpha$ but, like the $\beta$-phase in brass, it can be worked hot. Unlike $\beta$-phase in brass, it decomposes into a eutectoid of $\alpha + \gamma_2$ at 565 °C on slow cooling, which resembles pearlite in steel. The $\gamma_2$-constituent is very hard and brittle and its presence in small quantities increases the strength of the alloy; larger amounts embrittle the material, and for this reason aluminium content rarely exceeds 11%.

Three types of alloy are used:

(*1*) *Wrought $\alpha$ alloy* contains 5 to 7% aluminium, and is easily worked either hot or cold. In the cast condition the structure is cored, but a twinned structure is produced after working and annealing (similar to Fig. 71). Heat exchanger tubes are now made in $\alpha$ aluminium bronze.

Typical properties of an alloy containing 7% aluminium are:

| Cu 93; Al 7 | PS | TS | El | BH |
|---|---|---|---|---|
| Cast | — | 325 | 69 | — |
| Rolled, 40% reduction | 608 | 620 | 18 | 195 |
| Annealed 650 °C | 110 | 430 | 71 | 76 |

(*2*) *10% $\alpha$–$\beta$ alloy* is used as castings and also in the hot worked condition. As cast it consists of free $\alpha$ and a eutectoid of $\alpha + \gamma_2$ (or a type of $\beta$) in the form of a Widmanstätten structure (Fig. 214). Unlike the 10% tin–bronze which forms disseminated islands of eutectoid between the cored dendrites due to the wide solidification range and slow diffusion, the 10% aluminium bronze has a narrow freezing range, and the eutectoid is stable and cannot be removed by annealing. The relative size and distribution of the α and

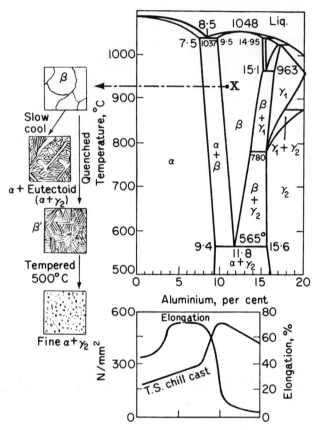

FIG. 213 Equilibrium diagram of copper–aluminium alloys with associated mechanical properties. Diagrammatic micro-structures are for alloy X after different thermal treatment

eutectoid greatly influence the mechanical properties of the alloy, which are affected by heat-treatment as in the case of steel.

On heating at 570°C the eutectoid (analogous to pearlite) forms the β solid solution (austenite analogue) which, as the temperature rises, dissolves the free α-constituent. The solution of α is complete at temperature X (Fig. 213) just in the β zone. If the alloy is quenched from temperature X the separation of the α-phase is partly prevented and we get an acicular structure (Fig. 216), strongly resembling martensite in steel, and the alloy is hardened. Control of the correct hardening temperature is rendered difficult by the steepness of the solubility lines; a small change in composition causes a wide variation in temperature.

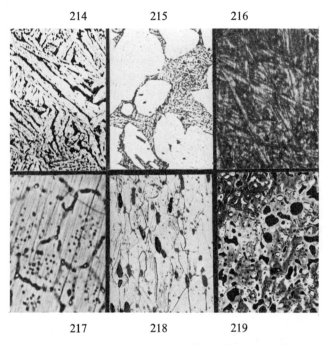

Fig. 214 10% aluminium bronze as cast; $\alpha$ light, fine eutectoid or $\beta_1$ (dark); etched acid ferric chloride ($\times$ 150)

Fig. 215 As Fig. 214, heated to 900°C, furnace cooled ('Self-annealed') coarse $\alpha + \gamma_2$ eutectoid in matrix of $\alpha$ (light) ($\times$ 150)

Fig. 216 As Fig. 215, but quenched in water. Martensitic structure of $\beta$-incipient separation of $\alpha$ ($\times$ 150)

Fig. 217 70/30 brass, 'burnt': incipient fusion at boundaries; unetched ($\times$ 80)

Fig. 218 Leaded 64/36 brass, cold drawn, for free cutting. Lead (black) elongated twinned grains with strain bands ($\times$ 80)

Fig. 219 Plastic bronze (10% Pb). Lead (black) cored matrix

If the alloy is tempered at 350–560°C small particles of $\gamma_2$-constituent form throughout the mass and the structure is somewhat similar to sorbite in steel (Fig. 213), i.e. $\alpha + \beta \rightarrow \alpha + (\alpha + \gamma_2)$.

The infrequently applied commercial heat-treatment involves quenching from 850° to 900°C followed by tempering for about 2 hours at 550–650°C. Such treatment is less effective on castings than on wrought material.

The α + β aluminium bronze is liable to develop a weak and brittle condition, known as 'self-annealing', when slow cooled from the β-region. This seems to be due to the formation of a coarse eutectoid (Fig. 215). One of the most effective means of overcoming the trouble in large castings is by adding 1 to 3% of iron to the alloy, thus producing much finer structures.

The mechanical properties of an alloy containing 10% aluminium are:

TABLE 60

| Condition | PS | TS | El | RA |
|---|---|---|---|---|
| Cast | 170 | 480 | 20 | 21 |
| 'Self-annealed' | 215 | 386 | 4 | 5 |
| Q 900°C | 385 | 385 | 2 | 4 |
| Q 900°C, T 550°C | 280 | 585 | 13 | 16 |
| Hot worked | 120 | 510 | 28 | 30 |
| Hot worked Q 900°C, T 55°C | 155 | 633 | 48 | — |

The fatigue limit ($5 \times 10^7$) cycles is about $\pm 215$ and $\pm 155$ N/mm² in air and water respectively.

*(3) Complex aluminium bronzes.* In complex aluminium bronzes the common additional elements are iron, nickel and manganese. Up to about 5% manganese or 3% iron can be added before a separate constituent appears.*

The 10/5/5 alloy contains a Ni–Fe–Al complex (ordered b.c.c.) which precipitates from β in the form of rosettes which give better mechanical properties than the acicular form precipitated from α in the 9% aluminium alloy. Quenching from 1000°C retains β with a hardness of 280 which increases to 420 after tempering at 400°C and is associated with a large increase in proof stress but no change in micro-structure.

Typical compositions of complex bronzes are given in Table 61. The castings are also covered by BS 1400, AB1 and AB2.

TABLE 61

|  | Al | Fe | Ni | Mn | TS | Elongation |
|---|---|---|---|---|---|---|
| Hot worked |  |  |  |  |  |  |
| DTD 164 | 10·0 | 1·5 | 2·0 | — | 587–695 | 18–30 |
| DTD 197 | 10·0 | 5·0 | 5·0 | 0–2·5 | 740–770 | 15–25 |
| Castings: |  |  |  |  |  |  |
| DTD 174A | 9·0 | 2·5 | 0–3·5 | 0–4·0 | 490–585 | 20–40 |
| DTD 412 | 10·0 | 4·5 | 4·5 | 0–2·5 | 615–695 | 12–20 |

*For details of these alloys see *J. Inst. Metals*, 1951–52, *80*, 419.

The wrought alloys are fabricated by hot working processes and may be cold worked to a limited extent only. The castings are available in die-cast and sand-cast forms and are useful for pump-bodies, pickling cradles and gears, gear selector forks, non-sparking tools and fittings for marine service. The resistance to impingement attack and cavitation erosion makes the alloys suitable for propellers. The wear and abrasion resisting properties of the aluminium bronzes are outstanding.

In certain acidic solutions aluminium bronzes may be attacked in a preferential manner, the aluminium being removed from the eutectoid—called *de-aluminification*. This trouble is absent in another modification of aluminium bronze called Superston 40, containing 12 Mn, 8 Al, 2 Ni, 3 Fe, an α–β alloy with a tensile strength of 695 N/mm² and 40 Joules Izod. Strength can be further increased by raising the aluminium content. Easier

TABLE 62  STRESSES CORRESPONDING TO DEFINITE CREEP RATES, IN STRAIN UNITS PER DAY (VOCE)* FOR ALUMINIUM, TIN AND SILICON BRONZE

| Percentage Composition | | | Condition | Temperature of Test, °C | Stresses N/mm² | |
|---|---|---|---|---|---|---|
| | | | | | 5th day | 40th day |
| Al | Fe | Ni | | | $10^{-4}$ per day | $10^{-4}$ per day |
| 10·06 | — | — | Sand cast | 250 | 124 | — |
| 10·47 | 0·01 | — | 1 hour, 825 °C, AC | 250 | 139 | 162 |
| 10·47 | 0·01 | — | 4 weeks at 250 °C | 250 | 153 | 185 |
| 10·47 | 0·01 | — | 3 days, 500 °C, SC | 250 | 171 | 205 |
| 10·47 | 0·01 | — | 3 days, 500 °C, SC | 400 | c. 32 | 42 |
| 9·7 | 3·4 | — | Sand cast | 250 | 147 | — |
| 10·13 | 2·8 | — | 1 hour, 825 °C, cooled in air | 250 | 150 | 190 |
| 10·13 | 2·8 | — | 4 weeks at 250 °C | 250 | 156 | 195 |
| 10·13 | 2·8 | — | 3 days 500 °C, SC | 400 | 56 | 60 |
| 9·5 | 5·4 | 4·8 | Sand cast | 250 | 198 | — |
| 9·86 | 4·97 | 4·87 | Extruded | 250 | 202 | 236 |
| 9·86 | 4·97 | 4·87 | Extruded | 400 | 46 | 57 |
| Sn | Zn | P | | | | |
| 9·90 | — | 0·03 | Sand cast | 250 | 100 | 90 |
| 9·92 | 2·05 | 0·01 | Sand cast | 250 | 99 | 97 |
| Si | Mn | Fe | | | | |
| 3·02 | 0·96 | 0·025 | CDA 1 hour, 575 °C | 250 | — | 145 |
| 3·02 | 0·96 | 0·025 | CDA 1 hour, 575 °C | 400 | — | 83 |

AC = air cool   SC = slow cool   CDA = cold drawn and annealed
*Metallurgia*, November 1946

casting properties enable thinner sections to be designed than is possible with aluminium bronze. This alloy is also a cheap substitute for beryllium copper tools.

## High-temperature properties

Aluminium is often beneficial in preventing oxidation at elevated temperatures, but superheated steam contaminated with sulphur dioxide or chlorine attacks the surface layers of aluminium bronzes at elevated temperatures.* Preferential oxidation of aluminium to alumina occurs and the oxide remains embedded in a copper-rich matrix. The reaction is accompanied by a volume increase of more than 9% which leads to rupture of the surface layers. Under similar treatment cast 60/40 brass was not attacked. Under most conditions, however, aluminium bronzes possess a high resistance to oxidation.

For service at elevated temperatures the transformation of $\beta$ into $\alpha + \gamma_2$ is not detrimental, but actually improves creep properties. Tempering at 500 °C for 3 days reduces the initial high creep and is preferable to normal cooling which gives an $\alpha + \beta$ structure in spite of the fact that the tensile strength, ductility and notch bar values at room temperature are inferior. Creep properties of aluminium bronzes and other copper alloys are given in Table 62, from which it will be seen that the tin bronzes are much inferior to the aluminium bronzes and, while the silicon alloy appeared superior at 400 °C, it suffered from large initial extensions.

## Precipitation hardened alloys (see also Chap. 14)

*Copper–beryllium* Striking results are obtained by temper hardening an alloy containing 1·8% beryllium with up to 0·5% cobalt or nickel. Quenching from 800 °C produces a soft supersaturated solution which can be hardened by cold work, while temper hardening at 300–320 °C is possible on both the soft and the work-hardened alloy.

The exceptionally high tensile and fatigue strength is associated with electrical conductivity values between 15 and 32% of copper. The high elastic limit with an elastic modulus of roughly only two-thirds that of steel

TABLE 63   PROPERTIES OF 2% BERYLLIUM–COBALT–COPPER SHEET

|  | LP | TS | El |
|---|---|---|---|
| Ann 800 °C, Quenched | 4·6/71 | 31/479 | 45 |
| Do. and tempered 300 °C, 3 hr | 40/618 | 78/1205 | 6·3 |
| Quenched 800 °C, cold rolled and tempered 300 °C, 1 hr. | 48/741 | 86/1328 | 2 |

*Metallurgia*, July 1946

allows a large deflection for a given stress in springs, Bourdon tubes, corrugated diaphragms and flexible bellows. It has a high resistance to wear, corrosion and also to softening up to about 250 °C. It is used for hammers and other tools when non-magnetic and non-sparking properties are desired, and its high cost is warranted; also dies for castings.

## Cu Zn Ni Al alloys

The addition of nickel and aluminium to brass in suitable proportions yields alloys which possess quite marked age-hardening properties. These alloys are softened by quenching from 800–850 °C and hardened by tempering at 350–500 °C.

TABLE 64
Cu, 72·5; Zn, ZO; Ni, 6; Al, 1·5%

| Condition | VPN | 0·1 PS | TS | El |
|---|---|---|---|---|
| Quenched 850 °C | 80 | 85 | 340 | 60 |
| Do. and tempered 500 °C, 2 hr | 190 | 448 | 618 | 17 |
| Cold dawn (25%) | 165 | 370 | 494 | 30 |
| Do. and tempered 500 °C, 2 hr | 245 | 649 | 695 | 4 |

### Silicon bronzes

The silicon bronzes contain 1–5% silicon with small additions of manganese, iron and zinc. They possess excellent corrosion-resisting properties and for this reason are favoured for chemical plant construction. The electrical and thermal conductivity is low and this assists in welding operation. Particulars of a typical alloy are: Everdur, Cu 96, Si 3, Mn 1 with a strength of 355 N/mm$^2$ and 50% elongation.

### High-resistance alloy

In addition to 40% nickel alloy (p. 312) a copper alloy containing 10% manganese and 2% aluminium has a high electrical resistance (41 microhms/cm$^2$/cm), the temperature coefficient of which is nearly zero over the range 20° to 350 °C. In the annealed state it is ductile and may be severely cold worked to give a tensile strength of 772 N/mm$^2$. It has a good resistance to corrosion and its electrical properties do not change on long heating within the range indicated. It is therefore useful as a resistance material operating at a moderately elevated temperature.

*Copper–iron–aluminium alloy*

An alloy containing 45/45/10 Cu–Fe–Al has a cast duplex structure of dendrites of iron-rich $B_1$ phase in a matrix of copper-rich α-solid solution (~ 50/50). Extrusion and cold working produces a fibrous structure with good strength and ductility, e.g. 1080–1160 N/mm² and 2–8% elongation. The properties can be controlled by the aluminium content increasing from 4·5% to a point where the γ-phase of the binary Cu–Al system causes brittleness (~ 12%). The alloy has potential uses for corrosion resistant fasteners for the building and general engineering industries.

# 16 Aluminium, Magnesium and Light Alloys

Aluminium has five principal characteristic properties of interest to engineers:

(1) Lightness; its specific gravity is only 2·7 as compared with 7·8 for steel and 8·8 for copper. For this reason the metal is used extensively in the form of alloys, for aerospace and automobile parts for structures of all types. In the last case the extreme lightness of the alloys enables a much greater volume to be used for a given weight, with a resultant increase in rigidity. For high speed reciprocating parts, such as pistons and connecting-rods, the use of aluminium alloys leads to a better balance, reduced friction, lower bearing loads and results in a considerable saving in the power required to overcome the inertia of the moving parts.

(2) Electrical conductivity is about 60% of that of copper, but weight for weight it is a better conductor than copper and is used for overhead cables. Sometimes a thin steel core is necessary for extra strength.

(3) High thermal conductivity, e.g. heat exchanger components. In the case of pistons, the high conductivity enables higher compression ratios to be used.

(4) Its resistance to corrosion is made use of in chemical plant for handling concentrated nitric acid, and food industries for pans and hollow ware; while aluminium foil is used in packaging and to seal bottles; alloys are used for marine and building purposes. Aluminium paint is also widely used.

(5) It has a great affinity for oxygen, a property which enables the metal to be used as a deoxidant in steels; and as 'thermit' (intimate mixture of iron oxide and powdered aluminium), for welding rails, ships' sterns and for the manufacture of hardener alloys such as ferro-titanium.

Aluminium is non-magnetic, has a low neutron absorption and isotopes with a very low half life. It is used as a matrix to carry boron carbide for thermal neutron shielding applications.

Aluminium alloys have a coefficient of linear expansion which is about twice that of cast iron or steel and allowance must be made for this when aluminium alloy pistons work in a cast-iron cylinder or when light alloy cylinder heads are secured by steel bolts. This problem can be reduced by the use of hypereutectic aluminium–silicon alloys which have coefficients of thermal expansion similar to those of the highly alloyed steels. Young's modulus is 70–75 $\times$ 10$^3$ N/mm$^2$ and sections need to be deeper for equal deflection compared with steel.

The corrosion resistance of aluminium and most of its alloys is high. The various grades of pure aluminium are the most resistant, followed closely by the aluminium–magnesium and aluminium–manganese alloys. Next in order are aluminium–magnesium–silicon and aluminium–silicon alloys. The alloys containing copper are the least resistant to corrosion; but this can be improved, in sheet or plate form, by coating each side of the copper containing alloy core with a thin layer of commercial purity aluminium (or special alloys), thus giving a three-ply metal. With thin clad sheets in particular, solution heat-treatment times should be reduced to a minimum to avoid the diffusion of alloying elements from the core to the surface of the coating.

Contact between aluminium alloys and other metals in the presence of moisture may set up electrolytic corrosion. Aluminium alloys corrode rapidly when in moist contact with copper and its alloys, less rapidly in contact with ferrous materials, but it is sacrificially protected by zinc, cadmium and magnesium. Hence, zinc and cadmium plated fixings are often employed in contact with aluminium alloy components. Unless ingress of moisture can be prevented, aluminium alloys should be insulated from other metals by barrier layers such as barium chromate with a slow drying oil varnish medium (see DTD 369A) or suitable fibrous packing.

Mercury-containing compositions attack aluminium and its alloys severely.

Aluminium can be worked either hot or cold, but after recrystallisation no twins are evident such as are typical of copper and its alloys.

Aluminium owes its resistance to corrosion to the natural film of oxide which covers its surface, but for applications a greater resistance is required and this is obtained by artificially thickening the oxide film by dipping in a solution of chromates or preferably by an electrolytic process called *Anodic treatment*. This consists in subjecting the clean article to intensive oxidation, as the anode in a sulphuric or chromic acid bath. The sulphuric acid process is simple and is commonly used. The freshly formed film is porous and requires 'sealing' for maximum protection and this may be accomplished by immersion in boiling water or in dichromate, lanoline or silicone containing solutions. The porous film can be coloured by organic or inorganic pigments for decorative purposes and the article can be chemically or electrochemically brightened before anodising for 'bright trim' applications, e.g. automobile hub caps, wheel trims etc.

A hard thick anodic film is used for many engineering applications for improving the wear resistance (Ministry of Defence Specification DEF 151).

*Extraction.* Aluminium is made by the electrolysis of alumina dissolved in molten cryolite. Major purifying treatment has to be carried out on the ore but metal of high purity (99·99% pure) can now be obtained by a further special electrolytic refining process. The main impurities are iron and silicon.

*Forms of material available*

Aluminium and its alloys are available in a variety of cast and wrought forms and conditions of heat treatment. The common wrought forms are forgings, solid, hollow and tube extrusions, sheet, plate, strip, foil and wire, whereas castings are available in sand, pressure and gravity die cast forms. In the post war years, the rapidly expanding market for aluminium alloys in the non-aerospace field has resulted in an increasing range of alloys and complexity of shapes to meet the more specialised requirements of users. Similarly, in the aircraft field, there has been a rapid technological advance in design and performance for supersonic military jet aircraft and commercial aircraft such as the Concorde SST. Thus, the technical developments of existing alloys, of new alloys and manufacturing techniques have had to keep pace with the more stringent service operating requirements.

The materials available are supplied in the main to British Standard Specifications which control the chemical compositions, minimum guaranteed tensile strengths, and release testing procedures. The alloys for commercial usage are covered by General Engineering Specification BS 1470–75 for wrought products and BS 1490 for cast materials. For special applications, such as for bright trim and free machining alloys, Supplementary Specifications are issued, e.g. BS 4300 weld tube. Aerospace alloys are covered by either BSL or DTD Specifications but although the materials may not differ in some cases to those supplied to the corresponding BSGE specifications, the test release requirements covered by BS 2 L100 are considerably more stringent. Material may be supplied, also, under the particular manufacturer's alloy designation, e.g. Hiduminium 48, a weldable high strength Al–Zn–Mg alloy where there is no covering BS Specification.

In the revised British Specifications the familiar W and WP designations for solution treated and artificially aged material are now TB and TF (BSGE specifications only) and the temper designations $\frac{1}{2}$H or H for half and fully hard rolled material are now H4 and H8, the full new range of tempers ranging from H1–H8, representing a scale of increasing strength. A complete list of the various designations is given in BS 1470–75 and 1490 (ingots and castings).

## Manufacturing techniques

*Melting.* Aluminium and its alloys, whether eventually to be used for castings or wrought products are melted in a wide variety of electrically heated, gas or oil-fired furnaces, the selection being dependent on quality and economic considerations. The majority of the alloying elements are added by means of key or hardener alloys. The melts are treated at various stages with proprietary fluxes appropriate to the type of alloy being processed. Aluminium alloys readily absorb hydrogen and the presence of water vapour in the furnace atmosphere is particularly detrimental. Accordingly, the melt is degassed with nitrogen, chlorine, or tablets containing hexachlorethane, etc. Treatment with chlorine is the most efficient and has the additional advantage of cleaning the melt of oxide films. The cleanliness aspect is particularly important in the case of wrought products, and great care is taken to remove the oxide inclusions. Many methods have been developed for doing this including fluxing, filtering, filter-degassing and chlorine treatment. Finally, the melts are grain refined with proprietary preparations, many of which contain titanium boride.

In the case of material for wrought products, the melts are nowadays invariably semi-continuously cast into rolling slab or billets for forging or extrusion. This casting technique is preferred due to the uniformity, grain size and general quality of the product.

*Castings* are made in sand moulds, shell moulds and permanent metal moulds filled under gravity or pressure. The economic number of castings increases in this order. Design can have a material effect in the case of manufacture of castings and sections as shown in Fig. 220.

*Wrought forms* are made by hot rolling from 400–500 °C to 250–300 °C and cold rolling is used for thin materials and foils. Extrusion at 400–500 °C produces a variety of shapes for direct use or for drawing or forging.

## Metallurgical considerations

### Non-alloyed aluminium

The various grades of aluminium which are commercially available in the various wrought forms are 99·99, 99·8, 99·5 and 99% purity. Ranges of tensile properties are given in Table 65.

TABLE 65 PROPERTIES OF 99·5% PURITY ALUMINIUM SHEET
(LONG TRANSVERSE VALUES)

| Temper | | TS $N/mm^2$ | | El % | BH |
|---|---|---|---|---|---|
| | | min | max | min | |
| Soft | (0) | 55 | 95 | 22–32 | 21 |
| Half Hard | (H4) | 100 | 135 | 4–8 | 32 |
| Hard | (H8) | 135 | — | 3–4 | 39 |

As distinct from more conventional alloying, commercially pure aluminium can be strengthened by dispersion hardening with fine flake aluminium oxide (6·5–13% oxide). The fine dispersion of aluminium oxide produces a high resistance to recrystallisation and has the effect of restricting the reduction in tensile properties and creep resistance as the temperature is raised to 550 °C. In addition, the mechanical properties return to the original when the soaking temperature is reduced to normal. This material which was marketed as SAP (Hiduminium 100) found only limited application and is no longer commercially available.

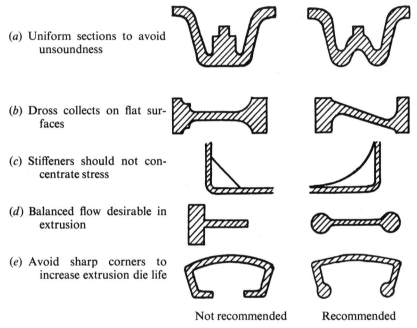

(a) Uniform sections to avoid unsoundness

(b) Dross collects on flat surfaces

(c) Stiffeners should not concentrate stress

(d) Balanced flow desirable in extrusion

(e) Avoid sharp corners to increase extrusion die life

Not recommended     Recommended

FIG. 220 Effect of design on ease of manufacture of aluminium alloy castings (a, b, c) and extruded sections (d, e)

*Aluminium alloys*

For engineering purposes numerous alloys have been developed which possess improved properties. The major alloying elements are: Cu, Mg, Zn, Mn, Si, Ni, Fe and Ti although Co, Sb, Zr, Ag, etc., are sometimes added for special purpose alloys.

The various alloys may be best considered under the two classifications of cast and wrought, the latter can again be subdivided into heat-treatable and work-hardened materials. The heat-treatment of aluminium alloys consists generally of three stages.

(i) *Solution treatment.* The alloy is heated to the maximum practicable temperature (450 and 550 °C) below that at which the particular low melting point phases will liquate (called 'overheating') and maintained there for a sufficient time to ensure complete solution of the appropriate solute elements.

(ii) *Quenching.* The solution treated material is then cooled as quickly as possible to retain the solute elements in super saturated solid solution. Quenching into water at ambient temperature is most commonly employed but this can induce internal stress in particular in complex shaped forgings or extrusions which may give rise to distortion on machining or introduce problems during service. In this case, quenching into boiling water, oil or molten salt is used. Modifying the cooling rate can have profound effects on certain critical mechanical properties in certain alloys. In the case of extrusions, rolled products and some regular shaped forgings, stretching (1·5–2·5%) is often applied after quenching to relieve the residual stresses. Thermal stress relief treatments cannot be applied to heat-treatable alloys for the components, would have to be heated to too high a temperature which would cause appreciable loss in strength.

(iii) *Ageing.* After quenching, certain alloys will harden significantly if left at ambient temperature (natural ageing). For higher strength, however, the alloys are heated for predetermined times at specially chosen temperatures, usually between 110° and 215 °C.

*Wrought alloys—work-hardenable*

These alloys meet the demand for stress-bearing panels used for bodies in both land and air transport. They are stronger than pure aluminium and in the appropriate temper they have suitable properties for forming without the complication and expense of hardening by heat-treatment. Extra hardness can be developed only by cold deformation; softening occurs at 350–400 °C. Details of typical alloys are given in Table 66.

All the alloys given can be welded (see p. 413). The aluminium–magnesium alloys possess a high resistance to corrosion, especially to sea-water and marine atmospheres, and their chief use is where this property is of value. They sometimes contain up to 1% manganese and up to 0·5% chromium.

About 15% magnesium is soluble in aluminium at 450 °C, and about 1·5% at room temperature. Consequently in commercial alloys a considerable portion of the magnesium is precipitated as β-phase as separate islands when formed at 300 °C, but as grain boundary films when formed by heating at 70–100 °C. These β-films render the 5 and 7% magnesium alloys susceptible to intercrystalline corrosion, which is accentuated by stress or cold work (see also p. 414). The resistance to corrosion of the material can be restored by reheating, followed by a controlled rate of cooling.

TABLE 66   WROUGHT ALLOYS—NOT HEAT-TREATABLE (SHEET FORM— LONG TRANSVERSE VALUES)

| BSGE Material Designation | Alloy Type | Temper | 0.2% PS N/mm² min | TS N/mm² min | TS N/mm² max | El % on 50 mm |
|---|---|---|---|---|---|---|
| NS3 | 1¼% Mn | 0 | — | 90 | 130 | 20–25 |
|  | 1¼% Mn | H4 | — | 140 | 175 | 3–7 |
|  |  | H8 | — | 175 | — | 2–4 |
| NS4 | 2¼% Mg | 0 | 60 | 160 | 200 | 2–0 |
|  | 2¼% Mg | H6 | 175 | 225 | 275 | 3–5 |
| NS5 | 3½% Mg | 0 | 85 | 215 | 275 | 12–18 |
|  | 3½% Mg | H4 | 225 | 275 | 325 | 4–6 |
| NS8 | 5% Mg | 0 | 125 | 275 | 350 | 12–16 |
|  | 5% Mg | H2 | 235 | 310 | 375 | 5–10 |

## Wrought alloys – heat-treatable

*The aluminium–magnesium silicide alloys*

This is one of the most commonly used families of aluminium alloys. The alloys have medium strength, good formability, satisfactory corrosion resistance and adequate weldability. They are used for most applications except where elevated temperature service is not required where there are more suitable aluminium alloys.

These alloys contain small quantities of magnesium and silicon, the amounts and proportions are deliberately different in the various alloys to promote the most important required service criterion. Sometimes they are present in the correct proportions to give the intermetallic compound $Mg_2Si$ and sometimes to give a slight excess of either silicon or magnesium.

The alloys are used for a wide variety of applications including architectural sections, structural sections, container bodies, roadway balustrading, railway bogies, etc.

They are used in the naturally aged or fully heat-treated conditions, the latter consisting of solution treatment at 520–540 °C, cold water quenching and artificial ageing at 160–180 °C.

There is one version of the alloy (E91E) in which the impurity levels are carefully controlled and which has an electrical conductivity only about 10% lower than pure aluminium, hence its use for overhead electric conductors and other electrical equipment.

The alloy most commonly used for structural applications (H30) has the composition Al 0.7%, Mg 1.0% Si, 0.6% Mn and properties are given below:

| H30 | 0.2% PS N/mm² min | TS N/mm² min | El % |
|---|---|---|---|
| Extruded Bars and sections fully HT (TF) (up to 150 mm) | 270 | 310 | 8 |
| Extruded only solution T (TB) (up to 150 mm) | 120 | 190 | 16 |
| Drawn Tube fully HT (TF) (to 6 mm) | 255 | 310 | 7 |

The age-hardening of Al–Cu–Mg–Si alloys is due to both copper and magnesium (see Figs. 62, 63). The main phases which precipitate during ageing are $CuAl_2$ and $CuAl_2Mg$ but in the correctly heat-treated alloy they are so fine as not to be visible under the optical microscope. A typical structure of the fully heat-treated alloy is shown in Fig. 221. $Mg_2Si$ is present in the micro-structure but this is mainly not dissolved during the solution treatment at 500 °C and therefore takes no part in the subsequent ageing. The quench sensitivity of the alloy is due to the copper in solution at 500 °C tending to precipitate during quenching on the alpha (AlMnFeSi) intermetallic phase already present thus denuding the solid solution for subsequent ageing. Accordingly, the slower the rate of cooling the less copper will be available for age hardening.

Selected minimum guaranteed tensile properties for the H15 and BSL97 alloys are given in Table 67.

TABLE 67  PROPERTIES OF H15 4·3 Cu, 0·6 Mg, 0·8 Si, 0·75 Mn, L97 4·6 Cu, 1·3 Mg, 0·25 Si, 0·7 Mn

| Form and Condition | Specification | Size Range mm | Test Direction | 0·2% PS N/mm² | TS N/mm² | El % |
|---|---|---|---|---|---|---|
| Extruded TB | HE15 | 20–75 | Long | 250 | 390 | 11 |
| Extruded TF | HE15 | 20–75 | Long | 435 | 480 | 8 |
| Unclad Plate —TF | BSL97 | 6–12·5 | Long Trans | 280 | 430 | 10 |
| | | 25–40 | Long | 310 | 430 | 10 |
| | | | LT | 280 | 420 | 4 |
| | | | Short Trans | 260 | 380 | 35 |

*Aluminium–zinc–magnesium–copper alloys*

These have the highest strength of all aluminium alloys and are used mainly in aircraft structural applications in the form of clad sheet, forgings and extrusions. Strengths of the order of 600 N/mm² are achieved by certain of the alloys.

The alloys may be divided into those containing manganese alone and those with both manganese and chromium. One advantage of the chromium free alloys is that complex shaped die forgings can be quenched into boiling water after solution treatment to produce a low level of internal stress but the presence of chromium renders the alloy highly quench sensitive so necessitating more rapid quenching. However, the disadvantage of the cold water quenching is overcome in the form of extruded sections or rolled products by stretching before ageing to effect stress relief. The compositions and minimum properties of typical widely used chromium free (DTD 5024) and chromium containing (DTD 5074) alloys are given in Table 68.

TABLE 68

|  | $Zn\%$ | $Mg\%$ | $Cu\%$ | $Mn\%$ | $Cr\%$ |
|---|---|---|---|---|---|
| DTD5024 | 5·5 | 2·8 | 0·45 | 0·5 | — |
| DTD5074 | 5·8 | 2·5 | 1·4 | 0·15 | 0·2 |

| Form and Material | Test Sample | Size mm | 0·2% PS N/mm² | TS N/mm² | El % |
|---|---|---|---|---|---|
| Forgings made from extruded stock DTD5024 | m/c from extruded stock | Up to 20 | 450 | 490 | 6 |
|  |  | 20–150 | 480 | 540 | 6 |
|  |  | 150–250 | 430 | 490 | 4 |
| Extruded Bar DTD5074 | m/c from extruded bar | Up to 20 | 480 | 540 | 4 |
|  |  | 20–150 | 520 | 585 | 4 |

These alloys are solution treated at 460–470 °C as liquation can occur on heating rapidly to temperatures above 480 °C. They are usually artificially aged at 110–135 °C. A micro-structure typical of forged fully heat-treated DTD 5024 is given in Fig. 222. The age hardening phase is $MgZn_2$ but it is too fine to be resolved by optical microscopy.

The early history with these alloys presented problems due to stress corrosion, susceptibility, but nowadays treatments are available which minimise this difficulty either by quenching after solution treatment in molten salt at 180 °C or performing a second ageing at about 160–180 °C.

*The aluminium–copper–magnesium–silicon alloys*

This type of alloy although being practically the first heat-treatable alloy to be discovered (duralumin) still finds wide application for many general engineering and aircraft structural purposes in the form of forgings, extruded bars and sections, sheet, plate, tube and rivets. These alloys possess considerably higher strength than the aluminium–magnesium silicide alloys but have a much lower corrosion resistance, due to the high copper content. In the case of sheet the latter problem is overcome by cladding on both sides with 99·5% purity aluminium. The alloys have good fracture toughness and find usage for service at temperatures up to about 120 °C.

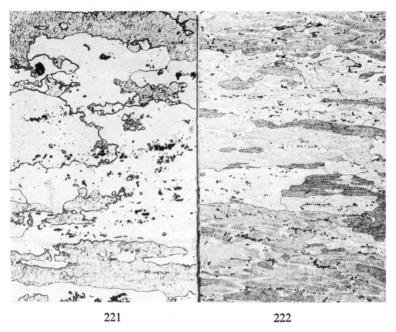

221 222

FIG. 221 Forged, fully heat-treated BS L97 etched Villela's reagent (× 100)
FIG. 222 Forged, fully heat-treated DTD 5024 etched in dilute acid (HCl, HNO$_3$, HF) (× 100)

Two of the commonly used alloys in this family are H15—Al 4·3%, Cu 0·6%, Mg 0·8% Si 0·75%, Mn and BS L97—Al 4·6%, Cu 1·3%, Mg 0·25%, Si 0·7% Mn.

The Al–Cu–Mg–Si–Mn alloy billets and slabs are hot worked at 400–470 °C to disperse the eutectic, break up the hard constituents and improve the properties, notably ductility. At temperatures significantly above 470 °C some of the alloys in this group become hot short; the BS L97 alloy has a much greater susceptibility in this respect than H15. Below 300 °C, the alloys are too brittle to work in the cast condition. After hot working, these alloys can be cold worked to an appreciable extent. They can be softened by annealing at 360 °C. The alloys show signs of overheating at about 510 °C and consequently are solution treated at 495–505 °C. If gross liquation occurs this results in a deterioration in mechanical properties. These materials are quench-sensitive and must be quenched, following solution treatment, into water at less than 40 °C, to attain the required level of mechanical properties on subsequent artificial ageing.

The alloy may be used in the naturally aged condition for after quenching the hardness and strength increase, rapidly at first and then more slowly up to a maximum value after about five days at ambient temperature. However, the alloy is more commonly used in the artificially aged condition—165 to 185 °C—when it is much stronger.

*Special purpose alloys*

A large number of complex alloys are available for special applications such as creep resistance, high-speed automatic machining, welding etc.

*Aluminium casting alloys*

As the majority of the aluminium casting alloys are used for all the various casting processes and several find service in both the as-cast and heat-treated conditions, it is most convenient to consider them in compositional groups rather than by casting processes. Details are given in Table 69.

The pressure die-casting alloys are not heat-treated as they blister badly during solution treatment. This is due to the air which is entrapped during casting, expanding and distorting the weak hot metal.

Some of the gravity castings are given an age hardening treatment without a separate solution treatment to provide a level of strength between that of the as-cast and fully heat-treated states. This procedure has an economic advantage over full heat-treatment and is possible due to a certain proportion of the precipitating elements being retained in solid solution by the rapid chilling in the mould. In other cases, where complete dimensional stability on machining is required some of the casting alloys are given a stabilising treatment at about 250 °C after solution treatment and quenching.

*Aluminium–silicon and aluminium–silicon–copper alloys*

The aluminium–silicon series of alloys is the most widely used for the production of all types of casting due to the excellent fluidity and casting characteristics of the molten metal.

In this alloy system, a eutectic is formed at $11 \cdot 7\%$ silicon (Fig. 223). One of the most used commercial alloys, LM6, is of approximately eutectic composition but others contain less silicon. As cast the properties are not particularly good due mainly to the coarse nature of the eutectic and any primary silicon present. The structure can be refined and the properties vastly improved by adding a 'modifying' agent to the melt e.g. metallic sodium (0·005–0·015%). This treatment produces a marked refinement of the eutectic and appears to displace the eutectic as shown in Fig. 223. The effect is equivalent to super-cooling the alloy. Fig. 228 shows the structure of an alloy sand cast normally consisting of coarse eutectic and primary silicon and Fig. 229 the effect of modification in producing dendrites of aluminium solid solution and very fine eutectic. Remelting tends to destroy the modification effect and further treatment is necessary.

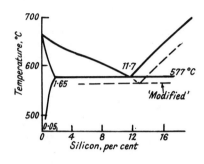

Fig. 223 Aluminium–silicon equilibrium diagram

The eutectic alloy LM6 (Al–12% Si) is not heat-treatable and has a good resistance to corrosion, although not as good as the 5% Mg alloy. In the molten state, it has a high degree of fluidity and its low shrinkage on solidification enables castings of intricate sections to be made dense and free from cracks. The alloy is also lighter than the Al–Cu alloys and is widely used for automobile castings, including water-cooled manifolds and jackets, motor housing etc. and for pump parts in the chemical and dye industries and for many domestic castings.

Other popular alloys are LM2 (10% Si–1½% Cu), LM24 (8½% Si–3½% Cu) and LM4 (5% Si–3% Cu) in which the silicon content is reduced and various amounts of copper are added. Reducing the silicon reduces the fluidity slightly and eventually at about 8% silicon modification is no longer essential. Adding the copper increases the strength and machinability but lowers the castability, corrosion resistance and ductility. Therefore, the alloys have been developed to give specific compromises of all these characteristics and economics. The composition of LM4 was originally formulated to permit a high proportion of secondary aluminium alloy to be used in its manufacture and it is used for vehicle castings, instrument cases and electrical equipment.

The addition of copper makes the aluminium–silicon alloys responsive to heat-treatment but the latter is only carried out in the case of LM4 as LM2 and LM24 are used for pressure die-casting.

Small amounts of magnesium are added to the aluminium–silicon alloys, for example, LM8 (Al–5·5%, Si–0·6% Mg) and LM25 (Al–7%, Si–0·3% Mg). These alloys are also heat-treatable and have similar strengths to the Al–Si–Cu alloys but have superior castability and corrosion resistance. However, they need tighter compositional control and secondary metal

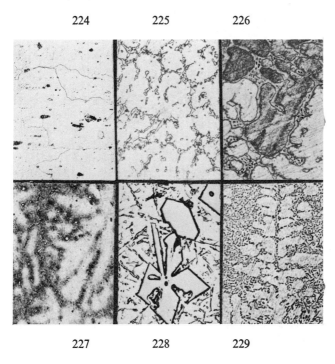

Aluminium alloys, etched—2% hydrofluoric acid

FIG. 224 Heat-treated duralumin, longitudinal. Solid solution with $FeAl_3$ (dark) and $CuAl_2$ or $MnAl_3$ (light)

FIG. 225 Heat-treated Y-alloy: particles of $CuAl_2$ and $NiAl_3$

FIG. 226 Cast 12% copper alloy: cored dendrites and eutectic of aluminium solid solution and $CuAl_2$

FIG. 227 Cast 2L5. Cu 2; Zn 13. Cored crystals with particles of $CuAl_2$ ($\times$ 100)

FIG. 228 13% silicon alloy cast: primary silicon and coarse eutectic Al + Si ($\times$ 100)

FIG. 229 Same as Fig. 228, but 'modified' with sodium. Primary aluminium and fine eutectic ($\times$ 100)

additions are not so easy to accommodate. LM8 is gradually being replaced by LM25 and these alloys are used for applications where corrosion resistance and good mechanical properties are service requirements.

A complex heat-treatable eutectic aluminium–silicon alloy which is used for elevated temperature applications where a low coefficient of thermal expansion and good bearing properties are essential, e.g. pistons, is LM13 (Al–12%, Si–1·6%, Ni–1%, Cu–1% Mg). Hypereutectic alloys with 20–25% Si have been investigated extensively due to their attractive characteristic of even lower thermal coefficients of thermal expansion but have found no great application as yet.

*Aluminium–magnesium alloys*

The cast aluminium–magnesium alloy, LM5 (Al–4·5%, Mg–0·5% Mn) is highly resistant to corrosion, anodises to give a good clear film, machines well and takes a high polish. It is superior to all other alloys in marine environment and hence finds particular application for moderately stressed marine craft fittings.

LM5 and the higher strength, heat-treatable alloy LM10 (Al–10% Mg) present some difficulty in the foundry due to:

(1) molten metal is readily oxidised (small beryllium additions are made to lessen this effect).
(2) degassing is difficult and the degassed metal picks up hydrogen on standing, which can lead to porosity in the castings unless particular care and control are exercised.

LM10 is used for applications where corrosion resistance is important and where high strength and good impact resistance are required. The alloy is usually solution treated at 425–435 °C and quenched into hot oil and used in this condition where it has a high degree of ductility. However, it naturally ages with time which means that the strength increases and the ductility progressively decreases. This ageing is accelerated under tropical service and the alloy can become somewhat embrittled and show a reduced stress corrosion resistance. These characteristics are largely avoided by reducing the magnesium content to 8% and adding $1\frac{1}{2}$% zinc which minimises the natural ageing propensity of the alloy, thus conferring a high degree of stability in these two properties. This modified alloy is covered by DTD5018.

The precipitating phase in LM10 is beta ($Mg_2Al_3$) and during exposure at tropical temperatures and above, this precipitates at the grain boundaries with detrimental effects on the mechanical properties. The addition of zinc modifies the composition of the phase, $Mg_3Zn_3Al_2$, and the mode and distribution of the precipitate.

## Aluminium–copper alloys

Aluminium–copper alloys form a eutectic containing 33% copper. Only one alloy, LM11, is now used to any extent and its usage is likely to decrease in future. LM11 contains 4·5% copper and is heat-treatable. It is mainly used for aircraft castings and for other highly stressed applications due to its good mechanical properties and shock resistance. The alloy has only a moderate corrosion resistance and its foundry characteristics are poor as it is very susceptible to hot tearing.

## Complex special purpose alloys

### Y-alloy (LM14)

Y-alloy obtains its name from the symbol used to identify the alloy in experiments at the National Physical Laboratory during the First World War. Its composition is Al–4%, Cu–2%, Ni–1·5% Mg.

Its special characteristic is its ability to retain a good strength at relatively high temperatures, a property which makes it an excellent piston material. It has a fairly good corrosion resistance. It is solution treated at 500–520°C, quenched into boiling water and naturally aged or precipitated at 100°C. For piston use, it is given a stabilising age at 200–250°C.

The micro-structure is rather complex and contains $NiAl_3$, Al–Cu–Ni complex, $Mg_2Si$, in addition to iron bearing constituents and a very small amount of $CuAl_2$ (Fig. 225).

### Hiduminium–RR350

Another complex casting alloy which was developed for elevated temperature usage is known commercially as Hiduminium–RR350. This material shows an advance on Y-alloy and has higher mechanical properties, notably creep resistance, at temperatures above 300°C for short times and the superiority is apparent at lower temperatures for extended service. It finds usage in aeroengine and other elevated temperature applications.

It is solution treated at 545°C, boiling water quenched and aged at 215°C. The composition is Al–5%, Cu–1½% Ni with small additions of Mn, Ti, Sb, Co and Zr.

### Premium quality castings

The minimum mechanical properties in the BS1490 series of casting alloy specifications apply to separately cast test bars and are not necessarily obtainable in actual castings. The demand for guaranteed properties in castings, initiated in the United States, has led to the concept of premium quality castings, made in sand or permanent moulds, for which minimum properties are specified throughout or in designated locations in castings. The very strict control on the founding of such castings, which must begin at the design stage, naturally results in higher production costs, but substitution for even more expensive forgings or parts fabricated from wrought products may be possible. The only alloy to be recognised commercially in the United Kingdom in this respect to date, is the aluminium—7% Si–0·3% Mg alloy, covered by DTD5028 and designated A356 in the United States. This is a version of LM25 in which the iron content is limited to 0·15% and the range of magnesium is 0·20 to 0·45%. In practice, even closer control of composition, combined with adequate melt cleaning, degassing, modification and temperature control is usually necessary to meet the specification requirements.

TABLE 69  ALUMINIUM CASTING ALLOYS

| BS 1490 LM | | Minimum Tensile Properties Test Bar | | | | Form Used |
|---|---|---|---|---|---|---|
| (M = as cast TB = sol treated only TF = fully HT) | | Tensile Strength N/mm² | | Elongation % | | S = Sand Casting PM = Permanent Mould PD = Pressure Die Cast |
| | | Sand | Chill | Sand | Chill | |
| 10 Si 1·5 Cu | 2M | — | 145 | — | — | PD |
| 5 Si 3 Cu | 4M | 140 | 150 | 2 | 2 | All forms |
| | 4TF | 235 | 275 | — | — | All forms |
| 45 Mg 0·5 Mn | 5M | 140 | 185 | 3 | 5 | S and PM |
| 12 Si | 6M | 160 | 185 | 5 | 7 | All forms |
| 5·5 Si 0·6 Mg | 8M TB | 160 | 235 | 2·5 | 5 | S and PM |
| | 8M TF | 235 | 275 | — | 2 | S and PM |
| 10 Mg | 10 TB | 275 | 310 | 8 | 12 | S and PM |
| | DTD 5018 | 275 | 310 | 5 | 10 | S and PM |
| 4·5 Cu | 11 TB | 215 | 260 | 7 | 13 | S |
| | 11 TF | 275 | 310 | 4 | 9 | S |
| 12 Si 1·6 Ni/Cu/Mg | 13 TF | 170 | 275 | — | — | S and PM |
| 4 Cu 2 Ni/1½ Mg | 14 TF | 215 | 275 | — | — | S and PM |
| 8½ Si 3½ Cu | 24 M | — | 175 | — | 1·5 | PD |
| 7 Si 0·3 Mg | 25 M | 125 | 160 | 2 | 3 | S and PM |
| | 25 TB | 160 | 235 | 2·5 | 5 | S and PM |
| | 25 TF | 235 | 275 | — | 2 | S and PM |
| | DTD 5028 TF | — | 275 | — | 5 | S and PM |

Note: DTD 5028—Properties guaranteed at designated locations in castings:
 0·2% PS—195 N/mm², TS 260 N/mm², 5% Elongation.
—Properties guaranteed in undesignated locations in castings:
 0·2% PS—175 N/mm², TS 225/N/mm², 3% Elongation.

## Mechanical properties

*Tensile properties at elevated temperature*

A number of alloys have been developed for service at high temperatures and Hiduminium–RR58 which is used in the form of sheet, plate, extrusions and forgings on the Concorde, is a typical example. Hiduminium–RR58 is a complex alloy containing Al–2·5%, Cu–1·5%, Mg–1·10%, Ni–1·1% Fe with small Mn and Ti additions. It is important to appreciate the effect of increasing test temperature on the strength of these alloys (Table 70).

TABLE 70  ELEVATED TEMPERATURE TENSILE TESTS ON FORGED BAR IN THREE ALUMINIUM ALLOYS AFTER SOAKING FOR 1000 HOURS AT THE TEST TEMPERATURE

| Alloy and Specification | Test Temperature °C | 0·2% PS N/mm$^2$ | TS N/mm$^2$ | Elongation % |
|---|---|---|---|---|
| Hiduminium–RR 58 to DTD 731 | Room Temperature | 375 | 455 | 9 |
| | 100 | 355 | 425 | 10 |
| | 150 | 335 | 385 | 15 |
| | 200 | 220 | 250 | 21 |
| Hiduminium 66 to BSL 77 | Room Temperature | 440 | 495 | 12 |
| | 100 | 430 | 475 | 12 |
| | 150 | 400 | 420 | 17 |
| | 200 | 180 | 235 | 29 |
| Hiduminium–RR 77 to DTD 5024 | Room Temperature | 460 | 520 | 12 |
| | 100 | 430 | 480 | 18 |
| | 150 | 235 | 295 | 28 |
| | 200 | 100 | 110 | 45 |

At sub-zero temperatures, the aluminium alloys do not show a sudden transition from ductile to brittle fracture at temperatures below −40°C.

## Fatigue strength

Unlike steels, aluminium alloys do not exhibit a clearly defined fatigue limit (see p. 25). It is customary, therefore, to quote an endurance limit (EL) which can range between $\pm 125$ to $\pm 215$ N/mm² at $10^8$ cycles.

For design purposes account must be taken of notch sensitivity. The effect of various stress concentrations may be assessed by carrying out tests on specimens containing grooves, notches or holes which are designed to produce a range of theoretical stress concentration factors ($K_T$). The notch sensitivity, q, may be obtained from the formula $q = \dfrac{K_F - 1}{K_T - 1}$ where $K_F = \dfrac{\text{(EL for plain specimens)}}{\text{(EL for notched specimens)}}$. If the stress concentration is fully effective then $K_F = K_T$ and $q = 1$ for highly notch sensitive material. If there is no difference between EL for plain and notched specimens then $K_F = 1$ and $q = 0$ (Table 71).

TABLE 71   ROTATING CANTILEVER FATIGUE TESTS ON FORGED BAR IN TWO ALUMINIUM ALLOYS ON PLAIN AND NOTCHED SPECIMENS ($K_T = 3 \cdot 0$)

| Alloy and Specification | Type of Specimen | Estimated Endurance Limit ($\pm$N/mm²) for: | | | |
|---|---|---|---|---|---|
| | | $10^5$ cycles | $10^6$ cycles | $10^7$ cycles | $10^8$ cycles |
| Hiduminium 66 to BSL 77 | Plain | 295 | 230 | 175 | 155 |
| | Notched | 150 | 110 | 85 | 70 |
| Hiduminium-RR 77 to DTD 5024 | Plain | 295 | 240 | 210 | 185 |
| | Notched | 130 | 95 | 75 | 70 |

At $10^8$ cycles for BSL 77  $K_F = 2 \cdot 2$ ∴ notch sensitivity value, q, = 0·7
and for DTD 5024,  $K_F = 2 \cdot 7$ ∴ notch sensitivity value, q, = 1·0

Where a component is designed for service under fatigue conditions, tests on the actual part should be made, if at all possible, and sharp corners or sudden changes in section should be avoided.

*Creep resistance.* Under elevated temperature service conditions, creep resistance is sometimes of prime importance. Although some tests are of short durations, in some cases it is necessary to determine strains produced in very long times (up to 100 000 hours in the case of generating plant and up to 30 000 hours for Concorde).

When selecting materials, it is vital to consider results for the form of alloy to be used since there are many factors such as grain size, heat treatment and cold work which have a pronounced effect upon creep resisting properties (Table 72).

TABLE 72  COMPARISON OF CREEP TEST RESULTS FOR HIDUMINIUM-RR 58 IN THE FORM OF CLAD SHEET AND FORGED BAR

| Test Temp. °C | Form | Estimated Stresses N/mm² to Produce 0·1% Total Plastic Strain in: | | |
|---|---|---|---|---|
| | | 100 hours | 1000 hours | 5000 hours |
| 100 | Clad Sheet | 285 | 260 | 235 |
| | Forged Bar | 320 | 305 | 295 |
| 150 | Clad Sheet | 215 | 175 | 135 |
| | Forged Bar | 255 | 230 | 195 |
| 200 | Clad Sheet | 110 | 55 | 20 |
| | Forged Bar | 165 | 125 | 75 |

*Fracture toughness.* The $K_{IC}$ values vary from 150 to 500 N/mm²$\sqrt{\text{cm}}$, depending on composition, form and the direction of test. It should be borne in mind that a high value for $K_{IC}$ does not necessarily mean that one material is superior to another unless the level of proof stress is considered since the quantity $\frac{(K_{IC})}{(\eta y)}$ reflects the performance of an alloy in terms of the crack length which can be tolerated before catastrophic failure.

*Stress corrosion resistance.* Stress corrosion cracking, (p. 120) may be an occasional problem with the high strength aluminium alloys such as are used extensively, for example, in the aircraft industry, e.g. DTD 5024 and BSL 77. The corrosive environment need not be severe, the normal atmosphere being sufficient in some cases depending on the stress levels but the problem is more likely to be encountered in marine environments. The resistance of a component is lowest in the short transverse direction, i.e. at right angles to the direction of grain flow.

Many of the alloys can be rendered less susceptible, or immune, to stress corrosion cracking by special ageing heat treatments which usually take the form of a treatment at two temperatures, e.g. DTD 5104. Variation in the rate of quenching from the solution heat-treatment temperature also has a profound influence upon the behaviour of the alloys, and one of the most successful techniques employed to confer immunity on an Al–Zn–Mg–Cu (DTD 5024) alloy is to quench from the solution heat-treatment temperature of 460°C, into molten salt at 180°C, following this with a single stage ageing treatment (DTD 5094). Slow quenching rates have an added advantage in that the residual stresses left in the material are lower than those resulting from quenching into cold water.

In addition to these techniques, the risk of stress corrosion cracking can be eliminated by insulating the material from the environment by anodising and painting.

## Ancillary processing techniques

*The joining of aluminium.* (For solders and brazing alloys see Chap. 20.) Alloys suitable for brazing include commercial purity Al, the Al–Mg–Si alloys, Al–1¼% Mn and Al–Mg alloys containing less than 2% Mg.

### Welding

Welded aluminium has found applications over a wide field including pressure vessels, pipelines, furniture, containers, armoured vehicles, aerospace and the shipbuilding industry (see Chap. 20).

Inert gas shielded arc welding has eliminated the need for flux and provides the concentrated heat source to offset the high thermal conductivity of aluminium. It has largely replaced the older metal arc and gas welding processes. The most commonly used filler materials are Al–5% Mg (NG6) and Al–5% Si (NG21).

The heat treatable wrought alloys inevitably lose strength on welding, the weld bead often being stronger than the adjacent heat affected zone. Heat-treatment after welding will improve properties but this is often impractical. Therefore, the loss in strength should be allowed for in design. The Al–Zn–Mg alloys are unique in this series in that, after welding, they continue to increase in strength at room temperature, producing the strongest as welded aluminium alloy joints, for example, Hiduminium 48 containing 4·5% Zn, 2·5% Mg and small additions of Mn, Zr and Ti.

*Finishing.* Many types of surface finish either decorative or protective can be applied to aluminium and its alloys depending on the surface condition of the original product and the end use involved. Mechanical processes such as scurfing, grinding, polishing, scratch brushing, blasting and barrelling can be used to produce an infinite variety of textures. Chemical etching is used frequently to give certain desired effects and freshly formed anodic films can be coloured by immersion in dye baths. Certain aluminium–silicon alloys give attractive grey colours when anodised in sulphuric acid, and by the use of other more complex electrolytes, colours in the gold, brown and black range can be obtained on a number of aluminium alloys without the use of dyestuffs, and are used for architectural applications. Relatively thick and 'hard' anodic films to enhance the wear resistance of aluminium alloys for engineering purposes are obtained usually by anodising at low temperatures and high current densities in sulphuric acid or sulphuric/oxalic acid electrolytes.

High purity aluminium and specially developed aluminium alloys, usually of the aluminium–magnesium type, can be given a high degree of specular reflectivity by chemical or electrochemical treatment in suitable solution. The highly reflective surface is made permanent by the subsequent application of a thin transparent anodic film. Bright trim on motor cars and domestic appliances are typical applications of this type of finish.

Aluminium and its alloys can be electroplated with a variety of metals for decorative or engineering purposes. After suitable pre-treatment, thick or 'hard' chromium plate can be applied directly to reduce the coefficient of friction and improve the wear resistance of aluminium alloy surfaces. Thick coatings of nickel can also be deposited chemically and are useful where inaccessible parts require to be coated.

Another form of surface treatment called Chemical Conversion coating, is a process whereby a film integral with aluminium alloy is produced by immersion or spraying with chemical solutions. The many present-day proprietary solutions may contain chromates, fluorides or phosphates or mixtures thereof, and the film formed gives a measure of protection from corrosion to the alloy, but the main application of conversion coatings is to provide an essentially inert keying surface to which paint or other organic coatings can be applied.

*Forming*

Pressings and deep drawn parts require generous radii for the initial draws and interstage annealing is applied if necessary.

## Magnesium and its alloys

There are two commercially successful methods for the production of magnesium:

(1) Electrolysis of fused $MgCl_2$. This uses anhydrous $MgCl_2$ prepared by the direct chlorination of magnesium oxide, the latter obtained by benefication of oxide or carbonate ores or from sea-water.

(2) Direct reduction of the oxide or carbonate by means of ferro–silicon at a temperature (1200 °C) above the vaporisation point of magnesium and under vacuum. The condensate is in the form of powder or small-sized crystals and requires melting for conversion to a saleable form.

Magnesium is the lightest commercial metal, its specific gravity (1·74) being only two-thirds that of aluminium. The metal itself is not sufficiently strong to permit its use in engineering applications. The properties are, however, improved by suitable alloying additions and the alloys in use find extensive applications in many engineering fields. The casting alloys have properties similar to those of the aluminium base casting alloys but the best magnesium base wrought alloys are not as strong as the high strength aluminium type. In strength to weight ratio magnesium castings are superior to those in aluminium alloys. The wrought magnesium alloys can profitably be used:

(*a*) for saving weight without sacrifice of stiffness, or
(*b*) for increasing stiffness without exceeding a given weight.

A further advantage of all magnesium alloys is their very great machinability. Certain simple precautions are essential to eliminate fire-risk, but these are well understood, and in turning and milling practice speeds of 300 metres per minute are in use.

In the molten state magnesium reacts with oxygen and nitrogen and to prevent burning during the melting of the metal or its alloys, specially developed fluxes are used—a fluid flux of low melting point being often used for the prevention of oxidation and local burning before melting is complete, followed by an 'inspissated' flux of stiffer consistency, capable of refining the melt by absorbing the fluid protective flux and oxide and nitride inclusions, and also of preventing surface oxidation during the later stages of the melting cycle. To prevent reaction between the molten alloy and the moisture in moulding sands, 'inhibitors'—sulphur, boric acid, ammonium bifluoride, etc.—are essential ingredients of the sand.

Magnesium has an hexagonal lattice structure and at temperatures below about 220 °C deformation by slip is possible only on the basal hexagonal planes, and this tends to limit the capacity for cold work. Above this temperature slip on 12 pyramidal planes is also possible and the metal and its alloys can be readily hot worked. The usual temperature range for alloys is 260–320 °C.

Three main groups of alloys are available. The binary magnesium–manganese alloy is suitable for all sheet-forming processes and is readily welded by argon arc processes. It possesses good corrosion resistance but is now obsolescent.

The magnesium–aluminium–zinc alloys (8–9% type) are suitable for sand, gravity and pressure die castings and also for extrusion (6% type) or press forging (8% type) or sheet (3% type). The cast alloys are amenable to heat-treatment and in the solution heat-treated state the strength and ductility are greatly increased (Table 73). These alloys are grain refined with consequent enhancement of the mechanical properties by carbon inoculation or superheating at 900 °C during melting cycle.

The third and latest group of alloy employs the intense grain refining effects of zirconium. These alloys are available in both cast and wrought forms, and in high strength and high temperature types. The high strength casting alloys contain zinc, sometimes with minor additions of rare earth metals or thorium and provide higher proof stress combined with better ductility and greater soundness. The high temperature casting alloys contain rare earth metals or thorium and are pressure tight and creep resistant to about 250 and 350 °C, respectively. The wrought zirconium alloys have higher tensile properties and are more readily worked than the other alloys. They are also more resistant to loss of properties on hot forming. Wrought alloys containing thorium, in some cases with manganese in place of zirconium, have been recently developed. For increased resistance to carbon dioxide Magnox containing 0·002–0·03% beryllium and 0·8 aluminium was developed for canning uranium at Calder Hall.

| Nominal Composition | | | Code | 0.1 PS N/mm² | TS N/mm² | FE N/mm² | El % | BH | Izod ft lbf | Izod J | Form | Use |
|---|---|---|---|---|---|---|---|---|---|---|---|---|
| Zn | Zr | | | | | | | | | | | |
| **Casting alloys** | | | | | | | | | | | | |
| 4.5 | 0.7 | | Z5Z | 130 | 230 | ±77 | 5 | 70 | 3 | 4 | HT | High strength alloy, useful up to 150°C |
| 4.0 | 0.7 | 1.2 RE | RZ5 | 120 | 200 | ±88 | 3 | 70 | | | HT | Pressure tight and crack resistant |
| 5.5 | 0.7 | 1.8 Th | TZ6 | 140 | 255 | ±77 | 5 | 70 | | | HT | Strong, castable, pressure tight |
| 2.2 | 0.6 | 2.7 RE | ZRE1 | 77 | 140 | ±70 | 3 | 55 | 1 | 1 | A | Castable, weldable, pressure tight, use up to 250°C |
| 2.2 | 0.7 | 3 Th | ZT1 | 77 | 185 | | 5 | 55 | 2 | 3 | HT | Creep resistant to 350°C |
| | 0.7 | 1.7 RE 2.5 Ag | MSR | 155 | 240 | ±108 | 4 | 80 | | | HT | Highest strength casting, used up to 250°C, pressure tight, weldable |
| 0.5 | 0.3 Mn | 8 Al | A8 | 70 / 70 | 140 / 200 | ±81 / ±84 | 2 / 6 | 55 / 55 | 1 / 4 | 1 / 5 | C / HT | General purpose alloy, good founding, ductility, strength and shock resistance |
| **Wrought alloys** | | | | | | | | | | | | |
| 3 | 0.6 | | ZW3 | 210 | 300 | ±120 | 8 | 70 | 20 | 27 | | High strength sheet extrusion and forging alloy, no stress corrosion |
| 1.3 | 0.6 | | ZW1 | 170 | 260 | ±117 | 8 | 70 | 16 | 19 | | Sheet extrusions, weldable, no stress corrosion |
| 5.5 | 0.6 | | ZW6 | 200 | 300 | ±90 | 10 | 73 | 18 | 24 | | High strength extrusion and forging alloy, not weldable |
| 2.0 | | 1 Mn | ZM21 | 135 | 230 | | 10 | 60 | | | | Med strength sheet and extrusion alloy, easily formed and welded, no stress corrosion |
| 6.0 | | 1 Mn | ZM61 | 300 | 315 | ±120 | 6 | | | | HT | High strength extrusion alloy |
| 1 | | 6 Al 0.3 Mn | AZM | 140 | 260 | ±120 | 7 | 65 | 15 | 20 | | General purpose alloy, gas and arc weldable, Gas and argon weldable, low strength general purpose alloy, good corrosion resistance |
| | | 1.5 Mn | AM 503 | 110 | 200 | ±75 | 5 | 50 | 8 | 11 | | |

RE = rare earth metals   Th = thorium   C = cast   HT = Heat-treated   A = annealed

In general none of the alloys required high temperature solution treatments and full properties are developed by a simple precipitation treatment or a preliminary anneal at operating temperature.

## Design

Notches and other stress raisers should be reduced to a minimum in view of the rather low impact values of some of the alloys.

The modulus of elasticity of magnesium alloys is only about 45 $KN/mm^2$ and consequently in comparison with similar parts in steel or aluminium it is necessary to stiffen sections if deflections are to remain constant.

This is preferably done by deepening and widening the general form of the casting in order to maintain reasonably *thin* walls. Stiffening ribs should have large fillets; flange faces should be generous in area and a large number of small bolts used to distribute the stress.

## Corrosion

All magnesium alloys are roughly as resistant to atmospheric attack as unprotected mild steel. The corrosion resistance of the Mg–Al alloy is improved, however, if the iron and nickel contents of the metal are reduced to a minimum and high purity alloys are now available. Magnesium is resistant to alkalis but is generally attacked by acids and chloride solutions.

Magnesium parts are frequently used in the unprotected state or with only a chemically produced chromate film. Cold acid chromate or black chrome manganese baths are frequently used. These are:

(a) *The acid chromate bath.*
   20–25 parts by volume nitric acid.
   100 parts by volume cold water.
   15% of the mixture by weight, potassium dichromate.

Immerse article for 10–120 seconds; wash in warm water.

(b) *The chrome manganese bath* (2 hr cold or 15 min, 70–80 °C).
   sodium or patassium dichromate         10% w/v
   crystalline manganese sulphate          5% w/v
   magnesium sulphate                      5% w/v

Treatment in the chromate baths (details set out in DTD 911) may be followed by the application of a zinc chromate primer and finally a suitable paint finish.

Hard anodising by the HAE or Dow 17 processes may also be carried out. MEL fluoride anodising followed by surface sealing with an epoxy resin and conventional painting gives maximum protection against corrosion.

# 17 Titanium and other New Metals

The development of supersonic flight and of nuclear energy is stimulating a demand for metals which were only laboratory curiosities before the Second World War. Consequently titanium, zirconium, beryllium, niobium and other metals are rapidly becoming available for the engineer, but developments are occurring faster than can be published in text books.

**Titanium**

Because of the difficulty of reducing the oxide, titanium is commonly made by reducing the tetrachloride with magnesium or sodium at a temperature of 800–900 °C under argon. The reduced sponge is formed into a consumable electrode which is arc melted in a water-cooled or liquid metal-cooled copper crucible maintained under a vacuum. Such expensive melting procedure is necessary because the molten metal reacts rapidly with refractories and absorbs oxygen, nitrogen and hydrogen which harden and completely embrittle the metal. Fortunately, at hot working temperatures, the diffusion of nitrogen and oxygen in titanium is relatively slow and, by careful procedure, these impurities can be confined to a thin surface layer which must be removed (Fig. 230). Hydrogen is also maintained below 0·003 % but furnace atmospheres and pickling can readily increase the hydrogen content which can penetrate deeply. The effect of hydrogen on the structure is shown in Fig. 233. The tendency of titanium to seizure or galling causes difficulty in extrusion, cold drawing and machining, but surface treatments such as anodising or coating with copper minimise the trouble. Rolling textures are easily developed, and their effects on mechanical properties are marked. With the preferred orientation in a tube, for example, the normal method of accommodating plastic deformation by twinning is only possible in compression as is shown by Fig. 231 from compression and Fig. 232 from tension side of a bent tube.

*Properties*

Titanium has a low density, is non-magnetic and its electrical and thermal conductivity and specific heat are near to those of stainless steel (Table 75). Titanium has proved somewhat disappointing as regards high temperature creep properties, but new alloys are showing considerable improvement.

Up to 882 °C titanium has a hexagonal structure—known as α—while

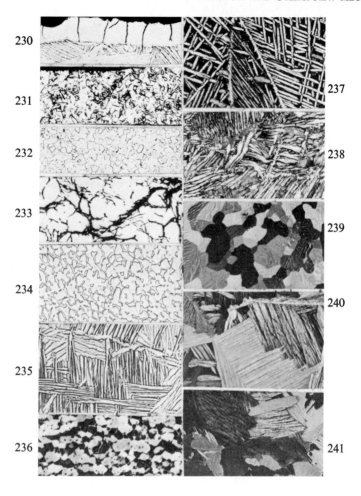

FIG. 230 Cracked brittle oxygen diffusion layer on surface of IMI alloy, Ti 317
FIG. 231 Twinning in compression side of bent tube ($\times$ 100)
FIG. 232 No twinning in tension side of bent tube ($\times$ 100)
FIG. 233 1 Al, 1 Mn alloy containing 0·012% hydrogen, annealed 600°C 1 hr shows $\alpha + \beta$ structure with hydride as branching network ($\times$ 500)
FIG. 234 4 Al, 4 Mn alloy (314A) annealed within $\alpha + \beta$ field at 800°C fine grained polygonal $\alpha + \beta$ ($\times$ 500)
FIG. 235 As Fig. 234 but fabricated in $\beta$ field 1050°C. Coarse basket weave structure ($\times$ 500)
FIG. 236 IMI alloy 679, rod worked in $\alpha + \beta$ field. Fine grained $\alpha + \beta +$ complex
FIG. 237 IMI alloy 679, worked in $\beta$ field. Widmansträtten $\alpha + \beta$ structure

# TITANIUM

above this temperature it changes to a body-centred cubic form known as β, which is decidedly more easily worked than α—titanium. Alloying elements affect this α–β change point and they can be classified roughly into four groups shown in Fig. 242.

(a) Elements completely soluble in both α and β, e.g. zirconium.
(b) Elements completely soluble in β but with limited solubility in α, e.g. molybdenum, niobium and vanadium.
(c) Elements with very limited solubility in α but extensive solubility in β and a eutectoid formation, e.g. iron, chromium, manganese.
(d) Elements partially soluble in α and β but with greater solubility in α, raising the change point and known as α stabilisers, e.g. aluminium and tin, also oxygen and nitrogen.

Beta stabilising elements, forming Groups b and c alloys, cause the α–β change to be lowered and spread over a wide range of temperature. In practice the difference between group b and c is not great because the eutectoid constituents in group c are generally slow to form and are suppressed when alloys cool at normal rates. The alloys then behave as if the boundaries of the α + β range continued unbroken to the lowest temperatures.

The behaviour of titanium alloys in passing through the β → α change has certain features in common with that of steel passing through the γ → α change, i.e. martensitic and pearlitic types.

A sample *rapidly* cooled from the beta field changes to alpha by an almost instantaneous reaction which produces a martensitic-like structure shown in Fig. 241. The alpha is supersaturated with the alloy addition because of the diffusionless change and is often referred to as α' or alpha prime. Unlike steel, however, the martensite is not hard and of little practical value because the solute is substitutional instead of interstitial (as carbon) and far less distortion is produced. With increasing additions of alloying elements the temperature of the transformation falls and increasing amounts of β are retained at room temperature. Minimum additions for complete retention of beta phase after quenching are 6·5 Mn, 4 Fe, 8 Cr, 7 Co, 8 Ni, 13 Cu. The retained β is sometimes not very stable.

---

FIG. 238  5 Al, 2½ Sn (IMI Ti 317 or RCA–110AT) alloy
  As rolled.  Distorted basket weave α structure  (× 100)
FIG. 239  As Fig. 238 but annealed 900 °C, 1 hr. AC Equiaxed α
FIG. 240  As Fig. 238 but annealed 1100 °C (i.e. β field) AC. Coarse basket weave alpha transformed from beta
FIG. 241  As Fig. 238 but annealed 1100 °C water quenched martensitic α with some transformed α at grain boundaries

Tempering at 350–400 °C sometimes causes β to form a transition phase called ω, which causes excessive brittleness. Sufficient ageing time is essential to form α, thus β → ω + β → α + β.

If alloys are cooled *slowly* from the β range, α separates from β by a nucleation and growth mechanism, forming a grain boundary network of α with transcrystalline lamellae within the grain (Figs. 235, 240). The residual β may be retained or may change to martensite-like form of α.

In practice the α + β titanium alloys are generally solution treated and aged. The simpler alloys, for example Ti–6Al–4V are also used after simple annealing to relieve stresses and to form small-rounded islands of β. Heating above the β transformation temperature often results in loss in ductility and unless carefully controlled produces an undesirable coarse α + β Widmanstätten structure (Figs. 235 and 237). Good properties are obtained by forging the material in the temperature range in which fine α separates from β (as Figs. 234 and 236).

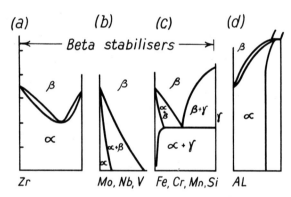

FIG. 242 Types of equilibrium diagrams for titanium alloys

The above discussion applies to β stabilising elements, but aluminium stabilises α, decreases density and improves properties and resistance to oxidation at elevated temperatures. Thus there are three main groups of alloy using combinations of alloying addition: (*a*) the all-alpha alloys are strong and maintain their strength at high temperatures but are difficult to fabricate, (*b*) the all-beta alloys which are less strong, easier to work and rather unstable at elevated temperatures but little used commercially yet, (*c*) the alpha–beta alloys with intermediate properties. The Ti–2½Cu alloy (IMI Ti 230) is in a class of its own. It has an alpha matrix hardened by a fine dispersed intermetallic compound and can be fabricated in the solution treated condition and then aged.

Commercial titanium of varying strengths is obtained by selection of raw material with varying impurity content. Titanium alloys are used for

TABLE 74 TYPICAL PROPERTIES OF TITANIUM AND ITS ALLOYS

| Alloy | | °C β→α | Treatment °C | 0.1% PS N/mm² | TS N/mm² | El | RA | Uses |
|---|---|---|---|---|---|---|---|---|
| Commercial Ti | IMI 115 | 890 | SR 300–500 | 232 | 417 | 30 | 50 | WSF |
| | IMI 130 | 890 | An 650–700 | 309 | 525 | 28 | | WSF |
| | IMI 155 | 890 | An | 540 | 695 | 25 | 50 | WSF |
| 2½Cu | IMI 230 | 895 | An 805 | 463 | 618 | 25 | 40 | WSF |
| | | | ST 805 WQ + Aged 8 hr | 587 | 740 | 15 | 35 | |
| | | | 400 + 8 hr 475 | | | | | |
| 5Al–2½Sn | IMI 317 | 1025 | An 800–900 | 772 | 880 | 16 | 32 | CW |
| 6Al–4V | IMI 318 | 995 | An 700 | 957 | 988 | 15 | 45 | WF |
| | | | ST 845–960 WQ + Aged 2 hr 480–600 | 1019 | 1143 | 14 | 45 | |
| 4Al–4Mo–2Sn–0·5Si | IMI 550 | 975 | ST 900 AC + Aged 24 hr 500 | 1000 | 1158 | 10 | 25 | F |
| 11Sn–2¼Al–5Zr–1Mo–0·2Si | IMI 679 | 950 | ST 900 AC + Aged 24 hr 500 | 1000 | 1112 | 18 | 40 | C |
| 11Sn–2¼Al–4Mo–0·2Si | IMI 680 | 935 | ST 915 AC + 700 AC* | 957 | 1158 | 19 | 42 | F |
| | | | ST 805–840 AC + 24 hr 500 | 1189 | 1313 | 15 | 40 | |
| 6Al–5Zr–0·5Mo–0·25Si | IMI 685 | 1025 | ST 1050 OQ + 24 hr 550 | 865 | 988 | 10 | 25 | C |

Young's Modulus = 106–124 kN/mm²  
Applications: W = Welding and forming, F = Forging, C = Creep, S = Sheet.  
Treatments: An = Anneal, SR = Stress relieve, ST = Solution treatment, AC = Air-cool, WQ = Water quench, OQ = Oil quench.

*This treatment gives improved fracture toughness.

compressor blades, discs, casings, engine forgings, bolts and in the chemical industry. The micro-structures of various titanium alloys are shown in Figs. 234–241. *Corrosion resistance* of titanium is remarkably good especially in sea-water and against crevice attack. Further, the repair or growth of the protective oxide film in a wide range of media can be stimulated by connecting the titanium to a positive source of direct current. In this way titanium may be suitable for very difficult chemical conditions.

## Metals for nuclear energy

Before dealing individually with the various metals used in the nuclear field it will be helpful to outline a few of the essential principles involved.

In Chap. 5 it was stated that the chemical properties of an element are closely related to the number of its valency electrons. Many elements, however, consist of a mixture of *isotopes*, i.e. each atom having the same number of valency electrons, hence same chemical properties, with a nucleus containing the same number of protons but different numbers of neutrons and therefore different atomic weights. Iron has four isotopes, containing 26 protons and 28, 30, 31 and 32 neutrons giving masses of 54, 56, 57, 58, written $Fe^{54}$, $Fe^{56}$, etc.

Natural uranium consists essentially of two isotopes, $U^{238}$, 99·3% and $U^{235}$, 0·7%. Isotope $U^{235}$ is fissile, that is if it captures another neutron it can split into two roughly equal fission particles, with the liberation of 2–3 neutrons and considerable heat. Under suitable conditions these newly released neutrons can produce a chain reaction.

### *Moderator*

These newly liberated neutrons possess a very high speed (2MeV kinetic energy) and the probability of being captured and producing more fission is unlikely unless the fuel is greatly enriched in $U^{235}$. It is necessary therefore to slow down the neutrons by repeated collision with atoms of low atomic weight such as hydrogen (as $H_2O$), deuterium (as heavy water $D_2O$) and carbon (graphite)—called *moderators*. The slow speed neutron has an energy equal to the thermal vibrations of the atoms and the reactor using such a moderator is called a *thermal reactor*. The slow neutron can be captured (1) by $U^{235}$ with fission, (2) by the moderator and (3) by $U^{238}$ to form plutonium $Pu^{239}$, or it may escape from the boundary of the reactor (Fig. 243(*a*)).

### *Fuel*

A concentrated fuel, such as *plutonium*, can be used without a moderator in a so-called *fast reactor*, e.g. at Dounreay. Plutonium is very reactive,

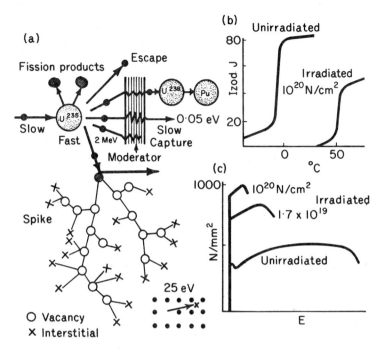

FIG. 243 (a) Neutron fission and reactions taking place with fast neutrons; escape capture, retardation in moderator and knock damage or spike of radiation damage. MeV = million electron volts; (b) effect of radiation on Izod; (c) effect of radiation on tensile curves for steel (ASTM A212B)

easily oxidised and highly toxic. It exists in six allotropic forms, the low temperature phases being very complex.

Uranium is very reactive, easily oxidised and exists in three allotropic forms. The alpha phase which is stable up to 760 °C is anisotropic (Fig. 244). Uranium is used as a fuel in most thermal reactors, but it has three disadvantages, growth, wrinkling and swelling.

Under neutron bombardment (below 400 °C) a single crystal of uranium lengthens [010] direction, shortens in the [100] direction and remains unchanged in the [001] direction. A rod rolled at a low temperature has a high [010] preferred orientation and therefore increases greatly in length on irradiation. This is known as *radiation growth*. A coarse random grained rod becomes increasingly *wrinkled* (150–200 °C), therefore it is necessary to use a rod with a random arrangement of fine crystals. *Thermal cycling growth* is similar and occurs when a uranium rod, but not a single crystal, is repeatedly heated and cooled (15–600 °C) (Fig. 245). Cycling through the α–β phase change can also lead to serious alterations of shape and density.

*Swelling* is another trouble experienced when uranium is irradiated above

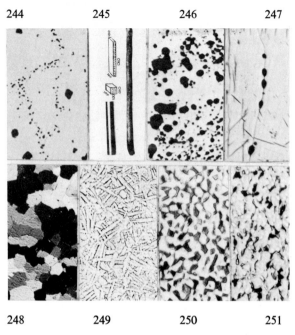

FIG. 244 Uranium showing carbide inclusions and networks of $U_6Fe$ and $UAl_2$
FIG. 245 Growth of uranium caused by thermal cycling or neutron bombardment
FIG. 246 Rare gas porosity in irradiated uranium
FIG. 247 Cavitation of Magnox strained 0·8% in 1000 hrs at 200°C ($\times$ 100)
FIG. 248 Zircaloy 2 or IMI Zirconium 20 cold rolled sheet annealed 750°C 30 mins. Polarised light $\times$ 200 shows fine precipitate of intermetallic compounds
FIG. 249 Zircaloy 2 or IMI Zirconium 20 rod extruded at 1000°C. Acicular ($\times$ 100)
FIG. 250 Zirconium 30 rod heated and water quenched. Alpha (white) and decomposed beta ($\times$ 500)
FIG. 251 Zirconium 30 annealed 810°C 30 min. alpha matrix with light grey areas of decomposed β and black intermetallic compounds of $Zr_2Cu$, and $ZrMo_3$

1000°C and is due to the formation of bubbles in the lattice of gaseous fission products, e.g. xenon and krypton (Fig. 246).

Some of these difficulties can be minimised by the use of uranium oxides, carbides and carbide or oxide cermets, which are isotopic and retain their hardness at much higher temperatures.

Thorium is another possible fuel and is free from phase changes below 1480 °C.

*Fuel canning materials*

The can positions the fuel correctly in the reactor, prevents corrosion by the coolant media and prevents escape of radioactive fission products. The canning material must have certain properties as follows:

(1) It must not absorb many neutrons, i.e. have a low neutron absorption cross-section, measured in Barns.

| Material | Cd | B | Hf | Ag | Ta | W | Cr | Ti |
|---|---|---|---|---|---|---|---|---|
| Barns | 2550 | 755 | 105 | 62 | 21 | 19·2 | 7·9 | 5·6 |
| Material | Ni | Mo | Nb | Al | Zr | Mg | Be | C |
| Barns | 4·6 | 2·5 | 1·1 | 0·23 | 0·18 | 0·06 | 0·01 | 0·003 |

(2) It must be compatible with both fuel and coolant. Aluminium, for example, reacts with uranium at temperatures as low as 200 °C, forming $UAl_4$ and $UAl_3$ compounds, and a diffusion barrier is necessary. Resistance to alloying with solid uranium and attack by molten uranium decreases in order W Ta Ni Zr Ti Mo.
(3) The material must have a reasonable strength at service temperature to prevent distortion and swelling of the fuel, or sufficient ductility to accommodate changes in dimensions of the fuel without failure.
(4) It must have reasonable thermal conductivity.

For thermal reactors aluminium, magnesium (p. 347), zirconium and beryllium or their alloys are preferred because of their low neutron absorption cross-section. However, Civil AGR cans are 0·015 in thick double vacuum-melted 20 Cr, 25 Ni, 1 Nb steel. At low rates of strain Magnox (p. 348) shows a tendency to cavitation, i.e. formation of holes at the grain boundaries. These lead to reduced ductility and to leak paths (Fig. 247).

*Zirconium*

The extraction, fabrication and general metallurgy of zirconium are somewhat similar to those of titanium. Zirconium minerals contain 0·5–2% hafnium which is a strong absorber of neutrons and must therefore be removed. Oxygen and nitrogen diffuses more rapidly in zirconium than in titanium, and this increases the difficulties in heat-treatment and scale removal. The main use of zirconium has been for cladding fuel elements and for structural components in water-cooled systems, hence attention has been paid to increasing its corrosion resistance. Traces of nitrogen, carbon, aluminium and titanium appear to promote breakaway corrosion (p. 143). For use in water-cooled reactors some of the harmful effects of these impurities are counteracted in Zircaloy 2 containing 1·5 Zn, 0·1 Fe, 0·05 Ni, 0·1 Cr, typical structures of which are shown in Figs. 248, 249. Zirconium has a relatively poor resistance to $CO_2$ at elevated temperatures, but improvement is obtained by the addition of 0·5 Cu, 0·5 Mo with an increase in tensile strength to 510 $N/mm^2$ and improved creep resistance at 450°C. This alloy, IMI Zirconium 30, useful in gas cooled reactors. (For weldability see p. 414.)

*Beryllium*

The cast metal is usually coarse grained and brittle and powder metallurgy techniques are employed to produce billets for working. Special safeguards must be taken to prevent toxic hazards, especially inhalation of fine dust.

Sheet material has a highly developed texture wherein the basal planes are close to the plane of the sheet. Such material can exhibit 30–40% elongation under conditions of *uniaxial* strain in any direction in the plane of the sheet. In the thickness direction the ductility is low and a *biaxial* tension leads to fracture. Alloying has not been successful so far in improving ductility because of the ready formation of compounds. Beryllium has sufficient isotropic ductility at 485–600°C, to permit hot forming.

At temperatures over about 760°C pure beryllium oxidises fairly rapidly and the metal is usually encased in mild steel for the working operations. It is corroded by traces of water in $CO_2$.

*Niobium* has a high melting point and good strength, ductility and corrosion resistance especially to liquid sodium coolants and excellent compatibility with uranium. Its oxidation resistance above 400°C is indifferent, but is greatly improved by alloying. Impurities such as oxygen, nitrogen, hydrogen and carbon have a detrimental effect on workability.

*Control material*

The chain reaction is controlled by inserting high neutron absorber rods such as boron, hafnium silver and cadmium. Calder Hall control rods consist of boron steels with a critical amount of aluminium to allow hot working, and sheathed in a low cobalt (0·001 %) austenitic stainless steel to prevent corrosion. Typical compositions are:

|  | C | Si | Mn | Al | B |
|---|---|---|---|---|---|
| Extruded tubes | 0·1 | 0·5 | 0·28 | 0·66 | 3·8 |
| Cast tubes | 0·12 | 0·6 | 0·31 | 0·29 | 5·0 |

*Pressure vessel*

The reactor core is contained in a steel vessel which will stand up to the pressure, temperature and irradiation. The steel used must (1) not be susceptible to brittle fracture during erection (p. 15); (2) have sufficient creep strength; (3) be compatible with coolant; (4) be weldable. So far aluminium-killed fine-grained steel has been used possessing a low ductile-brittle transition.

*Irradiation effects*

Neutron irradiation raises the ductile-brittle transition temperatures considerably. The effect of a given amount of irradiation increases as the temperature is raised from 15 to about 180 °C, and decreases to a negligible amount at 300 °C. In metals and alloys irradiation also produces an increase in hardness, tensile strength and a marked increase in yield strength, but it reduces work-hardening capability to the point where necking occurs with practically no uniform elongation. Annealed metals exhibit larger property changes than those in the initially cold worked state. Irradiation often acts like 'stirring' and promotes equilibrium or homogeneity in metastable alloys.

Radiation hardening can be eliminated by annealing at 300–450 °C for ferritic steels, copper and nickel, but a higher temperature is required for austenitic stainless steels.

Radiation hardening appears to be due to the introduction of dispersed point defects (p. 93) into the lattice which act like alloying in obstructing the movement of dislocations. These point defects are produced by so-

TABLE 75 PROPERTIES OF NEWER METALS

| | Beryllium | Hafnium | Niobium | Titanium | Vanadium | Zirconium |
|---|---|---|---|---|---|---|
| Crystal structure | c.p.h. | c.p.h.<br>b.c.c. | b.c.c. | c.p.h.<br>b.c.c. | b.c.c. | c.p.h.<br>b.c.c. |
| Mpt °C | 1283 | 2130 | 2468 | 1660 | 1860 | 1852 |
| Density, g/cm$^3$ | 1·85 | 13·36 | 8·6 | 4·51 | 6·1 | 6·51 |
| Thermal cond, cal/sec/cm/°C | 0·36 | 0·056 | 0·125 | 0·041 | 0·074 | 0·05 |
| Linear exp,/°C × 10$^{-6}$ | 11·54 | 5·9 | 6·89 | 8·5 | 8·3 | 6·0 |
| Youngs Mod, kN/mm$^2$ | 300 | 96 | 100 | 100 | 140 | 95 |
| TS N/mm$^2$ | 325 | 340 | 278 | 309 | 448 | 417 |
| El | 2·8 | 35 | 49 | 25 | 35 | 30 |
| VHN | 109 | 180 | 40 | 120 | 165 | 140 |

called *knock-on damage*. This occurs when a *fast* neutron collides with an atom knocking it off its lattice site so violently that it, in turn, knocks a whole cascade of other atoms into interstitial positions and leaving vacancies in the centre, i.e. a *spike* (Fig. 241). Annealing allows the vacancies and interstitials to annihilate each other and so reform the original lattice.

The effects of radiation damage in uranium have already been discussed. The properties of some of the newer metals are given in Table 75.

## Molybdenum

Attractive properties of molybdenum are its high melting point, 2640 °C, relatively low density (10·22) and high Young's modulus, 318 kN/mm². It has poor low temperature ductility and oxidation resistance—$MoO_3$ sublimes above 500 °C. It resists mineral acids and is used for valves; also stirrers for glass manufactures, boring tools where rigidity is required and as a spray undercoat in building of shafts.

Typical room temperature properties of 16 mm rod are:

|  | TS<br>N/mm² | YP | El | RA |
|---|---|---|---|---|
| As rolled | 700 | 546 | 40 | 61 |
| Stress relieved 980 °C | 672 | 566 | 42 | 62 |
| Recrystallised 1180 °C | 472 | 388 | 42 | 38 |

Three alloys of importance are: Mo–0·5 Ti; Mo–0·5 Ti 0·08 Zr (TZM) Mo–30W. TZM is only 15% stronger than unalloyed molybdenum at room temperature but more than four times as strong at 1320 °C (recry. temp. 1430) and finds uses as dies.

## Platinum metals

*Platinum*   f.c.c; sp gr 21·45; MPt 176 °C; E mod 172 kN/mm²; VPN 40.

It has outstanding corrosion resistance to acids and many fused salts but attacked at high temperatures by Si, B, Pb, Bi and S under reducing conditions. Its strength can be trebled by alloying with 10% Rh, Ir or Ru or dispersion hardened with thoria. It is used as crucible lining for optical glass. In various forms it is used as a catalyst.

*Palladium*   f.c.c; sp gr 12·02; MPt 1552 °C; E mod 115 kN/mm²; VPN 39.

It has the unique property of enormous solubility of Hydrogen (1000 times its own vol at RT) which it expels at 100–200 °C. An alloy 77 Pd 25 Ag is used as a foil filter to obtain ultra pure hydrogen and in fuel cells. Corrosion resistance is inferior to that of Pt.

*Rhodium*   f.c.c; sp gr 12·41; MPt 1960 °C; E mod 318 kN/mm² VPN 100.

It has high reflectivity and resistance to tarnishing, low electrical resistance, hence it is suitable for contacts and decorative purposes. Electrodeposited hardness is 900 VPN.

*Ruthenium*   h.c.p; sp gr 12·45; MPt 2310 °C; E mod 417 kN/mm²; VPN 200–350.

It is usually sintered but can be hot worked at 1200–1500 °C. It has a high elastic modulus and damping capacity and low contact resistance even at elevated temperature. Electro-deposited Ru (VPN 900–1300) resists wear by paper and textiles.

*Iridium*   f.c.c; sp gr 22·65; MPt 2443; E mod 516 kN/mm$^2$ VPN 220.

It is resistant to halogens and unattacked by molten Pb, Bi, Ga and Hg. Heaviest metal.

*Osmium*   h.c.p; sp gr 22·61; MPt 3050 °C; E mod 556 kN/mm$^2$; VPN 370–670.

Tetroxide is highly volatile and toxic. Metal is practically unworkable and is used for tipping pen nibs and instrument pivots.

*Nitinol or memory alloy, Ni–Ti*

An experimental intermetallic compound, 55 Ni 45 Ti, has some interesting properties with strength up to 900 N/mm$^2$ and elongation of 10% and Izod 38 Joules. Resistance to chemical attack is excellent. Given proper conditions Nitinol objects can be given a desired shape which is fixed by a 'memory' heating at about 480 °C. On cooling, the article can be deformed to another shape, but the original form can be restored by heating at $-50$ to 150 °C. It depends on the alloy's unique reversible stress-induced martensitic transformation. If strain is limited to about 8% a highly efficient recovery is possible when heat is applied. Nitinol exerts considerable force during recovery.

## Fibre reinforced materials

Newer fields of engineering, particularly aerospace, are demanding materials of higher specific strength and modulus. The strong materials, which are also light and stiff, are ceramics—covalent solids—such as glass, graphite alumina, boron nitride and silicon carbide. These materials also have high melting points and low expansion coefficients; their strength, however, is difficult to realise practically because of their sensitivity to minute surface cracks (e.g. 2/1000 mm) which are nearly always present or form if the surface is not protected. These difficulties can largely be overcome by using the ceramic as fibres bound together in a matrix which must (1) not damage fibre, (2) transmit stress to the fibre and (3) control cracks in the composite. These requirements are met by polymers and soft metals. Such composites occur naturally as bamboo and horn.

If a composite is stressed parallel to the fibres, the matrix yields plastically and under equal strain the stress within the fibres is enormously greater than in the matrix. If stress is high enough to break fibres, the softness of the matrix hinders the propagation of the crack, and in any case the fibres do not fail in one plane. When a fibre breaks the two ends attempt to pull away from each other but are prevented from doing so by the matrix

TABLE 76  PROPERTIES OF FIBRES

| Material | Density g/cm² | Dia mm | TS N/mm² | Young's Modulus kN/mm² | Specific Tensile Strength N/mm² | Specific Modulus kN/mm² |
|---|---|---|---|---|---|---|
| Steel | 7·75 | 0·076–0·18 | 2760–4130 | 206 | 356–532 | 26·6 |
| Boron | 2·6 | 0·1 | 2410–3440 | 378–412 | 925–1330 | 145–159 |
| RAE Carbon fibre | | | | | | |
| type I | 1·92 | 0·0076 | 1380–1720 | 310–378 | 794–989 | 178–217 |
| type II | 1·77 | 0·0076 | 1720–2060 | 172–206 | 955–1142 | 95·5–114·2 |
| S Glass as drawn | 2·49 | 0·005–0·01 | 4820 | 86 | 193·5 | 34·6 |
| S Glass in practical structure | 2·49 | 0·005–0·01 | 2410–2760 | 86 | 969–1108 | 34·6 |

which adheres to the fibre (shear force per unit area). As the fibre attempts to relax, the flow of the matrix parallel to the stress counteracts the tendency. Shear forces come into play and gradually build stress back into the broken fibre as depicted in Fig. 252. Stress is transferred to the fibre from both ends and if the fibre has a breaking stress of $\sigma_f$, the critical length of a fibre $l_c$ below which fibres crack, pull out of the matrix, is given by

$$l_c = \frac{\sigma_f r}{\tau} \quad \begin{array}{l} r = \text{radius of fibre} \\ \tau = \text{shear strength of bond} \end{array}$$

Provided the fibres exceed this critical length, they need not be continuous, and consequently composites can be made with short lengths of fibre.

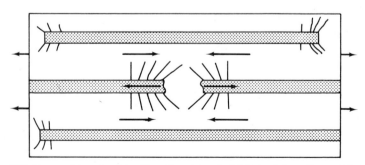

FIG. 252 Represents a broken fibre the ends of which tend to pull apart but are prevented by the shear forces in the matrix

Glass fibre reinforced plastics first came into use in aircraft during the Second World War. Glass can readily be drawn into fibres and put into a matrix of unsaturated polyester resin. Large pieces can be built up by the application of successive layers. Glass reinforced plastics suffer from two disadvantages: firstly, the elastic modulus is low, and sufficient rigidity cannot be obtained in a structure; and secondly, the resin matrix burns, chars or flows at about 200 °C. These disadvantages can be reduced by using high temperature polymide resin with pure $SiO_2$ fibres. A special type of glass fibre can be used with cement bond to form flexible type of concrete.

Concrete is also reinforced with steel fibres (0·25 mm dia mild steel wire chopped into 25 mm lengths), known as Wirand. It has a flexural strength of 14 N/mm² at 14 days as against high strength concrete of 4 N/mm² at 28 days.

Stiffer materials are whiskers (few microns dia) % $Al_2O_3$, SiC and $Si_3N_4$, which are now available although expensive, and means have been developed of introducing them into resin and metallic materials. Whiskers are single crystals, free from dislocations and their strength approaches the ideal theoretical value.

Rolls Royce has used silica fibres coated with aluminium and consolidated.

Carbon fibres have also been coated with metal such as tungsten, which allows metallic matrices to be used (e.g. Al).

A cheap carbon material has been developed by RAE Farnborough from commercially drawn acrylic fibres by subjecting them to various treatments and then graphitising to produce long threads of graphite oriented with the basal plane parallel to stress oxide. Two types are in commercial production; Type I—high modulus, and Type II—high strength (see Table 76). These fibres have the strength of steel twice the modulus and one quarter the density. The fibres are produced as tows containing 10 000 filaments 0·076 mm dia and in lengths between 300 and 1·2 metres, and are normally surface treated to give improved inter-laminar shear strengths. The components can be made by (1) moulding a 'mat', (2) winding on a former for cylindrical articles requiring orientation of fibre, (3) grouping shaped laminations which are cured in a die.

Metallic fibres such as patented steel, stainless steel, tungsten and molybdenum wires can also be used in a metal matrix such as aluminium and titanium. The combination of specific strength and fracture toughness developed in the component can be better than that of either component. Composites in the form of sheet laminates offer interesting possibilities. Fabrication of these materials is by rolling (often involving encapsulation to avoid oxidation) or by casting or vacuum infiltration of the matrix around the wires. Interactions between the fibre and the matrix can cause loss of strength by formation of brittle intermetallic components. Stainless-steel wire reinforced aluminium has exhibited increased fatigue life and the interfaces tend to deflect cracks.

Another technique is to use unidirectional freezing of a eutectic alloy so that the components develop parallel micro-constituents.

Welding of components is virtually impossible and even the drilling of holes for bolts is risky while forming and shaping is not successful, consequently material and component fabrication is essentially one operation.

The economic pressure to use short fibres is stimulating research into methods of grading and aligning short fibres to form mats which can be impregnated with metal or polymers.

## Super-plasticity

A number of two phase materials have been observed to exhibit large elongations (up to 2000%) without fracture, and such behaviour has been called super-plasticity. In the second half of the 1960s a quickening interest has been taken in the phenomena, and evaluation trials carried out.

In a tensile test the relationship between flow stress $\sigma$ and the strain rate $\varepsilon$ is given by $\sigma = K\varepsilon^m$ where $K$ = constant and $m$ = index of strain rate sensitivity.

With normal metals 'm' is low ($\sim 0.2$–$0.3$) i.e. they are insensitive to changes in strain rate $\varepsilon$. When a local neck starts to form, thus locally increasing the strain rate, the effect on flow is minimal, and necking continues up to fracture.

When $m = 1$, the flow stress is proportional to strain rate and the material behaves like a Newtonian viscous fluid, such as hot glass. Super-plastic metals have high 'm' values ($0.5$–$0.8$) and when a local neck begins to form, the increase in strain rate immediately raises the flow stress, and necking is stopped and uniform elongation continues. Super-plasticity only occurs above $0.3$–$0.4$ $T_m$ K ($T_m$ is melting point). A further characteristic of super-plastic flow is that once it is initiated the flow stress required to maintain it is extremely low—very much lower than that required in conventional hot-working processes at low strain rates, but higher at higher strain rates.

The main structural requirement is an extremely fine grain size, less than 5 microns, and this fine structure must be maintained at the working temperature. This usually necessitates a two phase structure to restrict grain growth by mechanical interference. Eutectics or eutectoids provide convenient forms of fine grain material, but exception ductilities have been produced in fine grained Mg and Ni. Super-plastic flow is possible in some alloys, e.g. Ti, and in ceramics by stressing under cyclic temperature conditions at a phase-transformation temperature. A number of super-plastic alloys are given in Table 77.

The Zn–Al alloy (Prestal) has received much attention, and it can be

TABLE 77  SUPER-PLASTIC ALLOYS

| Base Metal | Alloy % | Super-plastic temp °C | Grain size micrometres | 'm' strain rate sensitivity |
|---|---|---|---|---|
| Zn | 22 Al | 200–260 | 1–2 | 0·5 |
| Al | 33 Cu | 440–530 | 1–2 | 0·9 |
| Co | 10 Al | 1200 | 0·4 | 0·3 |
| Cu | 10–12 Al | 500 | | 0·5 |
| Fe | 26 Cr 6·5 Ni | 870–980 | 2 | 0·5 |
| Mg | 6 Zn 0·6 Zr | 270–310 | 0·5 | 0·5 |
| Pb | 20 Sn | 20 | 3 | 0·5 |

shaped by methods normally used for plastics, e.g. vacuum forming complex shapes into refractory concrete or metal moulds, adequately vented, and capital costs are low. Blow forming and drawing out like hot glass (die-less drawing) is also possible with some alloys. Although the deformation rate must be low, the loads are low and the technique used in polymer and glass technology may eventually be used to replace expensive and complex forming technology.

# 18 Miscellaneous Non-ferrous Metals and Alloys

**Bearing metals**

In a properly lubricated bearing, a film of oil separates the moving parts and the composition of the bearings is of no great importance. In starting and stopping, however, and in severe running conditions this film cannot be maintained and lubrication depends largely on the 'boundary' oil film. The capacity to retain this surface layer of oil is an important feature of a bearing metal.

Until recently it was considered that alloys consisting of one uniform constituent rarely satisfied the requirements of a bearing metal, but that an alloy consisting of a hard constituent embedded in a soft matrix is required. Several exceptions are now known.

The soft matrix gives the ability to yield under excessive local pressures (such as arise in misalignment of the shaft) and prevents local seizing and enables some abrasive in the lubricant to be tolerated. The choice of the matrix depends on the average pressure on the bearing. If the load is heavy and the speed low a bronze matrix will be used. Where the load is light a lead–antimony eutectic is sufficiently strong. For general purposes tin-base bearings are common. Compatability is important—see Appendix II.

In practice bearing alloys fall into three main groups:

(i) white metal type, consisting of hard particles embedded in a softer matrix;
(ii) leaded bronze type, consisting of a hard matrix through which a softer metal is distributed. During sliding the soft metal is smeared over the matrix, where it functions as a thin metallic lubricant film. Al–Sn and PTFE fall in this group.
(iii) single metal or single-phase alloy.

Apart from suitable mechanical, frictional, corrosion-resisting and wear-resisting properties, a very desirable property of a wide range of bearing alloys is that one of its constituents should possess a relatively low melting-point. This will prevent excessive seizure since the high local temperatures developed under severe conditions of running will readily cause a local softening or melting of the low-melting constituent. Bearing metals should be capable of forming a solid soap with normal lubricants and should form a strongly bonded oxide surface film which minimises pressure welding.

# BEARING METALS

*White metal alloys* can be divided into two classes, the lead base and tin base, and the latter are known as Babbitt metals.

From Fig. 253 it will be seen that tin dissolves 8% of antimony at 246°C, but this decreases to 3·5% at ordinary temperatures with the result that SnSb particles are precipitated in the cored crystals. In many bearing metals, the antimony content is about 10% and the hard compound SnSb crystallises first in the form of cuboids, which, being lighter than the melt, tend to float to the surface, as shown in Fig. 254. This results in the bearing being too hard at the top and too soft at the bottom. This trouble can be reduced by quick cooling, but a more successful method consists in adding copper to the alloy.

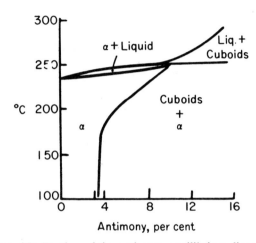

FIG. 253 Portion of tin–antimony equilibrium diagram

When copper is added to tin a hard needle-like constituent is produced which forms a eutectic with 0·75% copper. With 3% of copper the liquidus is raised and primary needles of the $Cu_6Sn_5$ occur with eutectic.

If both copper and antimony are present the $Cu_6Sn_5$ needles form first, and interlock in a network which prevents the relative movement of the SnSb cuboids which form later, and this results in a uniform structure (Fig. 255).

The structure is affected by the casting temperature and the rate of cooling.

Lead is added to the tin base alloys for cheapness, but it reduces the maximum pressure which can be used. The plain lead–antimony alloy (Magnolia) is also used as a bearing for light pressures. The hard constituent is primary antimony (Fig. 36) or SnSb in a matrix of a Pb–Sb–Sn ternary eutectic. Calcium and barium have each been added to lead to form bearing metals. Typical compositions together with the Brinell hardness and the 0·1 % PS under compressive stresses are given in Table 78. For tail shaft bearings of ships an alloy is used containing 68·5 Sn, 30 Zn, 1·5 Cu. Its structure consists of $CuZn_3$ and zinc in a matrix of tin–zinc eutectic. It corrodes preferentially to the shaft.

White metals are used on a bronze shell and under very severe working conditions may suffer from a peculiar form of fatigue cracking. After a period of service fine cracks appear on the surface of the bearing and form a network like a crazy-pavement. Later, pieces of the bearing may become detached.

254    255

FIG. 254 10% antimony in tin: SnSb cuboids in tin-rich solid solution  ($\times$ 80)
FIG. 255 Tin with Sb 10; Cu 3. Cuboids and star-like copper–tin constituent in ground mass of eutectic (tin − antimony solution + $Cu_6Sn_5$)  ($\times$ 80)

This difficulty is minimised by:

(a) using a high fatigue strength alloy such as a tin-base alloy free from lead, containing Sb, 7; Cu, 3, with sometimes additions of cadmium and nickel;
(b) paying maximum attention to the bonding of the lining to the supporting sheet;
(c) reducing the thickness of the lining—0·05 to 0·25 mm—is common nowadays, but this demands increased attention to the size of the hard particles.

*Trimetal* bearings are now widely used for lorry diesel engine crankshaft bearings. These consist of a steel backing, a 0·37 mm interlayer of copper–lead and a 0·037 mm overlay of lead–tin alloy. The interlayer provides tolerable properties if the overlay wears through.

TABLE 78  WHITE METALS

| BS No. | Sb | Cu | Sn | Pb | HV5 at 20°C | 0·1 PS N/mm² | |
|---|---|---|---|---|---|---|---|
| 3332/2 | 9 | 4·2 | 86·8 | | 29 | 3·4 | High compressive strength, high speeds and heavy loads |
| 3332/1 | 7·5 | 3·2 | 89·3 | | 27 | 2·5 | Strip lining for autos. High strength and ductility |
| 3332/3 | 10 | 5 | 81 | 4 | 32 | 4·4 | Cheaper but less ductile. Main bearings |
| 3332/4 | 12 | 3·5 | 75 | 9·5 | 31 | 4·5 | High speeds, heavy steady loads |
| 3332/6 | 10 | 3 | 59 | 28 | 27 | 4·1 | Dynamos, locos, shells and solid ingot |
| 3332/7 | 13·5 | 0·7 | 11·3 | 74·5 | 26 | 3·7 | Solid die castings. Segregates in centrifugal and sand casting |
| 3332/8 | 16 | 0·5 | 5 | 78·5 | 22 | 3·5 | Cheaper form of 3332/7. Med. pressure and speed |
| | 15 | 1As | 10 | 74 | 17 | 1·6 | *Strip lined bimetal for autos. Good fatigue |

*Strip is annealed 4 hrs 200 °C before shaping

## Bearing bronzes

For high bearing loads bronzes are used instead of the white metals, and these alloys have been discussed in Chap. 15. The 10% tin phosphor bronze is widely used, while the 15% tin bronze is used for locomotive slide valves and bearings for turntables. From 5 to 15% of lead is sometimes added for applications where lubrication or alignment is liable to be somewhat unsatisfactory (Fig. 219). The lead squeezes over the surface and prevents seizure.

A copper–lead bearing alloy which is widely used contains 25–30% lead and up to 1% tin to prevent attack of the lead by acidic oils. Thin overlays of lead–tin or lead–indium are also used. This alloy has a high thermal conductivity and is capable of carrying about 20% higher loading at higher speeds than lead or tin base white metals, and is therefore used for high-duty aeroplane, automobile and diesel engine crankshaft bearings. Severe edge loading can cause failure. Special melting and casting technique is required for these alloys in order to get a uniform distribution of lead particles. Steel strip is now lined with 60/40 copper–lead alloy without lead segregation trouble by powder metallurgy techniques.

Bearings are also made by moulding mixtures of powdered copper, tin and graphite under pressures of about 270 N/mm² and afterwards sintering the bushes in a reducing atmosphere. The graphite renders the bearing self-lubricating. Oil impregnation is also used in bearings for electric clocks, vacuum cleaners, etc.

*Silver Bearings*

For heavy loads, as in aircraft bearings, silver is also used. The silver is plated with lead to reduce the risk of seizure, followed by a thin coating of indium to increase corrosion resistance of lead to acidic oil.

*Aluminium alloys*

Low tin alloys containing 6·5 Sn, 1 Cu, 1 Ni have a high fatigue strength but their hardness and low embeddability necessitates the use of fully hardened shafts and close attention to oil filtration. Their high thermal expansion tends to cause loose bushes. Raising the tin to 20% with small proportions of copper, iron, silicon and manganese improves bearing properties, but it is necessary to break up to some extent the continuous networks of tin by cold rolling and recrystallisation. To function as a strong bearing, however, the aluminium alloy must be pressure welded to a steel strip and softened at 350°C without forming excessive brittle aluminium–iron compound. Such reticular aluminium bearings can operate at much higher loads and temperatures than white metal on a soft shaft without the journal wear that is experienced when copper–lead is used. Their fatigue strength is higher than that of 70/30 copper–lead. Applications include big end and main automobile bearings.

*Dry and anti-corrosive bearings*

To avoid contamination of food, textiles, etc., and for inaccessible positions, dry bearings are invaluable. One type consists of molybdenum sulphide bonded with an organic binder to a phosphated steel. A second type depends on using a material with negligible affinity for the shaft material. Polytetrafluoroethylene (PTFE) is extremely inert and has a low coefficient of friction and thermal conductivity, but is mechanically weak. A vast improvement in wearing properties is obtained by impregnating sintered porous bronze (a conductive path) with PTFE, to which is added a small amount of lead, e.g. Glacier DU. With a steel back such a bearing can operate under remarkably high loads and high or low speeds from $-200$ to 250°C. A thin layer of PTFE bonded to steel, or PTFE mixed with various fillers and machinable to size (e.g. Glacier DQ1 = 20 graphite + 20 bronze; Glacier DQ3 = 37·5 $Pb_3O_4$ + 2·5 bronze), are alternative forms of bearing. Glacier DX fills the gap between fully lubricated and fully dry bearings and consists of copper plated steel, sintered porous bronze interlayers impregnated with acetal co-polymer with a final layer (0·25 mm) of acetal co-polymer and is used for steering and suspension bearings in automobiles.

Carbon and nylon can be used under corrosive conditions. Nylon is sometimes reinforced with up to 80% spherical bronze particles.

### Zinc die-casting alloys

The main uses of zinc are in the form of the metal itself, zinc protective coatings on other metals—galvanising, sherardising, metal spraying—and alloys in which zinc is not the chief component, e.g. copper alloys. Zinc base pressure die-casting alloys, however, are now being increasingly used since the early troubles of embrittlement and growth have been overcome by the use of very pure zinc (99·99% pure). Small quantities of lead, cadmium and tin render the alloy susceptible to intercrystalline corrosion, the products of which cause the swelling of a casting. Lead and cadmium are therefore specified to be not more than 0·003% and tin not more than 0·001%. A test for intercrystalline corrosion consists in suspending samples in a closed container above water maintained at 95°C for a period of 10 days. On removal, the specimens are measured for growth. If corrosion has occurred, bending produces surface cracks.

Particulars of two alloys in common use are given in Table 79 of which about 98% of production is Grade A. These alloys undergo a shrinkage, completed in about 4 weeks' ageing, followed by a slight expansion. Where close dimensional tolerances are necessary the alloys can be substantially stabilised by an anneal at 100°C for 6 hours. Soldering of these alloys is not recommended.

TABLE 79  COMPOSITION AND PROPERTIES OF ZINC ALLOY DIE CASTINGS

|  | BS 1004/A | | | BS 1004/B | | |
|---|---|---|---|---|---|---|
| Al % | 3·9–4·3 max | | | 3·9–4·3 max | | |
| Cu % | 0·03 max | | | 0·75–1·25 max | | |
| Mg % | 0·03–0·06 max | | | 0·03–0·06 max | | |
| Zn %* | remainder | | | remainder | | |
| Temperature | TS $N/mm^2$ | El | Izod J | TS $N/mm^2$ | El | Izod J |
| −20°C | 301 | 4 | 5 | 340 | 3 | 5 |
| 0°C | 284 | 6 | 31 | 334 | 6 | 55 |
| 21°C | 283 | 10 | 58 | 327 | 7 | 65 |
| 95°C | 196 | 30 | 34 | 239 | 23 | 58 |
| Aged 7 yrs, 21°C | 207 | 16 | 55 | 242 | 13 | 57 |

*Other impurities: Fe, 0·075; Pb, 0·003; Cd, 0·003; Sn, 0·001% max

For improved creep strength at 100–150 °C a new zinc alloy (ILZRO14) is, 1-1·5 Cu, 0·15–0·25 Ti, 0·1–0·2 Cr, 0·01–0·04 Al. It must be cast in a cold chamber die casting machine. A sand casting alloy (ILZRO 12) contains 12 Al, 1 Cu.

Other die-cast alloys are lead and tin base, brass, aluminium bronze and aluminium silicon alloys.

There are two types of die castings, both being made in permanent metal moulds, and this practice results in finer structure and superior properties as compared with castings made in sand moulds.

In *gravity die* casting the molten metal flows into the mould under the action of gravity and the operations are similar to those in ordinary foundry practice. Usually the cores, to produce recessed portions, are of steel, though sand cores are used to make undercut portions. In *pressure die casting* the metal is injected into the mould cavity under pressure; the die is operated and kept securely closed during casting by hydraulic lock or by a toggle mechanism; cores are inserted and extracted semi-automatically and the solidified die casting is pressed away from the moving die block by ejectors. Thin sections down to 0·8 mm or even less can be cast, and the pressure die casting possesses good surface appearance and a close accuracy. Internal porosity tends to be found in heavy sections, particularly if this is joined to thin sections rendering adequate feeding difficult. The various operations are arranged to be semi-automatic or, in the case of small and medium-sized zinc alloy castings of suitable design, fully automatic. A quantity of 10 000 or more pressure die castings per annum is usually reckoned to be the economic minimum except in cases where the considerable saving of expensive machining operations makes a high initial die cost justify the use of the process for even smaller quantities.

Zinc alloy pressure die castings are usually made on hot chamber machines, in which the molten metal container is part of the machine and the injection plunger, usually operated by pneumatic power, forces the molten zinc through a 'goose neck' channel into the die cavity. Aluminium and magnesium alloys are principally die cast on cold chamber machines in which the metal is melted in a separate crucible and ladled or automatically fed into the injection chamber. Such machines are operated by oil or water hydraulics.

Close co-operation between the designer and the die-caster is of great importance, since in many cases some trifling alteration in design may lead to a simplification of the die accompanied by a reduction in cost and improved quality and appearance of the casting. The following points should be noted:

(*a*) The section of the die casting should be as uniform as possible and as thin as practicable, bearing in mind the required strength of the

component. Where local strength is required this should be provided by ribbing instead of by all-over increases in section.
(b) The casting is planned by the die-caster to contract on to the moving die half, whence it is ejected.
(c) Cored holes should be grouped into a minimum number of planes, to save die cost and provide maximum speed of output.
(d) When possible the parting line should be straight, since this assists in attaining accuracy of the product. The parting line is best on a sharp edge of the casting in order to locate the 'flash' in position where it can be easily removed without leaving an unsightly mark. The parting line should not cross between two portions which need to be spaced accurately from one another, since the two halves of the die tend to separate slightly during the casting operation.
(e) Loose cores and sunken lettering should be avoided.

## Lead alloys

Lead has an extraordinary permanence in air, due to the film of carbonate or sulphate, but is attacked by very pure waters. It has the characteristic of flowing under very low loads and this leads to local necking, which causes local failures in applications such as water pipes, where a good general extension is required. The addition of $0.06\%$ of tellurium refines the structure and reduces the tendency to extend locally. The tellurium–lead now contains $0.06\%$ copper in addition to $0.04\%$ tellurium and is useful in chemical plant. Unfortunately, such fine-grained lead has a poor resistance to creep. The fatigue strength of lead is also very low and this leads to intercrystalline failures of sheathed cables, due to vibration. The fatigue strength is doubled by adding $0.25\%$ cadmium with $1.5\%$ tin. The properties of this alloy can be modified by heat-treatment. Another alloy in this class contains $0.005\%$ silver and $0.005\%$ copper and is an example of the considerable effect small alloying additions can have on the strength properties of a very pure metal. Its creep resistance, enabling higher water pressures to be withstood, in a pipe of given dimensions, is distinctly better than that of pure lead. Pure lead dispersion hardened with $1\frac{1}{2}\%$ vol PbO has greatly improved creep strength at $20\,^\circ\text{C}$ ($\times 3$). Mixtures of lead powder in polyethylene are used for radiation shielding (5–12 to 1).

### Fusible alloys

These low melting point alloys are used as links in fire controlling systems which melt if temperature rises. Alloys are increasingly used to hold work-pieces including ophthalmic lenses instead of pitch, setting punches in press tools and proof casting to determine accuracy of a die.

| Bi | Pb | Sn | Cd | MP°C |
|---|---|---|---|---|
| 52·5 | 32 | 15·5 | | 96 |
| 49·5 | 27·3 | 3·1 | 10·1 | 70 |
| 40 | 60 | | | 170 |

Alloys containing more than 55% Bi expand, 48–55% Bi exhibit little volume change on solidification. The 40/60 alloy is used for moulds and castings.

### Nickel alloys

*Malleable nickel* has a good resistance to caustic alkalis, ammonia, salt solutions and organic acids. It is therefore used in chemical engineering for autoclaves, pumps, valves, agitators and pump linings, and also in the foodstuffs industry. For cheapness nickel-clad steel is frequently used, which consists of a thin coating of nickel rolled on to a mild steel base. In the molten state nickel readily absorbs carbon ($Ni_3C$) which forms graphite on cooling, and oxygen (NiO) which embrittles the material, and sulphur which forms intercrystalline films of $Ni_3S_2$. To render nickel malleable the addition of a deoxidant such as magnesium, titanium or manganese is essential.

Nickel is also used extensively for electroplating and as an important alloying element in iron-, copper- and aluminium-base alloys. Alloys in which nickel is the base are monel (p. 312), nickel–chromium (p. 246), magnetic alloys (p. 294). Sparking plug electrodes (Mn, 0·6–4; Sn, 2–1). Low expansion alloys. For electronic cathodes, nickel alloys containing 0·05 Mg or 0·04 Mg, 0·2 Si or 3·5 W are available (see BS 3072–76). The Ni–30% Cu alloy is used to resist sea-water attack but castings often contain 1–4% Si which improves resistance to corrosion by water and steam.

*The complex nickel–molybdenum* alloys, more commonly used in America, have exceptional resistance to attack by hydrochloric and sulphuric acids over wide ranges of temperatures and concentration. Good strength is maintained at elevated temperature and for a time the alloys were used in American jet engines. Examples are Hastelloy B containing Mo, 30; Fe, 5; Ni, 65%; and Hastelloy C, Mo, 17; Fe, 5; Cr, 14; W, 5; Ni remainder. A nickel–silicon (9%) alloy known as Hastelloy D has an exceptional resistance to sulphuric acid up to boiling-point. The presence of ferric or cupric salts may cause rapid corrosion of Alloy B. For heat-treatment Alloy B should be annealed at 1150–1180°C; Alloy C at 1200–1230°C; stabilising or spheroidisation anneal at 1040–1065°C for B and 1110–1120°C for Alloy C reduces the ductility of the alloys but minimises the effects of heat during welding.

Cast 50/50 NiCr alloys have come into substantial use because of their outstanding oil-ash (V, S, Na compounds) corrosion in boiler and oil refinery plants.

Castings tended to creep and lose ductility in service and have poor workability. These difficulties are overcome in two new patented alloys cast in 657 Ni + 50 Cr, 1·5 Nb, 0·2 max (C & N), wrought in 589 Ni + 50 Cr, 1·5 Zr.

# 19 Polymers

Natural organic material—wood, leather, flax, cotton, rubber—have long been used by engineers, but the development of man-made organic polymers has opened up a wide field of materials which are both competitive and complementary to metals as constructional materials. Such polymers are light, resistant to corrosion, easy to fabricate and offer special properties such as rubbery elasticity or high hysteresis. From the engineer's point of view, probably the greatest limitation of plastics is the way they creep at room or slightly elevated temperatures under relatively light loads, but design criteria and testing differs from that used for metals and the bewildering profusion of general types and even the variations within a particular batch are complicating features. A good batch of material can be ruined by bad processing and sometimes by the addition of reworked scrap. Thus average values of properties may not represent the actual value obtained in the finished product, and testing of a prototype is sometimes the only solution. Some of these problems exist with metals but the degree is so much greater with polymers.

Polymers are composed of many thousands or millions of atoms joined together into giant molecules, which are composed of a large number of repeating units called monomers.

*Addition and condensation polymerisation*

The process of linking together of monomers is called polymerisation, and two forms are used. Firstly, in the *addition* polymerisation the monomer must have two bonds which are broken to create two points at which new bonds may be formed, under the influence of temperature, pressure and a catalyst. For example, the monomer ethylene gas ($C_2H_4$) can be polymerised to form polyethylene as follows:

$$\begin{array}{c} H \quad H \\ | \quad | \\ C = C \\ | \quad | \\ H \quad H \end{array} \rightarrow \begin{array}{c} H \quad H \\ | \quad | \\ -C - C - \\ | \quad | \\ H \quad H \end{array} \rightarrow \begin{array}{c} H \quad H \quad H \quad H \quad H \quad H \\ | \quad | \quad | \quad | \quad | \quad | \\ -C - C - C - C - C - C - \\ | \quad | \quad | \quad | \quad | \quad | \\ H \quad H \quad H \quad H \quad H \quad H \end{array}$$

Monomer                         Polymer

# ADDITION AND CONDENSATION POLYMERISATION

This linear polymer can also be converted to a *branched* polymer by removing a side group and replacing it with a chain, for example:

```
         H   H   H   H   H   H
         |   |   |   |   |   |
     — C — C — C — C — C — C —
         |   |   |   |   |   |
         H   H   H   |   H   H
                     |
                 H — C — H
                     |
                 H — C — H      ←cross link forming a
                     |              structure
                 H — C — H
```

Cross link branches may interconnect to form a space-network structure, Fig. 256(*d*).

*Co-polymers* can also be formed by joining different monomers in the same chain, different combinations are shown below where M and O represent different monomers:

| | |
|---|---|
| linear polymer | — M — M — M — M — M — M — |
| regular co-polymers | — M — O — M — O — M — O — |
| random co-polymer | — M — O — O — M — M — M — O — M — |
| block co-polymer | — O — M — M — M — M — O — O — O — O — M — |

A chain can be composed of vinylchloride and vinylacetate comparable to a solid solution in metals with a wide variation in properties, Buna—S rubbers are co-polymers of butadiene and styrene.

The second type of polymerisation is called *condensation* polymerisation, in which a reaction occurs between the basic monomers, with the elimination of some smaller groups of atoms, such as water, when the unit is added to the end of the chain. A familiar example, Bakelite, is made from phenol and formaldehyde.

*Classification of polymers.* Polymers may be broadly divided into two classes—thermosetting and thermoplastic. *Thermosetting* plastics, usually supplied as moulding powders or casting resins, only soften in the initial fabrication stage and then harden, by further polymerisation, to a rigid brittle material. These plastics consist of long straight or branched molecules with numerous links between them, e.g. Bakelite, epoxy resins. Scrap cannot be re-utilised.

On the other hand, a thermoplastics material may be readily moulded or extruded because of the absence of cross links, and are capable of repeated softening and rehardening on heating and cooling. Mechanical properties are sensitive to temperature and to sunlight, and exposure to temperature may cause thermal degradation, i.e. breakdown of some bonds.

## Polymer structure

Polymer molecules are several thousand ångstrom units long, so that in bulk polymers chain ends are relatively few and far between, and only produce second order effects. The molecular chains are held together in the bulk polymer by weak forces and the chain is flexible because its segments can rotate about carbon–carbon bonds in the chain backbone. One c–c bond is always at an angle of 109° to the next, the bond direction being able to rotate on the base of such a cone, e.g. Fig. 256(*a*), c on cone B, d on cone C. This gives rise to many folded configurations. Free end to end distant is $\alpha$ to $\sqrt{}$number of links. The properties of a bulk polymer depend much more on how its molecular chains are arranged in space than on the molecular structure.

Polymers can exist in three main physical states.

(1) *Melts and rubber-like state.* At sufficiently high temperatures some low molecular weight, uncross-linked polymers will become a viscous liquid allowing the amorphous chains to flow past each other readily. Other polymers will become an amorphous assembly of highly flexible molecules, wriggling and mobile, and are considered to be in a rubber-like state, Fig. 256(*b*).

Raw rubber is normally an amorphous tangle of long chain molecules. Under thermal agitation the tangles have a random arrangement but when stretched, the snarls begin to disentangle and the chains become oriented and the attraction between the parallel portions of chains causes the material to stiffen, Fig. 256(*c*). When the force is released, the strained bonds return to the random tangle arrangement.

(2) *Amorphous glassy state.* As the temperature is reduced, the flexibility of the molecules is reduced because segmental rotations about the single carbon–carbon bonds are inhibited. Under suitable cooling the polymer will change from the rubber-like state to the amorphous glassy state, which is brittle Fig. 256(*d*). The temperature at which this transition occurs is denoted by Tg. Amorphous polymers such as acrylic and cross-linked natural rubber have values of Tg 110°C and −70°C respectively. Therefore at room temperature an acrylic is in a glassy state and natural rubber in the rubber-like state.

(3) *Partially crystalline state.* If the cooling rate is sufficiently low and the molecules have a favourable structure, i.e. few side blocks, crystalline spherulites may form by the radial growth of ribbon crystals from nucleation centres—chains can be folded like a chinese cracker, Fig. 256(*e*), to build up a platelet crystal about 300Å thick, which in bundles form the spherulites (Figs. 256(*f*) and 257).

In addition to chain folding of a molecule, crystallisation can occur between different chains, Fig. 256(*g*), and also when an amorphous polymer is stretched and chains become aligned along the draw direction, Fig. 256(*h*).

POLYMER STRUCTURE 381

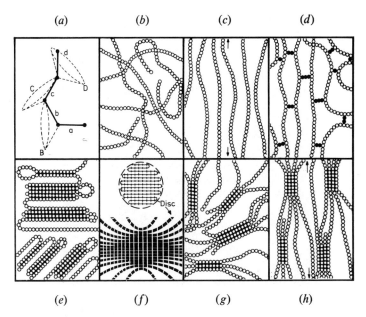

FIG. 256 (a) possible configuration of the c-c backbone of a polymer molecule
(b) amorphous polymer, random tangle of chains
(c) alignment of chain molecules during stretching
(d) cross linking of molecules; greater rigidity
(e) chain folding of single molecule to form crystallised region
(f) schematic representation of Spherulitic crystal
(g) laminar formation of different molecules to form crystals in amorphous matrix
(h) stretching causes oriented crystals from different molecules in amorphous matrix

A regular three dimensional atomic pattern or crystal can thus form by folding a single molecule repeatedly on itself or by bringing many different molecules close together in limited areas. Crystallisation causes a dense packing of molecules, thereby increasing the inter-molecular forces and producing higher strength, rigidity and brittleness. Crystalline polymers can have between 0–90% of their chains organised into regular regions, but there will always be some amorphous material present.

The variation of modulus with temperature for different types of polymers is shown schematically in Fig. 258. ABCDE represents a partially crystalline polymer. The region BC corresponds to the glass-rubber transition of the amorphous part of that polymer. In the region CD the crystallites act as large cross links between the now flexible molecules of the amorphous region, and the material therefore behaves as a stiff rubber. Crystallite

FIG. 257 Crystalline spherulites in amorphous matrix in polypropylene (*ICI Plastics Div*)

melting occurs over DE. An amorphous cross-linked polymer will behave as FGHJ. The region HJ represents the equilibrium rubbery modulus.

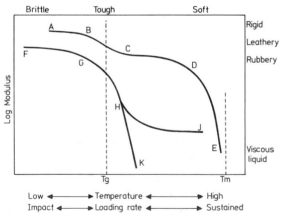

FIG. 258 The variation of modulus with temperature for different types of polymers. ABCDE represents a partially crystalline polymer, FGHJ an amorphous cross-linked polymer, FGHK an amorphous uncross-linked polymer, FG a thermoset

# Some common types of plastics

A thermoset will exhibit the behaviour of the region FG but the onset of softening coincides with the onset of thermal degradation. An amorphous uncross-linked polymer (FGHK) will flow as a viscous liquid at the high temperatures.

*Plasticisers.* A crystalline or amorphous polymer can be made more flexible by the addition of a foreign material of low molecular weight which can penetrate between the polymer molecules and reduce the attraction forces between them. Organic solvents, resins and even water are used.

*Strengthening of plastics.* The various methods of strengthening polymers and some commercial products are summarised below:

1. *Crystallisation.* Thermoplastics materials include polyethylene, polypropylene, polyoxy-methylene, poly vinyl alcohol, poly vinyl chloride, poly vinylidene chloride and polyamides, such as nylon 6 and 66.
2. *Cross-linking* of chains has yielded hard rubbers (by vulcanising with S), thermosetting resins, polyesters, network polyepoxides, polyurethanes and resins and plastics formed by compounding formaldehyde with urea, melamine or phenol.
3. Increasing the *rigidity of the carbon backbone* by attaching bulky groups to the chain to reduce bending, e.g. polystyrene, polymethylmethacrylate, polycarbonates, polyesters.
4. Mixing monomers to form *Co-polymers*, e.g. acrylonitrite-butadiene (ABS), styrene-butadiene (SBR).

Combination of these principles will lead to new high strength, high temperature plastics.

## Some common types of plastics

The common name is followed by the chemical name and physical state in the following summary. Many of the polymers are supplied in a range of grades.

*Thermoplastics*

*ABS*, acrylonitrile—butadiene—styrene, amorphous

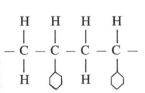

This co-polymer is tough, hard, rigid and can be plated and has excellent low temperature and chemical resistance. An alloy of PVC/ABS is self fire extinguishing.
Used for cabinets, car grilles.

*Acetals*, Delrin, formaldehyde polymers, partly crystalline.

```
  CH₂   CH₂   CH₂
 /   \ /   \ /
      O     O
```

*Co-polymers* (Kemetal) have better resistance to alkalis and hot water. Tough, rigid, resistant to moisture, impact and fatigue, and can replace metals in many instances. Used for handles, instrument housing, bearings, unlubricated gears, light duty springs and plumbing fittings. Temperature ranges: $-40\,°C$ to $120\,°C$ but acetyls will burn.

```
    H   CH₃   H   CH₃
    |    |    |    |
  – C  – C  – C  – C  –
    |    |    |    |
    H    C    H    C
         ⫽         ⫽
       O  OCH₃   O  OCH₃
```

*Acrylic* (perspex) polymethyl-methacrylate, amorphous glassy crystal transparency but softer than glass; low water absorption, good formability, resistant to sunlight, weather and to most chemicals: high dielectric strength. Used for edge lighting, lenses, glazing, sanitary wear. Can be shaped at $150–170\,°C$. Diakon is a powdered form low mol. wt. for injection moulding.

```
    F    F    F    F
    |    |    |    |
  – C  – C  – C  – C  –
    |    |    |    |
    F    F    F    F
```

*PTFE*, polytetrafluoroethene, partially crystalline.
It is characterised by chemical resistance, very low friction, excellent dielectric properties and wide working temperature range $250$ to $-196\,°C$. It is used for bearings (with fillers of bronze or graphite), electrical insulators, non-stick surfaces; fabricated by sintering of powder above $327\,°C$, since its viscosity is too high for normal plastic techniques.

```
    H    H    H    H
    |    |    |    |
  – C  – C  – C  – C  –
    |    |    |    |
    H    Cl   H    Cl
```

*PVC*, Vinyl chloride and Vinyl chloride/vinyl acetate co-polymers.
A versatile range of materials with properties ranging from rigid pipes and gramophone records to soft flexible sheeting for rain coats. They are mixed with heat and light stabilisers and for many applications, plasticisers, fillers and pigments.

*Nylons*, polyamides, partially crystallised, linear chains consisting of 4–11 methylene groups separated by an amide group.
The three common types from a wide variety are Type 6, 610 and 66, one difference being the cross-linking between molecules; Type 66 has the greatest cross-link bonds and therefore

the highest melting point, 264 °C, rigidity and strength; Type 610 the lowest, 222 °C. The frequency at which amide groups ($-NH.CO-$) occur along the chain affects the water absorption and chemical properties. Type 610 has the widest spacing and absorbs less water. Nylons are tough, strong, hygroscopic with low friction. Strength, rigidity and dimensional stabilities are increased by adding glass fibre and friction reduced by graphite and $MoS_2$. They are used in light engineering, especially bearings, gears, power tool housings and extensively as oriented fibre.

*Polyethylene polythene*, partially spherulitic, with short and long chain branching.

Two basic types are low density polythene made by high pressure ICI process and high density polythene made by low pressure process with a catalyst. Many grades are available with densities ranging from 0·913 to 0·940 g/cm³ (closer packing and highly crystalline) and consist of pure polymers of ethylene, mixtures of polymers, with addition of pigments and antioxidants; and flexible co-polymers of ethylene and vinyl acetate (EVA). Various polythenes are often distinguished by the melt flow index at 190 °C, which is a measure of its melt viscosity which is related both to processability and to mechanical properties (BS 2782 Method 105c). Surface hardness and elastic modulus depend mainly on crystallinity; electrical, thermal and optical properties depend mainly on chemical constitution of the polymer. It is cheap, tough, inert, in electrical insulator particularly for HF current; has good resistance to water and chemicals, but is sometimes susceptible to environmental stress cracking by alcohols, metallic soaps, silicon fluid, etc. This can be prevented by incorporating butyl rubbers in the polythene. Thermal stress cracking is also possible in articles containing high internal stresses. Uses include containers, pipes, packaging, cable covering.

ethylene:
$$-\underset{\underset{H}{|}}{\overset{\overset{H}{|}}{C}}-\underset{\underset{H}{|}}{\overset{\overset{H}{|}}{C}}-$$

propylene:
$$-\underset{\underset{H}{|}}{\overset{\overset{H}{|}}{C}}-\underset{\underset{H}{|}}{\overset{\overset{CH_3}{|}}{C}}-$$

*Polypropylene propathene*, partially crystalline —similar to ethylene: but the molecule of propylene is asymmetrical, consequently during polymerisation three basic types of chain structure are possible. If the $CH_3$ groups are on the *same* side of the carbon backbone, the polymer is called *isostactic*, if on alternate sides, called *syndiotactic*, and if irregular, is called *atactic*. In the latter type no crystallisation is possible and it is rubbery in fact. Isotactic polypropylene crystallises to a hard, stiff material. The proportions of these types in a commercial plastic control the properties. It has a low density (0·9–0·91), low water absorption, is free from stress corrosion cracking and can thus be used in contact with detergents but is attacked by strong oxidising reagents. The addition of stabilisers are necessary to resist ultra-violet light and heat oxidation, especially in contact with copper alloys. Co-polymers of propylene and ethylene have greater resistance to shock under cold conditions without loss of rigidity and hardness at high temperatures but are more prone to creep. It is used for domestic appliances and automobile components, valves and pipes.

## *Polystyrene*

Styrene can be polymerised to give polystyrene varying somewhat in mechanical properties but varying widely in mouldability, clarity, tainting tendency and resistance to ageing and stress cracking. These properties can be further modified by additives, by co-polymerising and by blending with rubber, thus giving an immense variety of materials. The main groups are:

**PS** General purpose or unmodified polystyrene.
**TPS** Toughened polystyrene, commonly polymerised in the presence of rubber.
**EPS** Expanded or expandable polystyrene.
Co-polymers SAN (styrene acrylonitrite) also ABS.
They are used extensively for packaging, including vending cups, housewares, lavatory cisterns, tiles, refrigerator liners and insulation.

## Thermo-set plastics

*Alkyds*, amorphous glassy

These are a reaction product of an alcohol with an organic acid and are related chemically to polyester resin. It is usual to distinguish between diallyl ortho phthalate (DAP) and diallyl isophthalate (DAIP) compounds.

Large numbers of different resins are available and are extensively used for paints and moulded items which have excellent dimensional stability and electrical properties, low moisture absorption (0·05%), and will function continuously at $\sim 170\,°C$. They are used mainly in the electrical industry. They have lower hot strength and higher cost than phenolic mouldings.

*Polyesters*, amorphous glassy

They are made by reacting a dihydric alcohol, e.g. ethylene glycol with an unsaturated dibasic acid, e.g. fumaric or maleic anydride. This produces a long chain of highly reactive groups which can be cross linked by a monomer, e.g. styrene or diallyl phthalate (DAP). Thus there are a wide range of resins which can be tailor-made for specific duties and which are mainly used in the glass-reinforced industry for car bodies, boat hulls. Other uses include surface coating, casting, nut locking and flooring. They can be cold or hot cured without gas liberation or surplus liquid, but the exothermic reaction limits the mass that can be cured; 8% contraction occurs.

They are not fire extinguishing.

*Aminos*, amorphous glassy

Amino resins are made by condensing formaldehyde with urea (UF type) or melamine (MF).

Under heat and in the presence of a catalyst polymerisation is completed. Acids may be used to increase the curing.

They are excellent moulding materials and adhesives; used for laminates, tableware and electrical fittings. Mouldings have a glossy hard wearing surface but are brittle and slowly attacked by dilute acids and alkali. They are self-extinguishing. The resins may be obtained as hard solids which are modified with other resins to make surface coatings, or as liquid resins which are hardened by amines or acids for the manufacture of plastic tooling compounds, castings and flooring, or adhesives. They have excellent adhesive and electrical properties, chemical resistance, toughness and flexibility. They do not evolve volatiles during cure, shrinkage is low and dimensions change negligibly afterwards.

*Phenolics* (Bakelite), phenol formaldehyde, amorphous, glassy

They have good heat, electrical and chemical resistance, but strong alkali and oxidising agents will destroy them. Filters include wood, cotton cloth, mica, asbestos, graphite, glass fibre and nylon each conferring characteristic properties. They are used for electrical components, domestic appliances, varnishes, laminates. The storage life in liquid form is limited, and moulding and laminating operations require pressure during curing because volatiles are given off.

*Silicones* have a silicon–oxygen backbone—similar to sand, glass and mica—but attached to the Si atoms are organic side groups such as methyl, phenyl or vinyl, e.g.

$$O - \underset{CH_3}{\overset{CH_3}{Si}} - O - \underset{CH_3}{\overset{CH_3}{Si}} - O - \underset{CH_3}{\overset{CH_3}{Si}} - O$$

By altering the organic group and the length and complexity of the basic linkage silicones can be produced in the form of inert liquids, gums and resins, and by adding fillers, in the form of compounds, greases and rubbers. Silicones are mainly unaffected by temperature from $-75$ to $+250\,°C$, are resistant to moisture, weathering, oxidation and to living organisms; have low coefficient of friction and good dielectric properties, and are water repellant. They are used for insulation impregnation, varnishes, laminates, water proofing, car polishes and rubber seals and moulds.

*Rubber-like plastics*

SBR Styrene–butadiene co-polymer. General purpose rubber whose physical properties are slightly inferior to natural rubber. Used for tyres.

Butyl rubber, isobutylene–isoprene. It has a high hysteresis and low friction, low gas permeability and resistance to ageing. Used for tyre treads and inner tubes.

Nitrile rubber, butadiene–acrylonitrile co-polymer. It has good resistance to heat and petroleum liquids. Used for hose in cars.

*Less known polymers*

Other polymers are polysulphone, a rigid high strength ductile thermoplastics Poly (4–methyl–pentene–1) TPX, a low density (0·83) thermoplastic for applications demanding a combination of clarity and chemical resistance.

Polyurethane is available as flexible and rigid foams and as an elastomer. Polycarbonates, possessing good heat resistance, are produced by the condensation of bisphenol A with phosgene.

*Design*

Principles of design are somewhat similar to those met in casting metals, for example, undercuts and non-uniform thickness should be avoided and generous fillets and draft should be allowed. Large flat surfaces should be stiffened with ribs and where possible compound curved surfaces or sandwich construction should be used in glass fibre mouldings.

*Fabrication*

The relatively poor thermal conductivity of plastics influences the fabrication method, for example, heated screw plasticisers are common instead of heated billets as in metals. Extrusion, injection moulding, blow moulding, compression moulding, vacuum forming by evacuating the air between a sheet and the moulds (see superplasticity), calendering between rolls are common methods.

Fabrication can greatly influence properties for example. A slight variation in extruding conditions may reduce the dielectric strength of polyethylene to 1/20th of its ideal figure. On the other hand, fabrication may so orientate the structure to give improved properties.

Metal coating of plastics may be achieved by electro-plating (on a conducting film) vacuum depositing or spraying.

Welding of parts is possible by many processes, e.g. spin or friction welding, hot plate welding involving local heating of the two surfaces on a hot plate and then clamping together, hot gas welding with filler rod, induction and ultra sonic welding. Cementing is also common, e.g. for styrene, ABS, nylon and acrylic. More difficult to cement are polythenes, acetates, polycarbonate and fluocarbon resins. Snap fits are also useful.

*Mechanical properties* (see methods of testing plastics BS 2782)

At normal temperatures metals deform with an elastic–plastic behaviour, as shown in Fig. 12 and within the elastic region the unloading curve coincides with linear loading curve. Elastomers also exhibit elasticity even after considerable extension in that the material returns to its original shape on release of the load, but the two curves do not coincide and a hysteresis loop is formed, Fig. 259($a$), and this increases under high deformation. This hysteresis effect enables rubber to convert large amounts of strain energy into heat, and is useful as a shock absorber.

Cross-linking reduces viscous flow and the tensile behaviour of nylon is somewhat similar to that of mild steel in that it strains elastically until a yield point is reached and then deforms plastically up to the ultimate ten-

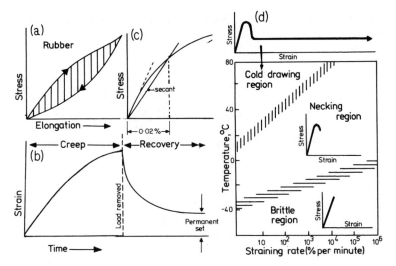

FIG. 259 (a) Elastic hysteresis in rubber
(b) Visco-elastic behaviour
(c) Stress-strain curves for a polymer and the secant used for calculating Young's modulus
(d) Effect of temperature and straining rate on the tensile behaviour of polypropylene with typical stress-strain curves

sile stress with high elongation of 50–150%, which however is much lower than for rubbers which have value of 400–600%.

In general, however, plastics display a combination of solid-like and liquid-like behaviour termed *visco-elastic*. A rapidly applied load evinces a partially elastic response because the contorted molecules can be instantaneously stretched; under a slowly applied load they glide slowly past one another in plastic flow, which is both time and temperature dependent and non-recoverable, i.e. somewhat similar to metals under creep conditions, p. 21.

Provided a thermoplastic has not yielded, original dimensions can be recovered when the load is removed, even after long periods of creep. Such recovery is time dependent, as shown in Fig. 259(b). If yielding has occurred, a permanent set results.

Since the stress-strain curve is not linear, an approximation must be used in the calculation of Young's modulus, such as the use of the secant between zero and 0·2% strain, Fig. 259(c). The design problem may be approached by postulating that a constant load be applied to a component for a specified lifetime at a maximum service temperature with a maximum allowable strain. The method requires the use of creep curves, as shown in Fig. 260(a), (b) and (c).

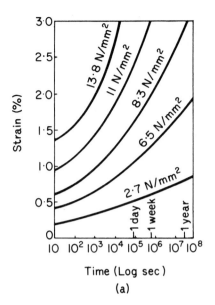

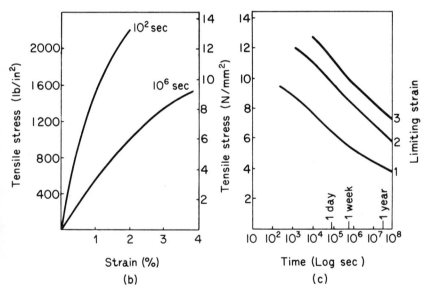

Fig. 260 (a) Propathene homopolymer creep intension at 20°C (density 0·909 g/cm)
(b) Isochronous stress-strain curve
(c) Isometric stress-time curve *(after ICI Ltd)*

## Low temperature service

Polymers are usually unsuitable for low temperature work as they are rapidly embrittled as the temperature falls to sub zero, Fig. 259(*d*). Typical strain curves are shown, and particularly interesting is the cold drawing type with large elongation, indicating some lack of work hardening.

## Fatigue of polymers

These behave in a variety of ways depending on the temperature, stress and cycling speed. Often it is the simple question of the propagation of existing micro-cracks, or induced surface defects from degradation processes. Unfortunately, particularly with amorphous polymers above the glass temperature Tg, high cycling rate produces sufficient heat in the body of the specimen, as a result of hysteresis effects, to soften it, and failure by severe deformation instead of fracture then occurs.

During fatigue on polymers a high degree of damping occurs, and it is highly difficult to obtain meaningful results on normal testing. The damping effect in these materials, particularly the elastomers, is made use of in anti-vibration mountings and sound damping materials.

TABLE 80  BASIC MECHANICAL PROPERTIES OF SOME PLASTICS

|                     | Young's Mod $N/mm^2$ | TS $N/mm^2$ | El % | Izod J |
|---|---|---|---|---|
| Polyethylene HD     | 55–104    | 21·4–38  | 50–600   | 1·5–1·2    |
| Polypropylene       | 90–138    | 29·6–39  | 50–600   | 0·6–8·0    |
| Polystyrene         | 275–410   | 34·5–83  | 1–2·5    | 0·2–0·5    |
| Poly vinyl chloride | 240       | 59       | 2–40     | 0·8        |
| PTFE                | 26–45     | 17–24    | 250–300  | 3·5–5      |
| Melamine (filled)   | 900       | 48–90    | 0·6–0·9  | 0·2–0·35   |
| Poly acetals        | 275       | 69       | 15–75    | 1·4–3·3    |
| Nylon 66            | 290       | 80       | 40–80    | 1·3        |
| Nylon glass filled  | 780       | 173      | 3        | 1·4        |
| Mild steel          | 20 600    | 350      | 30       | 30         |

Some typical mechanical properties are given in Table 80 but it is important to note that any quoted property is only valid for the conditions under which it was evaluated. Variations in temperature strain rate humidity produce considerable changes.

# 20 Metallurgical Aspects of Metal-joining

Metals are joined together by soldering and welding in its various forms and this involves many metallurgical considerations.

*Solders*

Numerous soft solders have been standardised (BS 219), for various purposes, but they consist essentially of tin and lead in varying proportions with occasionally the addition of antimony (1–3%). Antimonial alloys are unsuitable on brass because antimony and zinc together cause brittleness of joint. A eutectic is formed with 62% tin, and an alloy of this composition is known as a tinman's solder. It has the advantage of solidifying quickly without passing through a weak pasty stage. Plumbers' solder, on the other hand, consists of about 2 parts of lead and 1 part of tin and begins to freeze at about 240 °C, but remains pasty down to the eutectic temperature (183 °C). The long range of solidification enables the plumber to make his 'wiped' joint. Reaction of tin with the substrate promotes wetting. Copper reacts faster with tin than do iron or nickel and hence is more solderable. For higher operating temperatures than Sn–Pb alloys will withstand tin with 5% Sb (95A) or lead with 1·5% Ag are used.

There are a number of specialised alloys such as tin–lead–iridium–zinc to reduce embrittlement when soldering on to gold coatings; tin–52 iridium for soldering glass or glazed ceramics and tin–70 cadmium for systems where stray emfs due to thermoelectric effects must be kept to a minimum. Zinc chloride or a resin type of flux is necessary to dissolve any thin oxide films on the metal surfaces to be joined.

Soldering of aluminium is difficult due to the oxide film. The Al–Zn solder is melted on to the metal and the surface beneath it is scraped. The corrosion resistance of the joint is poor. An alloy containing 97·5 Pb 1·5 Ag 1 Sn has been suggested for nuclear energy applications.

*Brazing* is really a high temperature soldering operation using borax as a flux and one or other of the copper–zinc alloys (p. 307), but now extended to a wide range of filler materials shown in BS 1845:

| Type Al | Aluminium brazing alloys | (5 and 12% Si) | 530–630 °C |
| Type AG | Silver brazing alloys | (40–100 Ag) | 600–960 °C |
| Type CP | Copper phosphorous alloys | (CP1 = 5P 14 Ag; CP3 = 8P) | 645–800 °C |

| Type CU | Copper brazing metal | 1085°C |
| Type CZ | Brazing brasses | 860–980°C |
| Type NI | Nickel base alloys with P, Si and Cr | 875–1135°C |
| Type PD | Palladium bearing | 805–1235°C |
| Type AU | Gold bearing alloys | 905–1020°C |

Silver is added to copper to form silver solder with the addition of zinc, cadmium or tin to regulate the melting range and flow behaviour. Typical alloys are:

TABLE 81

| BS No. | Ag | Cu | Zn | Cd | Sn | Melt °C |
|---|---|---|---|---|---|---|
| AG 1 | 50 | 15 | 16 | 19 |    | 620–640 |
| AG 3 | 38 | 20 | 22 | 20 |    | 605–650 |
| AG 4 | 61 | 29 | 10 |    |    | 690–735 |
| AG 5 | 43 | 37 | 20 |    |    | 700–775 |
| AG 6 | 60 | 30 |    |    | 10 | 600–720 |
| AG 7 | 72 | 28 |    |    |    | 780–980 |

Copper containing 15 Ag, 5P (e.g. Silfos) melting at 700°C is also used.

For mass production, hydrogen furnace brazing offers many advantages. The hydrogen reduces the surface oxides, obviating the need for flux. Capillary attraction pulls the fluid copper into the close-fitting joint.

Vacuum brazing is also becoming a necessity for some alloys e.g. Nimonics. The electronics, nuclear power and aerospace industries have created a demand for a range of high temperature brazing alloys suitable for diverse materials such as used in a magnetron involving a dozen operations. This necessitates step-by-step brazing with successively lower melting point alloys. Palladium and gold based alloys are used. Palladium readily forms solid solutions with many metals hence good weldability. Joint strength may be far greater than the strength of the filler alloy since a strong base material stiffens up the joint, and joint design and brazing conditions have an influence.

Gold bearing alloys are mainly based on gold–copper and gold–nickel, both systems form continuous solid solutions characterised by a dip at about 80% Cu, giving a very narrow melting range which is associated with high fluidity and ability to fill gaps. The gold alloys resist corrosion and produce ductile joints without excessive alloying and are free from volatiles which make them suitable for operation in a vacuum at elevated temperatures. The Au–Ni alloys have greater high temperature strength and resistance to oxidation. Typical BS alloys are:

*Welding* can be divided into pressure and fusion processes. In fire-welding of wrought iron and carbon steel by the blacksmith, the parts which

# WELDING

TABLE 82  SOME PALLADIUM BEARING ALLOYS

| BS No. | Pd | Ag | Cu | Mn | Ni | Melt °C | Use |
|---|---|---|---|---|---|---|---|
| PD 1 | 5 | 68·4 | 26·6 | | | 805–810 | Low vapour pressure, stainless and Ni and Co super alloy |
| PD 5 | 20 | 52 | 28 | | | 875–900 | Good flow and penetration |
| PD 7 | 5 | 95 | | | | 970–1010 | |
| PD 9 | 20 | 75 | | 5 | | 1000–1120 | Creep resistance joints in Co, Ni, W and Mo alloys |
| PD 11 | 21 | | | 31 | 48 | 1110–1120 | Ni and Co super alloys, W, Mo, resistant to alkali metals and ammonia |
| PD 12 | 20 | | 55 | 10 | 15 | 1060–1110 | High strength, stainless and nickel based alloys |
| PD 14 | 60 | | | | 40 | 1235–1235 | High strength, low vapour pressure. Used for W, Mo, first-step joints, cermets |

have been heated below the solidus temperature are hammered and during the operation the surface scale is broken and crystals grow across the joint. The hot working produces a refined structure not obtained by fusion methods. The slag in wrought iron acts as a flux, but for steels sand is used, and this combines with iron oxide to form a fluid slag.

In electric resistance methods, such as butt, spot and seam welding, the metal near the joint is made part of a low voltage electric-circuit, and the heat developed renders the metal plastic or molten locally and the pressure applied to the parts produces a good joint, Fig. 261. In friction welding two parts are rotated or oscillated relative to each other to develop frictional heat and then pressure is applied. Fig. 263 shows a typical structure. It

TABLE 83

| Type | Au | Cu | Ni | Fe | Melt °C |
|---|---|---|---|---|---|
| Au 1 | 80 | 19 | | 1* | 905–910 |
| Au 3 | 37·5 | 62·5 | | | 980–1000 |
| Au 4 | 30 | 70 | | | 995–1020 |
| Au 5 | 82·5 | | 17·5 | | 950 |
| Au 6 | 75 | | 25 | | 950–990 |

*Iron is added to retard ordering transformation and volume changes

may be used for joining dissimilar metals. In explosive welding the charge brings two plates together and produces marked flow lines as shown in Fig. 262.

Much attention is being given to pressure welding, sometimes called 'Solid Phase' or recrystallisation welding. It depends on producing intimate contact between the two surfaces and this requires the removal of surface contaminants, usually by scratch brushing and then the breaking up of the oxide and scratched layer by plastic deformation of about 60% RA for full bond strength.

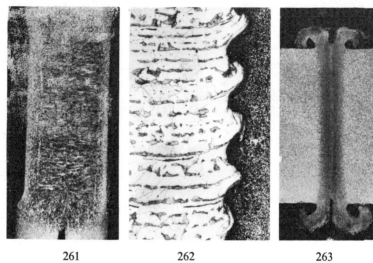

261      262      263
FIG. 261 Spot weld on mild steel ($\times 12$)
FIG. 262 Explosive weld between mild steel and brass ($\times 75$)
FIG. 263 Friction weld on mild steel ($\times 2$)

*Fusion welding* consists in melting the parent plate and also normally adding extra metal from a filler rod or coated metal electrode.

The oxyacetylene blowpipe, carbon arc or tungsten arc with hydrogen, helium or argon gas shield are especially suited for welding thin sheets. For welding heavy plates the metallic arc process using flux coated wires or carbon dioxide shield or the submerged arc using a fused granular flux or the electro-slag process are preferred.

The flux minimises the absorption of oxygen and nitrogen from the atmosphere; forms a fusible slag with any oxide present; reduces the volatilisation of alloying constituents in the article. In metal-arc welding electrodes the flux stabilises the arc, in many cases enabling ac supply to be used; at the same time it largely controls the weld contour and the penetration.

## Adhesive bonding

The use of adhesives for joining materials is gaining increasing recognition and has been particularly exploited in the aeronautical field. The main types of adhesive are:

(1) Air-drying, mainly thermoplastics as emulsions or solutions. Bond strength develops after removal of solvent.
(2) Fusible adhesives which on heating become liquid.
(3) Pressure-sensitive adhesives, mainly high viscosity liquids.
(4) Chemical reactive adhesives, mainly thermosetting liquids or solids activated by the addition of a hardener and/or heat.

One of the most successful bonding techniques is *Redux* bonding where the metal is given a coat of phenol formaldehyde, over which is scattered Polyvinyl formaldehyde powder. The prepared surfaces are brought together under pressure.

*The physical, mechanical, chemical and metallurgical properties* of the metals concerned largely govern the choice of welding process. Some of these properties are listed in Table 84. High thermal *conductivity*, e.g. copper and aluminium, means that the weld pool is cooled rapidly necessitating preheating, and the temperature of the surrounding plate is raised higher than in steel, i.e. wider heat affected zone. In resistance welding high *electrical conductivities* result in less heat for a given current, i.e. higher currents are required. Small additions of alloying elements dissolved in a high conductivity metal rapidly reduce the electrical conductivity (p. 286). *Thermal expansion* and contraction affect the distortion and cracking of welds. The non-ferrous metals and the austenitic stainless steels have higher coefficients of expansion than mild steel.

## Metallurgical effects of welding

Welding operations give rise to many metallurgical effects. The weld metal is essentially a *small casting*, with the inherent defects and characteristics of a casting. An appreciation of these properties can be easily attained by a study of the mechanism of crystallisation of metals and alloys described in Chap. 4.

The formation of cored columnar crystals at right angles to the parent metal can be seen in Fig. 264, which shows an electric weld consisting of two runs in Staybrite steel, a solid solution. The first run has been heated by the second and consequently diffusion of the alloying elements has occurred to a greater extent, as indicated by the less sharply defined structure. In ordinary steels the second run normalises the first layer and in making large welds consisting of many runs, an extra one is frequently made for this purpose and then machined off.

TABLE 84  PHYSICAL PROPERTIES OF METALS AND ALLOYS

| Material | Melting Range, °C | Specific Gravity, gm/cm³ 20°C | Coeff of Expansion °C × 10⁻⁶ 0–300°C | Thermal Conductivity Cal/cm²/cm/°C/sec Room Temp | Electrical Resistance Microhms/cm³ |
|---|---|---|---|---|---|
| Iron | 1537 | 7·87 | 13·5 | 0·190 | 10·0 |
| Copper (welding) | 1083 | 8·94 | 17·7 | 0·45 (pure = 0·92) | 1·7 |
| Nickel | 1455 | 8·90 | 14·4 | 0·145 | 7·8 |
| Silver | 960 | 10·5 | 19·1 | 1·00 | 1·5 |
| Aluminium | 660 | 2·70 | 25·6 | 0·520 | 2·8 |
| Magnesium | 651 | 1·74 | 27·9 | 0·370 | 4·6 |
| Cast-iron-low P | 1240–1120 | — | 11·6 | 0·148 | — |
| 0·5% carbon steel | 1500–1420 | 7·85 | 13·3 | 0·100 | — |
| 12% chromium steel | 1530–1510 | 7·70 | 11·4 | 0·046 | 55·0 |
| 27% chromium steel | 1510–1490 | — | 10·9 | 0·05 | 67·0 |
| 18/8 Cr–Ni–stainless | 1420–1395 | 7·92 | 18·1 | 0·033 | 73·0 |
| 80/20 Ni–Cr | 1420–1400 | 8·40 | 13·9 | 0·038 | 108·0 |
| Monel | 1350–1300 | 8·80 | 15·0 | 0·06 | 42·5 |
| Brass 70/30 | 950–925 | 8·53 | 19·8 | 0·290 | 6·9 |
| Bronze 7·5% tin 0·6% | 1030–850 | 8·73 | 18·2 | 0·148 | 10·5 |
| Aluminium bronze 93/7 | 1042–1040 | 7·6 | 17·8 | 0·192 | 12·1 |
| Everdur A | 1050–1000 | 8·54 | 18·0 | 0·112 | 26·0 |
| Aluminium silicon 13% | 577 | 2·66 | 21·0 | 0·35 | 5·4 |
| Aluminium magnesium 7% | 620–550 | 2·63 | 24 | 0·31 | 5·7 |

*Gas solubility* in liquid and solid metals, and gas reactions (p. 65), are important in controlling the porosity of a weld, e.g. hydrogen in aluminium reaction of hydrogen and copper oxide.

Metallic arc welds made with a bare wire are liable to *contamination by gases* from the atmosphere—oxygen and nitrogen. The nitrogen frequently appears as needles on certain planes in the crystals in the last run. At high magnification, a dark constituent is also found at the grain boundaries, which sometimes causes low impact strength especially after a water quenching treatment.

The form of the solubility curve of nitrogen in ferrite is very similar to that of carbon. Structures similar to martensite (termed nitro-martensite)

FIG. 264  Arc weld in Staybrite
Deposition by globules is indicated   ( × 60)

and to pearlite (termed braunite) being observed in the nitrogen iron system (see p. 170). On account of this, a weld made with a covered electrode exhibits a micro-structure with less pearlite and is cleaner than a corresponding one made with a bare electrode.

Fig. 265 illustrates the structure of the weld metal made with a bare wire electrode. Fine needles are seen together with trapped oxide and cavities. Manganese inhibits the formation of needles.

A study of formation of the Widmanstätten structure (p. 165) is of fundamental importance in understanding the structure of welds in mild steels.

*Slag inclusions* are frequently trapped in fusion welds due to joint or bead contour and the difficulty of melting the slag in subsequent runs. In

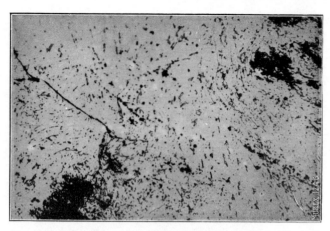

FIG. 265 Arc weld made with bare wire electrode (mild steel)
Oxide films and nitride needles (× 200)

metallic arc welds in mild steel microscopic inclusions are also present; their composition is different from, but related to, the type of electrode flux, Table 85.

TABLE 85  COMPOSITION AND PROPERTIES OF MILD STEEL ARC WELDS

| | Wire | | | Welds | | | | |
|---|---|---|---|---|---|---|---|---|
| | Normal Core | Sub-merged Arc | Bare Wire Weld | Deep Groove Weld Fe–Mn–Silicate Flux | Down-hand Fillet Viscous Rutile Flux | Positional Fluid Rutile Flux | Low Hydrogen Lime–Fluoride Flux | Sub-merged Arc |
| C% | 0·10 | 0·13 | 0·03 | 0·06 | 0·06 | 0·06 | 0·08 | 0·09 |
| Mn% | 0·40 | 0·92 | 0·04 | 0·30 | 0·40 | 0·40 | 0·70 | 1·0 |
| N% | 0·003 | 0·005 | 0·14 | 0·02 | 0·03 | 0·02 | 0·01 | 0·01 |
| O% | 0·012 | | 0·25 | 0·10 | 0·12 | 0·11 | 0·05 | 0·087 |
| H% | 1 | | 2 | 17 | 15 | 15 | 3 | 0·5 |
| Main inclusion | | | | (MnFe)O, 2FeOSiO$_2$ | MnOSiO$_2$ | MnOSiO$_2$ | 2MnOSiO$_2$ | MnO + MnOSiO$_2$ |
| TS N/mm² | | | 401 | 448 | 525 | 479 | 494 | 525 |
| E | | | 6 | 30 | 27 | 30 | 38 | 26 |
| Izod J | | | 8 | 89 | 61 | 68 | 123 | 34 |

Controlled amounts of nitrogen and manganese together with a dispersion of fine non-metallic inclusions and especially a fine grain size with high dislocation density enable mild steel arc welds to have strengths of 430–570 N/mm² with adequate ductility. See Table 85.

The *contour of welds* by forming 'notches' can affect both fatigue (p. 24) and low temperature properties (p. 15) of a structure.

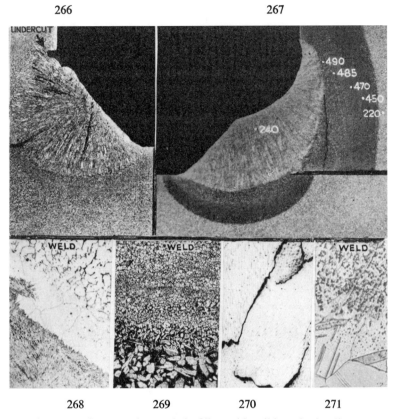

FIG. 266 Hot cracks around crystals in fillet weld; mild steel (× 3)
FIG. 267 Hard zone crack. Mild steel weld on high tensile steel (× 4)
FIG. 268 Special 18/8/3–Cr/Ni/Mo weld on armour plate. Ferrite islands in weld; martensite in plate (× 200)
FIG. 269 Aluminium bronze weld 10/5/5–Al/Ni/Fe on casting (× 200). Note fine grain size in weld
FIG. 270 7% aluminium bronze arc weld; intercrystalline cracks (× 100)
FIG. 271 8% tin bronze arc weld on 70/30 brass (× 200)

## Hot cracking of welds

Under constrained welding conditions the contractional strains sometimes cause intercrystalline cracks in the *hot* welds, the fractured surface being tinted with oxidation films (Fig. 266). High arc welding currents cause large columnar crystals to form with definite planes of weakness at the throat of the weld. Austenitic steels, e.g. 25 Cr, 20 Ni, which normally solidify with large *columnar crystals*, are prone to this trouble. In the 18/8 Cr–Ni alloys cracking is obviated by the addition of 2–4% molybdenum which refines the structure by forming a ferritic–austenitic structure (Fig. 268).

*Low melting-point compounds*, such as iron sulphide in steel, nickel sulphide in nickel, are a common cause of cracked welds (p. 169), especially in welding free-cutting steels. The addition of sufficient manganese largely overcomes the troubles in steel by forming a high melting-point sulphide.

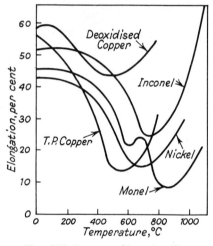

FIG. 272 Ranges of low ductility

*Ranges of low ductility* occur in some materials within certain temperature ranges as shown in Fig. 272. If welds are stressed at about 700°C cracks may develop.

*The effect of welding on the plate* is also important, since the operation causes temperature gradients in the plates, the limits of which are the melting-point of the metal and room temperature. In the case of steel a portion of the plate will be within the critical ranges.

If such a specimen is cooled slowly the ferrite which was taken into solution by the austenite is deposited at the boundaries of the small austenite crystals and is followed by the formation of pearlite. The areas of the latter are very much smaller than those originally in the specimen and each is surrounded by fine ferrite networks, which can be distinguished from the original skeleton of ferrite. This structure is illustrated in Fig. 273, and diagrammatically in Fig. 274. This treatment is also known as *under-annealing*.

Fig. 267 shows the hard zone formed in high tensile steel.

When the plates which have to be welded are in a cold-worked condition due either to slight cold-rolling or to deformation during fabrication, re-crystallisation effects arise as discussed in Chap. 7. At inclusions lamellar tearing can occur.

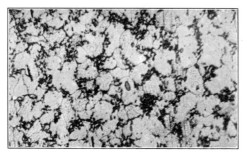

FIG. 273 Mild steel plate adjacent to arc weld. Shows refining of pearlite areas (× 200)

## Welding high tensile steels

Increasing use has been made of welding in the fabrication of low alloy steels with yield points up to 430 N/mm² (p. 227) and also of oil- or air-hardening steels, e.g. armour. While carbon and alloys render the welds more liable to hot cracks, the most important defect is likely to be cracks in the hardened zone of the plate, i.e. *hard metal cracks* which form at or near room temperature (Fig. 267).

The factors which influence this cracking are:

(a) *Constraint*, such as in rigid structures which sets up multi-axial stresses in the weld (p. 13). These stresses increase with thicker plates and are influenced by the design and the welding technique used.

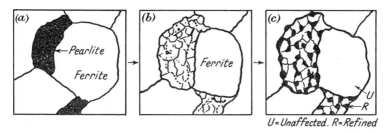

FIG. 274 Diagrammatic illustration of formation of structure shown in Fig. 273
(a) original; (b) between $A_1$ and $A_3$; (c) after cooling

(b) *Formation of a brittle martensite zone* adjacent to the weld. Its hardness is greater with increase of carbon contents in particular (p. 185), of alloy content (p. 195) and rate of cooling. To minimise cracking the end of the martensite reaction should be above 290 °C.

(c) *Hydrogen* dissolved in the weld bead and its diffusion into the parent metal. Atomic hydrogen is soluble in austenite at all temperatures but practically insoluble in cold ferrite or martensite. When the heat-affected zone transforms, therefore, atomic hydrogen is rejected and tends to collect in inclusions, pores and submicroscopic defects to form molecular hydrogen which cannot diffuse away and therefore builds up a high-pressure additive to the contraction stresses. Alternatively the hydrogen may reduce the cohesive strength of the steel (p. 13), thus enabling a brittle fracture to occur with little strain. Hair-line cracking in alloy steel ingots is a related phenomenon (p. 233).

One method of overcoming the trouble due to hydrogen has been the use of 18/8/3 Cr—Ni—Mo austenitic arc-welding electrodes, having a flux covering with the minimum of water containing substances. The austenite acts as a reservoir for the hydrogen which is moderately low (12 ml per 100 gm). The strength of the weld is about 700 N/mm$^2$ (Fig. 268 shows junction between a high tensile steel and the weld.)

For welding steels which do not harden higher than 450 VPN adjacent to the weld, special mild steel welds with low hydrogen content (less than 3 ml per 100 gm) are satisfactory. Cracking is unlikely to occur even with ordinary mild steel electrodes (H = 18 ml per 100 gm) if the mean hardness of the hard zone does not exceed 350 VPN. This hardness is reduced by:

  (i) The use of large size electrodes.
  (ii) Heavy beads.
  (iii) Continuous welding and
  (iv) Preheating to 100–200 °C which also assists in diffusion of hydrogen

The carbon equivalent value is sometimes specified (BS 4360) to fix the limit of carbon and alloy content in regard to welding.

$$CE = C + \frac{Mn}{6} + \frac{Cr + Mo + V}{5} + \frac{Ni + Cu}{15}$$

CE <0·14 no special precautions with rutile electrodes
0·41–0·45 use low hydrogen electrodes *or* preheat
>0·45 use low hydrogen electrodes + preheat

For critical steels it is helpful to precoat the surface of the joint with austenitic weld metal preferably of the 25–Cr 20–Ni type.

The stray arcing of an electrode on high tensile steels is liable to produce a localised hard zone from which can initiate a fatigue crack.

The deposition of mild steel over 18/8 austenitic welds is liable to produce cracks in the mild steel which is hardened by the pick-up of alloying elements.

The WI Controlled Thermal Severity (CTS) test is shown in Fig. 275. A square plate of the required material is fillet welded along two opposite sides to the base plate. The remaining two sides are then welded up with the electrodes to be tested, the second weld being laid down immediately after the first. The assembly is allowed to stand for a period of time, after which the welds are sectioned and examined for cracks. The severity of the test may be varied by altering the thickness of the plates, the hydrogen level in the test welds and the composition of the weld metal. The cooling rate is designated by means of a *Thermal Severity Number* (TSN). TSN 1

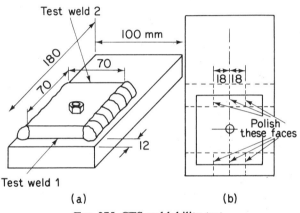

FIG. 275 CTS weldability test

is the thermal severity corresponding to heat flow along a single steel plate 6 mm thick. TSN 2 is obtained in butt weld between two 6 mm plates, while in a 6 mm tee joint, where there are three heat flow paths, the thermal severity number is 3. The TSN number is also increased in proportion to the plate thickness, so that a CTS testpiece in 12 mm plate would have TSN 6.

*Lamellar tearing* sometimes occurs in structural steels parallel to the surface of a plate beneath heavy fillet welds. The cracking usually follows planes of non-metallic inclusions within the plate giving a step-like fracture appearance in transverse sections.

## Hard facing

Welding is being increasingly used to hard face articles, especially for salvage purposes. The effect on the basis metal requires consideration;

materials such as mild steel, austenitic steel and copper alloys are not greatly affected by localised heating. High tensile steels, high speed steels, and cast iron are readily hardened and cracked by localised heat as discussed above. It is therefore necessary to preheat such materials to 150–550 °C.

Typical hard facing alloys are:

(1) *Austenitic* 14% manganese steel with a deposited hardness of 250 VPN which work-hardens to 400. Welds frequently contain 4% nickel and low carbon (0·5%) contents but still tend to crack and become porous. To avoid a brittle manganese alloy layer when welding on to ferritic parent metal a cushion layer of 18/8 austenitic steel should be used and also for joints in manganese steel.

(2) *'Cold' martensitic.* Arc welds contain 2–6% chromium with 0·5–1·0% molybdenum and 0·2–0·5% carbon giving hardness 350–650 VPN but which are readily softened on heating to 650 °C.

(3) *'Hot' martensitic.* Welds in this group contain large amounts of chromium, molybdenum, cobalt, tungsten and boron (p. 239) and retain or increase their hardness at red heat. They are suitable for applications, e.g. lathe tools, involving work at elevated temperatures.

(4) *Non-corrosive.* For applications involving corrosion such as acid valves, hardness must be combined with resistance to corrosion. High chromium contents in cobalt, nickel or iron base are useful, e.g. stellite, colmonoy. Complex aluminium bronze is useful for pressing dies.

(5) *Carbide.* Fused tungsten carbide particles imbedded in a mild-steel matrix are useful for oil-well drilling bits since a 'sandpaper' cutting effect is achieved.

(6) *Marfacing.* Maraging steel can be deposited on structural steels or on softer types of maraging steel. The as-deposited hardness of 350 is raised above 500 by heating to 480 °C.

## Welding cast iron

From Chap. 13 it will be clear that the term 'cast iron' covers a wide range of structures, the majority, however, consists of pearlite and graphite; low silicon irons with fine graphite flakes or spheroids are easier to weld than high silicon irons since large flakes do not form a foundation for a weld.

In the presence of high silicon contents pearlite is readily decomposed into ferrite and graphite at temperatures above 750 °C. Solution of the graphite near the melting-point usually offsets this decomposition and gives rise to hard martensitic zones as in welding high tensile steels. If the iron is low in silicon free cementite or white iron eutectic is formed on rapid cooling, forming hard spots. Because of the low ductility of cast iron the volume changes involved in welding promote liability to cracks even at distances from the weld in a rigid structure. To avoid cracks, hard welds and heat affected zone preheating to 400–600 °C is used followed by slow cooling. For this reason the oxy-acetylene process is often preferred for welding cast iron. The usual filler rods contain about C, 3·5; Si, 3·0; P, 0·6%; but alloy rods are available to match the austenitic cast irons. So-called bronze welding employs 60/40 brass which forms a joint at about 1000 °C, which minimises some of the welding difficulties.

FIG. 276 Cast iron weld using copper–nickel rod (× 100)

Note the cored dendritic structure in the weld

For arc welding electrodes are used with mild steel, monel, nickel and bronze core wires. Without preheat the mild steel welds dissolve carbon and become unmachinable due to the presence of martensite or white iron. A readily machinable *weld* deposit is obtained by the use of nickel alloy or bronze electrodes since hard carbon compounds are not formed in these alloys. In welding without preheat, however, a hard heat affected zone is still formed, independent of the electrode used. This zone can be minimised by the use of a 250 °C preheat.

In welding without preheat steep temperature gradients should be avoided by intermittent welds spread over the article and by the use of low currents which also assist in avoiding the undesirable dilution of bronze welds with iron. The structure of a weld made with a copper–nickel electrode is shown in Fig. 276. The dendritic structure of the weld is clearly evident.

18/8 austenitic steel electrodes also make satisfactory welds in some applications.

## Welding stainless steels

### Effects of oxygen

The formation of adherent oxide films has already been discussed in Chap. 8. From a rust-resisting point of view this property is of fundamental importance but unfortunately this characteristic adds difficulties when welding is undertaken, necessitating the use of a powerful flux.

### Carbon

The effect of carburising atmospheres on the steels is important, since the chromium readily forms chromium carbide and the alloy does not develop its full corrosion resistance. An excess of acetylene gives carburised welds which are white, flat and easy to make, but are unsatisfactory as regards corrosion resistance and mechanical properties.

*The chromium* steels become increasingly brittle as the chromium content exceeds 15% and if held at 450–550°C, but toughness is improved by the addition of nitrogen in amounts equal to 1/100 of the chromium content and also by raising the temperature to 120°C. A preheat of 200–400°C should be used to minimise the risk of cracking and hardening. The addition of niobium or titanium to combine with the carbon is useful in reducing the capacity for hardening of the plate, but is not effective in the weld, to soften which requires annealing at 770°C for the 14–18% Cr alloys and 870°C for the 25% Cr alloy. Where annealing of the weld is impossible the 18/8 austenitic electrodes are sometimes successfully employed, provided that service temperature fluctuations do not set up stresses due to the differences in coefficient of expansion; that galvanic corrosion is not serious in the particular environment and also that sulphur rich gases are absent.

The influence of welding on the micro-structure is shown diagrammatically in Fig. 277 which

(a) indicates the formation of a hard heat effected zone in stainless iron, and
(b) the formation in the higher chromium irons of coarse-grained metal which has a poor resistance to shock and cannot be cured by heat-treatment alone.

*The austenitic stainless steels* present no particular difficulty in welding. Fig. 277(c) shows the defect known as weld decay described on p. 256. This trouble is now overcome by the addition of titanium, niobium or silicon to the plate and niobium to the weld since titanium readily oxidises in arc welding. Fig. 264 shows the structure of a weld in Staybrite and Fig. 278 a weld in Weldanka in which ferrite is also present.

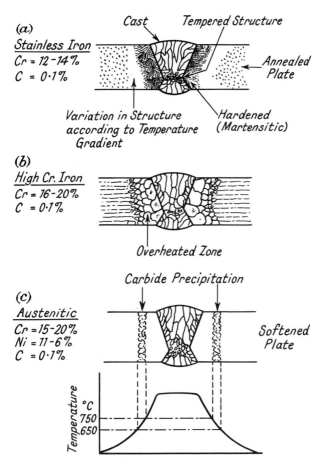

FIG. 277 Diagrammatic representation of effects of welding on stainless steels

In the presence of strong nitric acid titanium containing 18/8 steels after very high solution treatment (e.g. 1350 °C) often suffer rapid intergranular attack at the weld junction known as *knife edge* corrosion. Niobium stabilised steels are freer from this trouble.

The Schaeffler diagram shown in Fig. 279 indicates the phases formed for various stainless steels, using nickel and chromium equivalents. The effects of dilution and the amount of ferrite can be estimated from the diagram.

## Welding copper and its alloys

The welding of copper is rendered difficult by its high thermal conductivity which, depending on its purity, is five to nine times that of mild steel. This necessitates the use of preheating and conservation of heat by the use of asbestos sheets.

Adequate provision must be made for contraction during welding especially since copper has a low strength at elevated temperatures. Tacking should not be used.

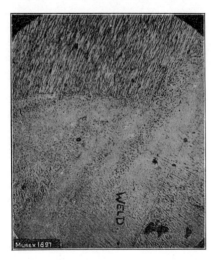

FIG. 278 Weld in Weldanka—17/10/1·5–Cr/Ni/Si

For welding, the type of copper has an important influence on the properties of the weld. Tough pitch copper is unsuitable since the dispersed oxide particles present in rolled material are replaced by intercrystalline films in the remelted copper, resulting in mechanical weakness, especially at elevated temperatures. When hydrogen is present, such as in the oxyacetylene process, reaction occurs with the oxygen (see p. 303), resulting in porosity. Hammering of the hot weld serves to break up the oxide network and to close the pores, but the trouble is overcome by the use of deoxidised copper plate with a filler rod containing 1% silver and 0·05% phosphorus. Where high electrical conductivity is essential oxygen-free high-conductivity copper should be used, but the welds may have lower properties than with deoxidised copper. Selenium or tellurium in excess of 0·01% causes porosity in the weld and in the basis metal near the weld. Copper filler can be used with argon arc.

Satisfactory metallic arc welds on copper can be made with 7% tin or aluminium bronze electrodes.

## Welding brasses

Owing to the low boiling-point of zinc (925 °C) its volatilisation is liable to cause serious porosity and unpleasant fumes unless oxidising conditions

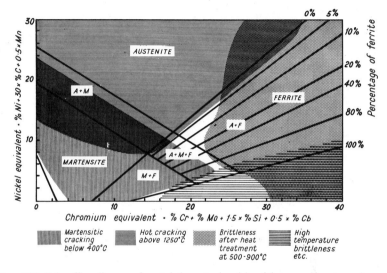

FIG. 279 Schaeffler diagram for stainless steel welds with approximate regions of defects depending on composition and phase balance

(*By courtesy of Bystram, Murex Welding Processes Ltd*)

prevail to form a protective film of zinc oxide. Although the volatilisation of zinc prevents the use of a brass wire for metallic arc welding, excellent welds can be made with either tin or aluminium–bronze electrodes. Fig. 271 shows the junction of a typical example. Some of the additional elements in the high tensile brasses affect their welding; lead promotes hot cracks but iron is harmless. Aluminium is troublesome on account of the refractory film of aluminium oxide making the metal sluggish and preventing coalescence of the drops. Annealing of welded jobs at 260–280 °C is frequently desirable to remove stress and hence risk of season cracking (p. 120).

## Welding bronzes and other alloys

### Tin bronze

The thermal conductivity of bronzes is similar to that of steel so that the heating condition should be similar also. The wide freezing ranges of the bronzes give rise to cracks at elevated temperatures and porosity due to shrinkage cavities. Care should be taken to prevent stressing the joint while it is hot and puddling or overheating should be avoided to minimise segregation and also the absorption of gases which result in porosity on solidification. A neutral or slightly oxidising atmosphere is desirable.

The ductility of bronzes containing more than 8% tin may be impaired by the presence of a brittle micro-constituent—$\delta$ phase (see p. 314). The presence of lead over 2% is also liable to lead to hot shortness.

### Silicon bronzes

The thermal conductivity is low and there is no trouble from oxidation of the melt. A silica film forms on the surface of the liquid metal which tends to prevent ingress of gas. The weakness of these alloys at temperatures near the melting-point leads to cracked welds when there is any restraint present.

Intercrystalline cracking in both deposited weld metal and plate material have been found, although this cracking phenomenon appears to be associated more with welds made by the oxyacetylene process, probably due to the deleterious effect of water vapour on Everdur at elevated temperatures. The argon or carbon-arc and electrical-resistance welding processes obviate the moisture vapour and welds made by these processes are found to be almost completely free from intercrystalline unsoundness.

### Aluminium bronze

The main difficulty in welding is due to the tenacious film of aluminium oxide which is not removed readily by the usual fluxes used for oxyacetylene welding of copper alloys. Consequently oxide films are frequently present in such welds. Satisfactory welds can, however, be made with metallic arc electrodes. In multi-run welds with the 7% alloy, however, a reduction in strength occurs due to intercrystalline cracks forming at about 700 °C under stress (Fig. 270). Bismuth in excess of 0·0003% appears to be the cause; this can be nullified by the presence of lithium. The structures of aluminium bronze welds are shown in Figs. 269 and 270.

### Nickel silver alloys

These can easily be joined by soft and hard soldering, brazing, oxyacetylene, arc and electric resistance welding.

## Nickel alloys

The nickel rich alloys have a low ductility at certain temperatures (Fig. 272) and welding under restraint should be avoided in order to prevent intercrystalline cracking. Gasses are easily absorbed and re-evolved to give pores. Specific deoxidisers, e.g. aluminium in monel and titanium in nickel, have a beneficial effect on the properties of the weld; while silicon can cause a cracking effect on arc welds. Too much carbon is liable to form graphite which breaks up the continuity of the structure. Iron ($> 3\%$) contamination of nickel-arc welds leads to porosity. Nickel and its alloys are very sensitive to lead and sulphur which lead to hot cracking. Sulphur enrichment of the surface of a nickel plate with resultant thickening of the grain boundaries can readily occur during preheating with gas burners or during heat-treatment in open gas-fired furnaces. Stray arcs on the surface of nickel can also cause star cracks.

## Light alloys

Many of the difficulties met in welding these materials are due to their inherent properties.

(1) High conductivity increases width of heat-affected zone and frequently necessitates preheating while high expansion influences distortion.
(2) Cleanliness is essential and fusion faces must be free from oxide, grease and moisture. In gas welding active fluxes are required and the removal of these hygroscopic fluxes from the welds is essential if corrosion is to be prevented. The densities of the metal and fluxes are so similar that care is necessary to avoid flux entrapment. A typical gas welding flux is (BS 1126) NiCl 0–30, KCl 0–60, KF 5–15, NaCl remainder. No fluxes are required in this inert gas arc welding process.
(3) Gas porosity arises from absorption of hydrogen by the molten metal (p. 65) from the welding flame or water vapour from the flux. In alloys containing 3–10% magnesium surface reaction with atmospheric moisture creates atomic hydrogen (H) which diffuses into the plate. At flaws and inclusions within the sheet undiffusible molecular hydrogen ($H_2$) forms and develops pressures (p. 131) sufficient to cause blisters and porosity in the heat-affected zone.

The presence of minor impurities such as lithium, barium, strontium, and particularly calcium, greatly intensify this trouble. Provided these impurities are absent, melt is degassed and annealing is in electric furnaces, alloys containing up to 7% magnesium are free from this defect.

While the welded 5% magnesium alloy is only slightly affected by stress-corrosion, the 7% magnesium alloy fails rapidly and heating the weld at 125 °C for 24 hours produces extremely rapid failure. The trouble is related to the amount and mode of precipitation of the β-phase. As welded, the weld junction contains cavities and separate large particles of β. After heat treatment at 125 °C continuous films of β are formed in the heat-affected zone. Prior plastic deformation has an important influence since unwelded and undeformed 7% magnesium alloy heated to 125 °C showed little stress corrosion damage.

The aluminium magnesium alloys are also sensitive to exaggerated grain growth when critically strained material is heated (p. 125). Cracks may thus be induced in the heat-affected zone of a joint.

Cracking with high purity aluminium and heavy welds is minimised if the iron exceeds the silicon content. The iron–silicon constituent also largely affects the resistance of aluminium to corrosion by nitric acid, since it occurs in the heat-affected zone and in the weld as intercrystalline films which are readily attacked. The trouble is reduced by the use of rods of higher purity, hot hammering to break up and disperse the impurity and metallic arc welding to produce a shallower heat-affected zone which is also beneficial in the case of magnesium-containing alloys.

Aluminium–silicon alloys can be readily welded.

Welds in duralumin-type alloys have low strength unless heat-treated subsequently. Gas welds often show brittleness and cracking is prevalent. It would appear that the copper-free $Mg_2Si$ type alloys are the most promising of the heat-treated alloys.

Some magnesium-base alloys can be gas welded but arc welding necessitates the use of a neutral gas shield such as helium or argon. Stress relief after welding is desirable to avoid stress corrosion cracking.

*Welding dissimilar metals* involves a consideration of the formation of brittle compounds in the alloy system. For example, aluminium cannot be welded to copper or steel for this reason. Copper, bronze and brass can be welded to mild steel with a tin-bronze arc electrode; stainless steel to mild steel with 18/8 electrodes. Friction welding is very suitable for welding dissimilar metals.

*Titanium and zirconium.* Argon arc welding is suitable for titanium and its α alloys, but satisfactory ductility is not readily obtained with α–β alloys. It is advisable to supply argon to back face. Zirconium requires even better protection and often necessitates welding inside a chamber filled with inert gas. Titanium can be satisfactorily resistance, pressure and flash-butt welded.

*Niobium* requires similar treatment, and trouble is often met with oxide film diffusing rapidly throughout the metal with consequent embrittlement.

*Beryllium* poses a number of problems because of its brittleness, which should be taken into account when designing the joints to avoid restraint; also because rapid oxidation and grain growth occurs at fusion temperature. Even 0·05% oxygen in argon causes unsoundness. Argon arc, electron beam and pressure welding can give satisfactory joints.

# 21 The Measurement of Temperature

The accurate measurement of temperature is becoming increasingly important in almost all branches of industry, but particularly in the heat-treatment of metals. There are many errors which may arise in the measurement of temperature, and the object of this chapter is to outline the construction of the common instruments and, in particular, to explain the underlying principles and the errors involved.

*Temperature scale*

The absolute fundamental temperature scale is based on the Second Law of thermodynamics applied to a 'perfect gas', for which

$$\text{pressure} \times \text{volume} = \text{constant} \times \text{temperature (K)}$$

The unit degree on the scale is defined arbitrarily as one-hundredth of the temperature interval between the freezing-point (273° abs) and the boiling-point (373° abs) of pure water. Although no perfect gas is available, hydrogen or nitrogen can be used in an instrument known as a constant volume gas thermometer and slight corrections made for the lack of perfection. The zero point is chosen as the lowest conceivable temperature corresponding to −273° on the Centigrade scale, which has its zero at the freezing-point of water.

The gas thermometer is a difficult instrument to use, and it must be regarded as a primary instrument which has been used to measure the freezing-points of a number of pure metals and substances which can be employed to calibrate secondary pyrometers of greater flexibility.

The International Temperature Scale (1948) is similar to the thermodynamic scale but uses the platinum-resistance thermometer and the disappearing-filament pyrometer. The unit interval is called the degree Celcius.

**Secondary pyrometers**

A number of instruments depending on various physical properties are:

(*a*) *Thermometers*—depending on the relative expansion of mercury in glass or in steel, in common use up to 300°C; the range may be extended to 500° by filling the stem with nitrogen under pressure.

(b) The heat content of hot bodies, e.g. *Simen's calorimetric* pyrometer.

(c) *Fusion.* Triangular pyramids, moulded from various mixtures of kaolin, lime, feldspar, magnesia, quartz, iron oxide and boric acid, used to indicate temperatures in steps between 600 and 2000 °C. Each cone has a definite composition and fuses at a certain temperature, indicated by bending and final collapse. Reducing atmospheres affect the end-point. Such cones are named Seger, Orton and Sentinel.

(d) *Colour extinction* (e.g. Pyroversum 600–1200 °C). The hot body is viewed through a graduated colour filter and the point on its scale at which the colour from the hot body is neutralised is noted. The objection to such colour pyrometers is that they depend on the sensitivity of the human eye to colour, and this varies not only with different observers but also with the same operator under different physiological conditions.

For indicating low temperatures such as are used in preheating parts prior to welding, special wax pencils—Tempilstiks—are useful. Somewhat related is the use of thermal-sensitive paints, which can be used to indicate temperature variations over a surface.

(e) *Electrical resistance pyrometer.* This depends on the variation of electrical resistance of a platinum wire with temperature. In the range below 600 °C it gives the greatest accuracy of all the secondary instruments, but it is delicate and bulky and for these reasons it is not used extensively in industrial installations.

(f) *Thermoelectric pyrometer.* This is now the most extensively used instrument and it will be described in detail. The basic principle of it is that when two dissimilar wires are joined to form a complete electric circuit and the two junctions maintained at different temperatures an electromotive force (emf) is set up due to the algebraic sum of:

(i) An emf developed between two different metals placed in contact, i.e. 'Seebeck' effect.*

(ii) An emf developed between the ends of a homogeneous wire when one end is heated, i.e. *Thompson effect.*

The magnitude of the emf therefore depends on the temperature difference of the junctions and the metals used. If one junction is held at a uniform temperature (cold junction) then the emf developed can be used to determine the temperature of the other junction (hot junction). This

*The Peltier effect is the opposite of the Seebeck effect in that a current, passed through a junction of two dissimilar metals, causes heat to be absorbed at one junction and emitted from the other. With multiple thermocouples of new alloys such as bismuth telluride (e.g. $Bi_2Te_3$–$Bi_2Se_3$), thermoelectric cooling is feasible for domestic refrigeration.

arrangement of conductors is called a thermocouple. The complete pyrometer consists of the following parts:

(a) Two dissimilar conductors in wire or rod form.
(b) Electrical insulation of the wires.
(c) Sheaths for protecting the wires from injurious gases.
(d) Provision for controlling the temperature of the cold junction.
(e) Instrument for measuring emf.

*Types of thermocouples*

Data for common couples are given in Table 86, and the construction is shown in Fig. 280.

TABLE 86  THERMOCOUPLES

| Couple Wires | Limit °C | Millivolts per 100°C approx. |
|---|---|---|
| Copper/constantan (Cu 60, Ni 40) | 400 | 4·3 |
| Iron/constantan | 800 | 5·3 |
| Chromel/alumel (Ni 90, Cr 10/Ni 98, Al 2) | 1100 | 4·1 |
| Platinum/platinum + 10 or 13% rhodium | 1450 | 0·96 |
| Platinum–rhodium 20% Platinum–rhodium 40% | 1800 | 0·35 |
| Tungsten/molybdenum | 2000 | 0·8 |
| Carbon tube/silicon carbide rod | 1800 | 30 |

The emf developed by the base metal couples is roughly four times that of the platinum couple and, owing to low costs, thick wires can be used resulting in lower resistance of the external circuit and longer life of the couple. Consequently a simple and robust millivoltmeter can be used. Against these advantages, however, the base metals cannot be made as homogeneous, nor is their life as great at the higher temperatures as the platinum group. Thick wires allow heat to be readily conducted away from the hot junction.

Nickel and platinum alloys are ruined by sulphurous gasses, but the attack can be largely prevented by inserting the end of the couple sheath in a jacket containing lime which absorbs sulphur. The emf of a platinum couple is also effected by reducing gases in the presence of any metal, carbon or refractory silicates due to the absorption of metal vapours, silicon or carbon. In a vacuum the Pt/Pt–13Rh couple deteriorates rapidly owing to contamination of the platinum wire with volatilised rhodium. This effect is less with Pt–1Rh/Pt–13Rh.

The tungsten–molybdenum couple, however, must be protected by a reducing atmosphere of dry hydrogen or hydrogen–nitrogen mixture. This couple should not be used below 1200 °C owing to an inversion in its emf—temperature curve.

*Hot junction*

The two dissimilar wires have to be electrically connected to form the hot junction and this can be done in several ways. For the constantan couples the ends of the wires can be cleaned, twisted together and silver soldered (Cu 48, Ag 48, Zn 4) using borax as a flux. The chromel–alumel couple can

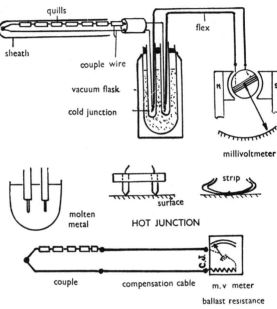

FIG. 280 A thermocouple pyrometer

be welded in an electric arc and the platinum wires in any oxyhydrogen flame. To get a couple sensitive to changes in temperature there should be no unnecessary mass of metal in the hot junction. The composition of the actual hot junction does not matter so long as the whole of it is at the same temperature. In certain immersion or surface instruments it is found unnecessary to weld the wires, the ends being left bare and contact made by the molten metal or the metal surface itself (Fig. 280).

*Insulation and protection of thermoelement*

Either one or both wires are electrically insulated by silica or fireclay capillary tubing and the whole is inserted in a sheath consisting of fused silica (used up to 1250 °C), glazed porcelain or alundum (Pythagorus) used up to 1700 °C. This sheath closed at one end should have high conductivity, low permeability to gasses and ability to withstand changes in temperature.

*Cold junction*

For laboratory use it is convenient to connect the couple wires to copper flex and each junction is enclosed by a glass tube closed at the end and fitted into a vacuum flask filled with oil or melting ice (Fig. 280). This method of controlling the cold junction temperature is not suitable for many industrial installations. It is frequently easier to transfer the cold junction from its proximity to the hot junction to a place where the surrounding temperature is lower and steadier. The thermocouple is joined, therefore, to what are called compensating leads, which are wires thermo-electrically interchangeable with the couple wires at room temperatures. Base metal couples are extended by wires of practically the same material as those employed in the couple itself, but in the form of stranded wires for flexibility. For platinum couples are used inexpensive wires of copper and nickel–copper alloy, which give the same emf as the couple over ordinary ranges of temperature.

In most industrial installations the compensating cables extend to the terminals on the indicator, but it must be remembered that variation in temperature at this point still needs a correction to be applied. With a millivoltmeter, the common method is to set the pointer to the actual cold junction temperature, and in certain instruments this is done automatically by a bi-metal spring. Since few couples yield an emf which is directly proportional to temperature it is *not possible to add the variation in cold junction temperature to the observed temperature.* The emf corresponding to this variation must be added to the observed emf.

## Measuring instruments

The emf is measured by either a millivoltmeter or a potentiometer and both methods can be adapted for recording.

*Millivoltmeter*

This consists essentially of a moving coil mounted between the poles of a permanent magnet. The coil can be suspended by metal fibres or more commonly supported by one or two jewelled pivots, the torque taken by coiled springs, and the movement indicated by a pointer.

The instrument should have a *high electrical resistance* relative to the external circuit and it should not vary greatly with temperature. While a resistance of 40 ohms may be sufficient for a base metal couple one of 500 ohms is usually desirable for a platinum couple. The reason for this is that the instrument uses current and measures an emf which depends on the total resistance of the circuit. Any variation in the external resistance will affect the reading and this may arise due to variation of the depth of insertion of the couple in the furnace since its resistance increases with its temperature (Fig. 282A). The high resistance instrument is usually made

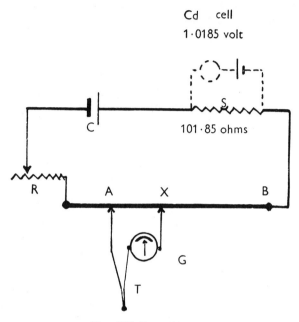

FIG. 281 Potentiometer

by adding a ballast resistance having a low temperature coefficient (manganin) which in turn necessitates more copper wire on the coil or a more sensitive movement or a stronger magnet. Such an instrument is of course more delicate than a low resistance instrument used for the same purpose.

The indicator should be *dead-beat*, i.e. the pointer should not oscillate many times before coming to rest. A heavy coil and pointer will tend to prevent quick damping. For consistency and accuracy the magnetic flux between the poles of the magnet must be constant and this will not be so if the instrument is mounted near cables carrying heavy currents or on steel stanchions, unless it is *magnetically shielded* by a soft iron cage or case. Finally the terminals should be shorted with heavy copper wire to damp the movement when the millivoltmeter is moved.

*Potentiometers* afford the most accurate method of measuring an emf

# MEASURING INSTRUMENTS

which is opposed by an equal emf to produce a balance. No current is used and consequently the usual changes in the external circuit have no effect on the readings.

Essentially a potentiometer is a number of resistance coils (A–B, Fig. 282) through which a steady current is passed from accumulator C. This

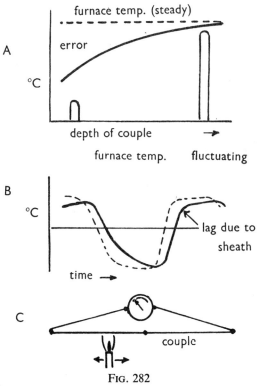

FIG. 282

A—Effect of depth of immersion of couple on temperature reading
B—Lag effect due to sheath when temperature fluctuates
C—Test for heterogeneity

current is kept constant by occasionally adjusting it by means of the rheostat R, so that the fall in potential through a fixed resistance, S, balances the emf of a standard cadmium cell $cd$ (1·0185 volts at 15°C). Then since in a simple circuit the fall in potential is proportional to the resistance, an unknown emf can be measured by balancing it against the voltage drop over the appropriate known resistance. One end of the couple is connected to A and the other end through a galvanometer to a suitable tap, X. The galvanometer indicates when a balance point is reached and this arrangement is known as the '*null*' method.

## Errors associated with thermocouples

Some of the errors which arise in the use of thermocouples are illustrated in Fig. 282. With a steady furnace temperature the reading will be affected by the *conduction* of *heat* along the sheath and wires and this will vary with the material of the sheath, thickness of the wires and depth of immersion of the couple in the furnace. There is also a very large time *lag* with heavily protected couples which introduces errors when the temperature is fluctuating. Other errors arise due to parasitic emfs arising from (1) heterogeneity in the couple wires which are in a temperature gradient, (2) electrical leakages into the couple circuit. Heterogeneity due to composition or cold work can be tested as shown Fig. 282c by connecting the two ends of the couple to a sensitive galvanometer and passing a small flame along the wires. No deflection should be produced except when the flame is at the junction unless the couple is contaminated.

*Calibration of a thermocouple* can be made by comparison with a standard couple of known accuracy or by measuring the emf produced at the boiling or freezing point of pure substances, such as boiling point of water, freezing point of tin, lead, zinc, aluminium, sodium chloride and copper (under charcoal).

The metal is melted in a small foundry crucible and the thermocouple, with sheath, is inserted in it so that the hot junction is surrounded with molten metal (Fig. 283A). While the metal cools readings of time are taken every millivolt drop. When the metal begins to solidify the pointer remains stationary and this reading corresponds to the fixed freezing temperature of the metal (at a given cold junction temperature).

The observed readings for a number of metals can be plotted against temperature to form a calibration curve for the couple (see Fig. 283c).

## Radiation and optical pyrometers

For measuring high temperatures it becomes necessary to use an instrument which need not be in contact with the hot body. Such pyrometers depend on the radiant energy emitted by the hot body and obey certain laws, which apply to a perfect 'black body'. A 'black body' may be defined as one absorbing all radiations falling on it, without loss by reflection or transmission. The radiation from the interior of a chamber at uniform temperature approaches closely to the ideal conditions. On the other hand, only a fraction of the theoretical energy is received from a body in the *open*, and this fraction (always less than unity) is called the *emissivity*, which depends upon (*a*) wave length, (*b*) temperature of surface, (*c*) character of surface. A few values are given in Table 87.

For bodies in the open corrections have to be made.

*The radiation pyrometers*, used under 'black body' conditions, are sub-

TABLE 87 EMISSIVITY OF SURFACES FOR RED LIGHT
(WAVE LENGTH = 0·00065 mm)

| | | | |
|---|---|---|---|
| Silver | 0·07 | Iron | 0·37 |
| Copper, solid | 0·11 | Iron oxide 800 °C | 0·98 |
| Copper, liquid | 0·15 | Iron oxide 1200 °C | 0·92 |
| Cuprous oxide | 0·70 | | |

ject to the Stefan–Boltzmann Law which may be stated as: 'an increase of 1% in the absolute temperature of the radiating body results in an increase of 4% in the energy emitted'. Stated mathematically the law is:

$$E = K(T^4 - T_0^4)$$

where $K = 1·34 \times 10^{-12}$ calories per square centimetre per second, E is total energy radiated by body at absolute temperature T to surroundings

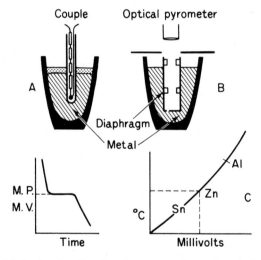

FIG. 283 Methods of calibrating thermocouples and optical pyrometers

at absolute temperature $T_0$. Since $T_0$ is small compared with T we can write

$$E = KT^4.$$

For bodies in the *open* we can use the formula to find true temperature T:

$$E = KeT^4 = KS^4 \text{ or } T = \frac{S}{\sqrt{e}},$$

where *e* is total emissivity, and S is apparent temperature indicated by pyrometer. For example, if the instrument reads 900 °C when sighted on iron in the open the true absolute temperature is

$$T = \frac{900 + 273}{\sqrt{0.37}} = 1928° \text{ abs. or } 1655°C$$

Most of the radiation pyrometers use either a lens or a mirror to concentrate the heat rays emitted by the hot body on to a small thermocouple, and the emf developed is measured by a calibrated millivoltmeter; ranges are usually 550° to 2000 °C, but there is really no upper limit.

Fixed focus and focusing instruments are used, as illustrated in Figs. 284A and 284B. In the Féry pyrometer the inclined mirrors (M) in front

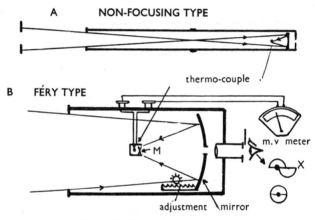

FIG. 284 Radiation pyrometers

of the thermocouple are used to indicate when correct focus has been obtained, namely, the image of the hot body appears broken in halves when out of focus.

Such radiation pyrometers require about 15 seconds to take a reading, but can be made recording in the same way as a thermoelectric pyrometer. It is essential that the image of the hot body be sufficiently large to cover the sensitive thermo-element, then the distance from the hot objective is immaterial. When measuring furnace temperatures the gasses should be turned off, since the higher temperature of the gasses may produce an error. Water vapour absorbs certain radiations and gives a low reading.

Improved designs are now available. One, which is suitable for measuring surface temperature quickly (5 sec), uses a hemispherical gilt reflector placed on the hot surface to provide black body conditions. To measure gas temperatures a suction pyrometer is desirable, consisting of a thermocouple surrounded by concentric radiation shields. The gas is aspirated past the couple at a high velocity.

The K and S pocket pyrometer, however, uses a bimetal spiral with a pointer and compensating spiral. The heat rays are concentrated by a lens on to the bimetal spiral which unfolds.

*Optical pyrometer*

In these pyrometers the *intensity of the light* from the hot body is compared with the intensity of light from some standard source, and both are matched in the instrument to *one* specific wavelength, usually red light

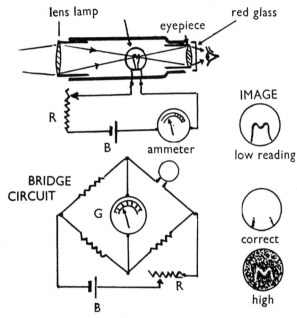

FIG. 285 Disappearing filament pyrometer

(0·00065 mm), and even a colour-blind man can do this. The commercial instruments differ in the optical devices used to make this comparison of the two intensities of light.

In the *Wanner* pyrometer a polarising device compares the red ray from the hot body with the ray of the same wave-length from an electric lamp the intensity of which is calibrated by an amyl acetate lamp.

In the disappearing filament pyrometer the intensity of a standard filament lamp is varied until it disappears against light from the hot body. Fig. 285 shows the construction of a typical instrument. The lamp is placed inside a telescope the focusing of which forms an image of the hot body in the plane of the filament. The filament and the image are viewed through the eyepiece of the telescope, adjusted to suit the operator's eye. An electric current is passed through the lamp filament so that it is just hot enough to disappear against the image of the hot body. This current can be measured by an ammeter, and, if previously calibrated, can be used to indicate the temperature of the hot body. In the simple electric circuit using an ammeter a large portion of the scale is useless, owing to the measurement of current used in heating the filament to 750 °C. In some instruments the voltage drop across the lamp is measured, but the most satisfactory method of using the whole of the scale without the 'set back' of the zero, is to make the lamp filament one of the arms of a Wheatstone Bridge and calibrate the galvanometer in degrees, with a zero reading corresponding to about 750 °C. The resistances in the circuit are so arranged that balance is obtained when the filament is at 750 °C. As the filament becomes hotter its increase in resistance throws the bridge out of balance and the galvanometer pointer moves over the scale.

To cut out colour differences a red glass filter is placed in front of the eyepiece. For temperatures above 1400 °C an absorbing screen is placed between the hot body and the filament so that the latter is never overheated, but this of course necessitates recalibration.

The principle on which the instrument is based is Wien's Law, which states that the intensity of any radiation (I) of wave-length ($\lambda$) emitted from a body at an absolute temperature (T) is given by:

$$I = c\lambda^{-5} e^{-\frac{K}{\lambda T}}$$

where $c$ and K are constants $e$ is the base of Naperian logarithms.

For *non-black body* conditions

$$\frac{1}{T} - \frac{1}{S} = \frac{\lambda \log E}{6232}$$

where S is apparent temperature and E is the emissivity of a non-transparent material at wave-length $\lambda$.

The reading obtained from an object in the open will depend upon its reflecting power and its surroundings. An oxide-free metal may reflect considerable sunlight into the instrument and give a false reading.

The optical and radition pyrometers can be calibrated against a thermocouple or the melting of a palladium wire in an electric furnace arranged to approximate a 'black body' by a series of diaphragms suitably disposed. Alternatively, the instruments can be focused on the bottom of a re-entrant tube in a crucible containing a pure metal and the freezing-point determined (Fig. 283B).

## Photo-electric

Increasing interest is shown in the use of photo-electric cells, some of which are sensitive to infra-red, for measuring temperature.

The *barrier-layer* selenium photo cell generates a current proportional to the light falling on it and a pyrometer using it can be robust and simple, but it can be damaged if it gets too warm. It is used widely on steel melting furnaces for temperatures above 1000 °C.

Vacuum *photo-emissive* cells (Ag–O–Cs or Sb–Cs surface) are sensitive but require electronic amplifiers. They are useful for measuring the temperature of hot-rolled steel strip down to about 900 °C.

The *photo-conductive* lead sulphide cell is suitable for high speed measurements down to 150 °C.

# Selected Bibliography

Alexander, W. O., *Metallurgical Achievements*, Pergamon.
Andrews, E. H., *Fracture in Polymers*, Oliver & Boyd.
Barratt, C. S., *Structure of Metals*, McGraw-Hill Book Co.
Batson, R. G. and Hyde, J. H., *Mechanical Testing*, Chapman & Hall.
Betteridge, W., *The Nimonic Alloys*, Arnold.
Biggs, W. O., *Brittle Fracture in Steel*, Macdonald & Evans.
Bullens, D. K., *Steel and its Heat Treatment*, John Wiley & Sons.
Cahn, R. W., *Physical Metallurgy*, North Holland.
Cottrell, A. H., *The Mechanical Properties of Matter*, Wiley.
Cottrell, A. H., *Theoretical Structural Metallurgy*, Arnold.
Cottrell, A. H., *Dislocations*, O.U.P.
Cullity, B. D., *Element of X-ray Diffraction*, Addison Westley.
Darken, L. S. and Gurry, R. W., *Physical Chemistry of Metals*, McGraw-Hill Book Co.
Darwin, G. E. and Buddery, J. H., *Beryllium*, Butterworth.
Dews, H. C., *The Metallurgy of Bronze*, Pitman.
Dieter, G. E., *Mechanical Metallurgy*, McGraw-Hill Book Co.
Eley, D. D., *Adhesion*, O.U.P.
Evans, U. R., *The Corrosion and Oxidation of Metals*, Arnold.
Gordon, M., *High Polymers*, Iliffe Books Ltd.
Greaves, R. H. and Wrighton, H., *Practical Microscopic Metallography*, Chapman & Hall.
Hedges, E. S., *Tin and its Alloys*, Arnold.
Hollomon, J. H. and Jaffe, L. D., *Ferrous Metallurgical Design*, Wiley.
Honeycomb, R. W. K., *The Plastic Deformation of Metals*, Arnold.
Hume-Rothery, W., *The Structure of Metals and Alloys*, Inst. of Metals Monograph No. 1.
Jastrzebski, Z. D., *Nature and Properties of Engineering Materials*, J. Wiley.
Jevons, J. D., *The Metallurgy of Deep Drawing and Pressing*, Chapman & Hall.
Johnson, W. and Mellor, P. B., *Plasticity for Engineers*, Van Nostrand.
Jones, W. D., *Fundamental Principles of Powder Metallurgy*, Arnold.
Kennedy, A. J., *Processes of Fatigue and Creep in Metal*, Oliver & Boyd.
Lancaster, J. F., *The Metallurgy of Welding, Brazing and Soldering*, Allen & Unwin.
Marsh, J. S., *Principles of Phase Diagrams*, McGraw-Hill Book Co.
McClintock, F. A. and Argon, A. S., *Mechanical Behaviour of Material*, Addison-Westley Pub. Co.
Miller, G. L., *Zirconium*, Butterworth.
Moneypenny, J. H. G., *Stainless Steels*, Chapman & Hall.
O'Neill, H., *Hardness of Metals and its Measurement*, Chapman & Hall.

Payson, P., *Tool Steels*, Wiley.
Rassweiler, G. M. and Grube, W. L., *Internal Stress and Fatigue in Metals*, Elsevier.
Read, W. T. Sr., *Dislocation in Crystals*, McGraw-Hill Book Co.
Rolfe, R. T., *Steels for the User*, Chapman & Hall.
Schwarzkopf and Kieffer, *Cemented Carbide*, Macmillan.
Sharp, H. S., *Engineering Materials*, Heywood.
Smallman, R. E. and Ashbee, K. H. G., *Modern Metallurgy*, Pergamon.
Smallman, R. E., *Modern Physical Metallurgy*, Butterworth.
Sully, A. H., *Creep*, Butterworth's Scientific Publications.
Tietz, T. E. and Wilson, J. W., *Behaviour and Properties of Refractory Metals*, Arnold.
Tipper, C. F., *The Brittle Fracture Story*, C.U.P.
Van Vlack, L. H., *Materials Science for Engineers*, Addison-Westley.
Zackay, V. K., *High Strength Materials*, Wiley.

'Effects of Neutron Irradiation on Metals and Alloys'. A. H. Cottrell, *Met. Reviews*, **1**, 479, 1956.
*Metals for Supersonic Aircraft and Missiles*, Amer. Soc. Metals.
*Metals Handbook*, Amer. Soc. Metals.
*Metals Reference Book*, C. Smithells. Butterworth.
*Recrystallisation Grain Growth and Texture*, Amer. Soc. Metals.
'Symposium on Hardenability of Steel', *Iron and Steel Inst. Spec. Rept.*, No. 36, 1946.
'Symposium on Internal Stresses in Metals and Alloys', Inst. of Metals, 1947.
*Temperature Measurement*, BS 1041.
*Welding Handbook*, Amer. Welding Soc.

# Appendix I  Periodic Table of Elements

| PERIOD | I | II | | | | | | GROUPS | | | | | III | IV | V | VI | VII | 0 |
|---|---|---|---|---|---|---|---|---|---|---|---|---|---|---|---|---|---|---|
| | 1 | 2 | | | Electropositive valency | | | | | | | | 3 | 4 | 3 | 2 | 1 | 0 |
| | | | | | Electronegative valency | | | | | | | | 3 | 4 | | | | |
| 1 | 1 H | | | | | | | | | | | | | | | | | 2 He |
| 2 | 3 Li | 4 Be | | | | | | | | | | | 5 B | 6 C | 7 N | 8 O | 9 F | 10 Ne |
| 3 | 11 Na | 12 Mg | | | Metals with variable valency i.e: transition groups | | | | | | | | 13 Al | 14 Si | 15 P | 16 S | 17 Cl | 18 A |
| 4 | 19 K | 20 Ca | 21 Sc | 22 Ti | 23 V | 24 Cr | 25 Mn | 26 Fe | 27 Co | 28 Ni | 29 Cu | 30 Zn | 31 Ga | 32 Ge | 33 As | 34 Se | 35 Br | 36 Kr |
| 5 | 37 Rb | 38 Sr | 39 Y | 40 Zr | 41 Nb | 42 Mo | 43 Tc | 44 Ru | 45 Rh | 46 Pd | 47 Ag | 48 Cd | 49 In | 50 Sn | 51 Sb | 52 Te | 53 I | 54 Xe |
| 6 | 55 Cs | 56 Ba | 57 La * | 72 Hf | 73 Ta | 74 W | 75 Re | 76 Os | 77 Ir | 78 Pt | 79 Au | 80 Hg | 81 Tl | 82 Pb | 83 Bi | 84 Po | 85 At | 86 Rn |
| 7 | 87 Fr | 88 Ra | 89 Ac † | (104) | (105) | (106) | (107) | (108) | | | | | | | | | | |

| 57 La | 58 Ce | 59 Pr | 60 Nd | 61 Pm | 62 Sm | 63 Eu | 64 Gd | 65 Tb | 66 Dy | 67 Ho | 68 Er | 69 Tm | 70 Yb | 71 Lu |
|---|---|---|---|---|---|---|---|---|---|---|---|---|---|---|
| 89 Ac | 90 Th | 91 Pa | 92 U | 93 Np | 94 Pu | 95 Am | 96 Cm | 97 Bk | 98 Cf | 99 E | 100 Fm | 101 Mv | 102 | (105) |

* Lanthanide or rare earth series
† Actinide or transuranic series

Non-metals are near to right-hand corner.
Metals near to line parallel dividing metals from non-metals (Al, Zn, Sn, Pb) have oxides which can behave as acids or bases, i.e. show some of the properties of non-metals.
Elements in same vertical column (i.e. same valency electrons) have similar properties, e.g. Be, Mg, Zn, Cd, or Cr, Mo, W.

# Appendix II   Metal Compatibility

The blacker the circles, the less compatible—and therefore the more suitable—that combination is for sliding applications

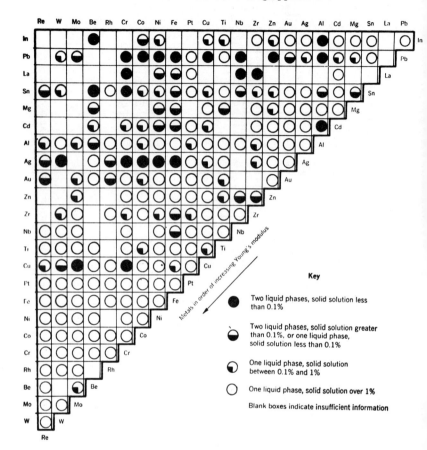

# Appendix III   Conversion Data

(a) Stress: tonf/in² to Newtons/mm² (or MN/m²)

| UK tonf/in² | N/mm² | UK tonf/in² | N/mm | UK tonf/in² | N/mm | UK tonf/in² | N/mm |
|---|---|---|---|---|---|---|---|
| 0  | 0·00  | 25 | 386·1 | 50 | 772·2  | 75  | 1158·3 |
| 1  | 15·44 | 26 | 401·5 | 51 | 787·6  | 76  | 1173·8 |
| 2  | 30·89 | 27 | 417·0 | 52 | 803·1  | 77  | 1189·2 |
| 3  | 46·33 | 28 | 432·4 | 53 | 818·5  | 78  | 1204·6 |
| 4  | 61·78 | 29 | 447·9 | 54 | 834·0  | 79  | 1220·1 |
| 5  | 77·22 | 30 | 463·3 | 55 | 849·4  | 80  | 1235·5 |
| 6  | 92·66 | 31 | 478·8 | 56 | 864·9  | 81  | 1251·0 |
| 7  | 108·1 | 32 | 494·2 | 57 | 880·3  | 82  | 1266·4 |
| 8  | 123·6 | 33 | 509·7 | 58 | 895·8  | 83  | 1281·9 |
| 9  | 140·0 | 34 | 525·1 | 59 | 911·2  | 84  | 1297·3 |
| 10 | 154·4 | 35 | 540·5 | 60 | 926·6  | 85  | 1312·8 |
| 11 | 169·9 | 36 | 556·0 | 61 | 942·1  | 86  | 1328·2 |
| 12 | 185·3 | 37 | 571·4 | 62 | 957·5  | 87  | 1343·6 |
| 13 | 200·8 | 38 | 586·9 | 63 | 973·0  | 88  | 1359·1 |
| 14 | 216·2 | 39 | 602·3 | 64 | 988·4  | 89  | 1374·5 |
| 15 | 231·7 | 40 | 617·8 | 65 | 1003·9 | 90  | 1390·0 |
| 16 | 247·1 | 41 | 633·2 | 66 | 1019·3 | 91  | 1405·4 |
| 17 | 262·5 | 42 | 648·6 | 67 | 1034·8 | 92  | 1420·9 |
| 18 | 280·0 | 43 | 664·1 | 68 | 1050·2 | 93  | 1436·3 |
| 19 | 293·4 | 44 | 679·5 | 69 | 1065·6 | 94  | 1451·8 |
| 20 | 308·9 | 45 | 695·0 | 70 | 1081·1 | 95  | 1467·2 |
| 21 | 324·3 | 46 | 710·4 | 71 | 1096·5 | 96  | 1482·6 |
| 22 | 339·8 | 47 | 725·9 | 72 | 1112·0 | 97  | 1498·1 |
| 23 | 355·2 | 48 | 741·3 | 73 | 1127·4 | 98  | 1513·5 |
| 24 | 370·7 | 49 | 756·7 | 74 | 1142·9 | 99  | 1529·0 |
|    |       |    |       |    |        | 100 | 1544·4 |

(b) Conversion Ft lbf to Joules

| Ft lbf | 1  | 2  | 3  | 4  | 5  | 6  | 7  | 8   | 9   |     |
|---|---|---|---|---|---|---|---|---|---|---|
| Joules | 1  | 3  | 4  | 5  | 7  | 8  | 10 | 11  | 12  |     |

| Ft lbf | 10 | 20 | 30 | 40 | 50 | 60 | 70 | 80  | 90  | 100 |
|---|---|---|---|---|---|---|---|---|---|---|
| Joules | 13 | 27 | 40 | 54 | 68 | 82 | 95 | 108 | 123 | 136 |

(c) BS 18 symbols in Tensile test

| Original gauge length | $L_o$ | Reduction of area % | $Z$ |
| Final gauge length | $L_u$ | Tensile strength | $R_m$ |
| Extensometer gauge length | $L_e$ | Upper yield stress | $R_{eH}$ |
| Cross section area | $S_o$ | Lower yield stress | $R_{eL}$ |
| Elongation after fracture | $A$ | Proof stress | $R_p$ |

(d) Comparison of elongation values on two old and new gauge lengths for steel are given below

| 4 $\sqrt{(Area)}$ | 8 | 10 | 12 | 14 | 15 | 17 | 18 | 20 | 22% |
|---|---|---|---|---|---|---|---|---|---|
| 5·65 $\sqrt{(Area)}$ | 5 | 7 | 8 | 10 | 11 | 12 | 13 | 15 | 17% |

(e) Prefixes denoting decimal multiples or sub-multiples

| Multiplication factor | Prefix | Symbol |
|---|---|---|
| $10^{12}$ | tera | T |
| $10^9$ | giga | G |
| $10^6$ | mega | M |
| $10^3$ | kilo | K |
| $10^2$ | hecto | h |
| $10$ | deca | da |
| $10^{-1}$ | deci | d |
| $10^{-2}$ | centi | c |
| $10^{-3}$ | milli | m |
| $10^{-6}$ | micro | $\mu$ |

(f) Conversion factors for everyday units

$$\begin{aligned}
1 \text{ in} &= 25\cdot4 \text{ mm} \\
1 \text{ sq in} &= 645 \text{ mm}^2 \\
1 \text{ lb} &= 0\cdot454 \text{ kg} \\
1 \text{ kg} &= 2\cdot2 \text{ lb} \\
1 \text{ lbf} &= 4\cdot45 \text{ N} \\
1 \text{ tonf} &= 9964 \text{ N} \\
1000 \text{ lbf/in}^2 &= 6\cdot89 \text{ N/mm}^2 \\
1 \text{ tonf/in}^2 &= 15\cdot44 \text{ N/mm}^2 \\
1 \text{ ångström} &= 10^{-8} \text{ cm} = 10^{-10} \text{ m} \\
1 \text{ micron} &= 10^{-6} \text{ m} \\
1 \text{ bar} &= 10^5 \text{ Nm}^{-2}
\end{aligned}$$

# Appendix IV  Comparison of En Nos. and Equivalent Values in BS 970

| En No | New No | Type | En No | New No | Type |
|---|---|---|---|---|---|
| 6 | 080 M 41 | 41 C | 48 | 527 A 60 | ¾ Cr |
| 8K | 080 M 46 | 46 C | 52 | 401 S 45 | Si Cr 3/8 |
| 11 | 526 M 60 | ¾ Cr | 54 | 331 S 40 | Ni Cr W 14/14/2¼ |
| 12 | 503 M 40 | 1 Ni | 56A | 410 S 21 | 13 Cr |
| 13 | 785 M 19 | 1½ Mn Ni Cr | 56AM | 416 S 29 | 13 Cr, S |
| 16 | 605 M 36 | 1½ Mn Mo | 56C | 420 S 37 | 13 Cr CO 24 |
| 16M | 606 M 36 | 1½ Mn Mo, 5S | 57 | 431 S 29 | 17 Cr 2½ Ni CO 15 |
| 17 | 608 M 38 | 1½ Mn Mo > Mo | 58A | 302 S 25 | Cr Ni 18/9 |
| 18 | 530 M 40 | 1 Cr | 58B & C | 321 S 12 | Cr Ni 18/9/Ti |
| 19A | 708 M 40 | 1 Cr Mo | | | |
| 19 | 709 M 40 | 1 Cr Mo > Mo | 58E | 304 S 15 | Cr Ni 18/9 |
| 23 | 653 M 31 | 3 Ni Cr | 58F & G | 347 S 17 | Cr Ni 18/9 Nb |
| 24 | 817 M 40 | 1½ Ni Cr Mo | | | |
| 25 | 826 M 31 ⎫ | 2½ Ni Cr Mo | 58H | 315 S 16 | Cr Ni Mo 17/10/1½ |
| 26 | 826 M 40 ⎭ | | 58J | 316 S 16 | Cr Ni Mo 17/11/2½ |
| 27 | 830 M 31 | 3 Ni Cr Mo | 58M | 325 S 21 | Cr Ni /8/9/Ti/S |
| 30B | 835 M 30 | 4 Ni Cr Mo | 59 | 443 S 65 | Cr Ni Si 20/1½/2 |
| 31 | 534 A 99 | 1½ Cr 1 C | 60 | 430 S 15 | 17 Cr O·1 C |
| 32A | 045 M 10 | C | 100 | 945 M 38 | 1½ Mn Ni Cr Mo |
| 32M | 212 M 14 | CS | 110 | 816 M 40 | 1½ Ni Cr Mo |
| 34 | 665 M 17 | 1¾ Ni Cr | 111 | 640 M 40 | 1¼ Ni Cr |
| 36A | 655 M 13 | 3¼ Ni Cr | 201 | 120 M 15 | C |
| 36C | 832 M 13 | 3½ Ni Cr Mo | 202 | 214 M 15 | C free cutting |
| 39A | 659 M 15 ⎫ | 4% Ni Cr | 206 | 523 A 14 | ½ Cr |
| 39B | 835 M 15 ⎭ | | 207 | 527 A 19 | ¾ Cr |
| 40 | 722 M 24 | 3 Cr Mo | 351 | 635 M 15 | ¾ Ni Cr |
| 40C | 897 M 39 | 3¼ Cr Mo V | 352 | 637 M 17 | 1% Ni Cr |
| 41A | 905 M 37 | 1½ Cr Al Mo | 353 | 815 M 17 | 1½ Ni Cr Mo |
| 42 | 070 A 72 | '72' Carbon | 355 | 822 M 17 | 2 Ni Cr Mo |
| 43 | 080 A 52 | '52' Carbon | 361 | 805 M 16 ⎫ | |
| 44 | 060 A 96 | '96' Carbon | 362 | 805 M 20 ⎬ ½ Ni Cr Mo | |
| 45 | 250 A 53 | Si/Mn | 363 | 805 M 25 ⎭ | |
| 47 | 735 A 50 | 1% Cr V | | | |

# Index

Abnormality of pearlite, 168, 208
Abrasive powders, 47
ABS, 384
Absolute temperature scale, 415
Acetal co-polymer, 383
Acetals, 384
Acicular cast iron, 280
Acid, Bessemer process, 156; corrosion by, 140; open-hearth steel, 156; steels, 155, 156
Acrylic fibre, 365
Acrylonitride–butadiene–styrene, 384
Activation energy, 96
Adhesive bonding, 397
Admiralty brass, 308; gun-metal, 315
Age-hardening, 96
Ageing, 44, 171
Agricultural tools, steel for, 173
Air hardening Ni–Cr steels, 231; softening of, 181
Airscrews, metal for, 209
Alcomax, 298
Alkyds, 387; aminos, 387
Allotropy or iron, 162
Alloy, case-hardening steels, 206, 207; cast irons, 178; tool and die steels, 236
Alloying elements, characteristics of, 212, 217
Alpha brass, 306
Aluminising, 261
Aluminium alloys, 331; bearing, 372; die-casting dies for, 238
Aluminium bronze, 319; creep of, 323; effect of additions on, 322; fatigue, 323; heat-treatment, 322
Aluminium–chromium–molybdenum steel, 209
Aluminium–copper alloys, 96, 310, 314; equilibrium diagram of, 96, 320
Aluminium in brass, 308, 311; in bronze, 319; in steel, 167, 208–9 properties of, 327
Aluminium–magnesium alloys, 333, 340
Aluminium–magnesium–manganese alloys, 333, 348
Aluminium–manganese alloy, 332
Aluminium–silicon alloys, 333, 337; equilibrium diagram of, 338
Aluminium, soldering of, 393
Aluminium–zinc alloys, 340
Aluminium–Zn–Mg–Cu alloys, 334
Amorphous glassy state, 380
Amorphous metal, 104
Amsler fatigue machine, 25
'Anka H', 246
Anneal, sub-critical, 175
Annealing, double, 181; effects of, after cold work, 121; full, 177; of air-hardening steels, 181; of aluminium alloys, 331-2; of brass, 126; of cast iron, 282; of cast steel, 179; of monel, 126; of steel, 126, 177; of tool steel, 181; twins, 116
Anode, 141
Anodic oxidation of aluminium, 328
Anti-friction alloys, 369
Antimony, in anti-friction alloys, 369; in bronze, 318; in copper, 302
Antimony–lead equilibrium diagram, 83
Appearance of micro-structure, 50
Applications, of alloy steels, 227; of carbon steels, 172
Armco iron, 154
Armour, welding of, 403
Arsenic in copper, 300
Aston process, 154
Atactic, 386
Atmospheric corrosion, 145
Atomic arrangements, 72
Ausforming, 225

# INDEX

Austempering, 199
Austenite, 163, 183; grain size, 167
Austenitic, cast irons, 280; manganese steel, 217; stainless steel, 246, 255; welds, 408
Autotempering, 189
Axes, steel for, 175
Axles, steel for, 185, 218, 231

Babbit metals, 369
Back reflection X-ray test, 77
Bacteria corrosion, 144, 145
Bainite, 183, 194
Bakelite, 388
Balanced steel, 66, 172
Ball-bearing steels, 220
Barba's law, 9
Barium carbonate, use of in case-hardening, 204
Barns, 359
Basic Bessemer steels, 156
Basic Siemens process, 156
Bearing bronzes, 318, 371
Bearing metals, 368; sintered, 372
Bending tests, 31, 32
Beryllium–copper alloys, 324
Beryllium, properties of, 360; welding of, 414
Bessemer process, 156
Beta-brass, 305
Betatron, 35; linear acceleration, 35
Bibliography, 429
Bismuth, in bronze, 318; in copper, 302
Black-body conditions, 422
Blackheart malleable castings, 283
Blast furnace, 152; diagram of, 153
Blisters, pickling, 131
Block slip, 104
Block wall, 295
Blow-holes in castings, 65
Boiler drums, steel for, 173
Bolt, macro of, 42; steel for, 234
Bonderising, 149
Boron, 224; on control rods, 360
Boundary, crystal, 50, 61, 114, 165

Boundary, oil film, 368
Bragg equation, 78
Branched polymer, 379
Brass, annealing of, 126, 307; cold-working of, 308; creep of, 306; effects of alloying elements on, 309; effects of zinc on, 304; equilibrium diagram for, 304; for brazing purposes, 393; hot working of, 308; properties of, 306, 307, 309; structures of, 305; welding of, 411
Braunite, 399
Brazing, 305, 393
Breakaway corrosion, 143
Brearley, 245
Brinell test, 3
British Standards Specifications, 2
Brittle fracture, 13, 44
Broaches, steel for, 173
Bronze, aluminium, 319; annealing of, 116, 117; effects of tin on, 313; for bearings, 318, 371; phosphor, 315; welding of, 410
Built-up edge, 134
Burning of steel, 179
Burnt brass, 308

Cadmium in copper, 291; in lead, 375
Calcium boride, 300
Calcium in lead, 369
Calibration of pyrometers, 422
Calorising, 260
Cap copper, 306
Carbides, formation of, 212, 214
Carbon–chromium steels, 237
Carbon, effect on welding stainless steel, 408
Carbon equivalent, 273
Carbonitriding, 210
Carbon steels, composition of, 172; structure of, 165; properties of, 163, 174
Carburising, effect on fatigue limit, 28; of steel, 204
Cartridge brass, 306
Case-hardening steel, failures of, 205

Case-hardening, boxes, 263; by nitrogen, 209; of steel, 203; steels, 206, 207
Casting of magnesium, 248
Castings, chill, 61; steel, 165; structure of, 62
Cast iron, 270; compositions of, 282; damping, 31; effects of alloying elements on, 278; effects of impurities on, 273; effects of rate of cooling on, 271; for nitriding, 210; growth of, 285; heat-treatment of, 282; melting and casting of, 278; microstructures of, 277, 279, 283; stress relief, 282; tensile strength of, 277, 285; transverse strength of, 31; welding of, 406
Cathode, 300
Caulking tools, steel for, 173; contraction, 322
Caustic embrittlement, 121
Cementite, 161
Centrifugal castings, 65
Cerium, 281
Cermets, 268
Changes in the solid state, 88
Characteristics of alloying elements, 217
Charpy test, 16
Chemical affinity effect, 76
Chill castings, 63
Chill crystals, 63
Chisels, hardening of, 186; steel for, 173, 236
Chromate treatment of magnesium, 350
Chromates, effect on corrosion, 143, 144, 148
Chrome lines, 220
Chromium-bronze, 291
Chromium, in permalloy, 299; in stainless steels, 247; steels, compositions of, 220, 230, 235, 237, 249, 297
Chromium–nickel alloys, 221
Chromium–vanadium steels, 223
Civilian transformation, 196
Classification, of alloys in steel, 215; of carbon steels, 172; of stainless steels, 245
Close anneal, 175
$CO_2$ process, 69
COD, 17
Coatings on iron, 146
Coaxing, 27
Cobalt, 224; in heat-resisting steels, 263, 268; in high-speed steel, 239; in magnets, 297; in permalloy, 299
Coefficients of expansion, 398
Coercive force, 297
Coherency, 98
Cohesive stress, 14, 115
Coinage bronze, 315
Cold drawing, 131
Cold-drawn steels, effect of annealing on, 178
Cold extrusion punch, steel for, 237
Cold forging, 129
Cold junction, 419
Cold work, 104, 120; effect on mechanical properties, 7, 10, 30, 120; mechanism of, 106
Colmonoy, 406
Colony, eutectic, 84
Columnar crystals, 63
Compatibility of metals, 368
Compensating leads, 419
Complex brasses, 309
Composite materials, 365
Composites, 365
Compound stresses, 13
Concrete, 365
Condenser tubes, 145, 308, 312
Conductivity, of aluminium, 327, 332, 398; of copper, 300, 398
Connecting-rod, steel for, 231
Constantan, 312, 417
Constituents, identification of, 47, 50; in hardened steel, 182
Constitution of metallic systems, 82
Consumable electrode furnace, 159
Contamination of couples, test for, 422
Continuous annealing, 176
Continuous casting, 65
Continuous cooling curves, 199

INDEX 439

Contraction effects, 64
Controlled transformation 18/8, 226, 250; rolling, 227
Conversion of Brinell hardness to tensile strength, 3, 83, 86, 192, 423
Cooling curves, 59
Cooling rate, on cast iron, 271; on steel, 191
Copper alloys, 303; strength at elevated temperature, 323
Copper–aluminium alloys, 319; equilibrium diagram of, 96; heat-treatment of, 97, 319, 320
Copper–beryllium alloys, 324
Copper–chromium, 291
Copper, composition of, 300
Copper, conductivity of, 291; effect of selenium, 411; effect of tellurium, 411; in steel, 224, 227, 246, 299; in white metals, 249
Copper–manganese–nickel, 325
Copper–nickel alloys, 312
Copper–nickel–aluminium alloys, 325
Copper–nickel equilibrium diagram, 86
Copper–nickel–silicon, 292
Copper, properties of, 398
Copper–silver equilibrium diagram, 87
Copper, soldering of, 393
Copper–tin equilibrium diagram, 89, 314
Copper, welding of, 410
Copper–zinc alloys, 304
Copper–zinc–nickel alloys, 311
Cored crystals, 87, 116
Corner segregation, 68, 144
Corrosion, breakaway, 144; by acids, 140; by atmosphere, 145; by corrosion, 120; by soil, 145; by waters, 143; fatigue, 30; fretting, 151; knife-edge, 409; mechanism of, 140; of aluminium, 148, 328; of lead, 147; of magnesium, 350; of tin, 215; of welds, 409; of zinc, 142, 146
Corten, steel, 227
Crack opening displacement, 16
Cracks, 44; detection of, 33

Crankcases, metal for, 338
Crankshafts, steel for, 211, 218
Crater wear, 243
Creep, 21, 211; bronze, 323; cupro-nickel, 312; magnesium alloy, 349; slip, 91; steel, 255, 263; tests, 24
Critical cooling velocity, 186, 198
Critical illumination, 56
Critical points, 162; effect of quenching on, 190
Cross slip, 110
Cryogenic iron, 282
Crystal boundaries, 52, 61, 114, 167
Crystallisation of metals, 59
Crucible process, 155
CTS test, 405
Cu–Fe–Al alloy, 326
Cupping test, 32
Cupro nickel low temperature properties, 21
Curie temperature, 295
Cutlery stainless steel, 245, 249, 254
Cutting speeds, 134
Cyanide case-hardening, 204


Damping capacity, 31
De-aluminification, 323
Deep drawing, 132
Defects, classification of, 44
Deformation, effect on properties, 7, 10, 28, 120; mechanism of, 104
Degassing bronze, 317; of aluminium alloys, 330
Dendritic crystals, 60
Densities of metals, 398
Deoxidised copper, 300, 410
Deoxidisers, 66, 156, 316, 376
Depth of hardening, 200
Design, effect on corrosion, 146; effect of Izod, 16; effect on heat-treatment, 190; effect on die castings, 373; on aluminium, 331; on magnesium, 350
Dezincification of brass, 145
Diagrams of thermal equilibrium, 82

Delrin, 384
Diamond, 73
Diamond dust, 47
Die blocks, steel for, 179, 239
Die castings, 373
Die materials, 94
Diffusion of elements, 127
Directional properties, 117
Disappearing-filament pyrometer, 425
Discard in ingots, 65
Dislocation, 91; climb, 110, 112
Dispersion hardening, 100
Divorced pearlite, 176
Double annealing of steel, 181
Double reduced steel, 177
Drawability, 119
Drawing processes, 132
Drills, steel for, 173
Drop forging, 129
DTD Specifications, 2
Ductile fracture, 11
Ductility, 5, 11
Duralumin, ageing of, 96; forging temperature of, 127
Durville casting process, 319
Dynamic strain ageing, 227
Dynapat, 129

Ears on pressings, 118
EDD, 171
Effervescing steel, 66
Elastic limit, 6
Elastic modulus, 6, 328
Elastic–Plastic, 7
Electric furnace, processes, 115, 157
Electric welding, 397
Electrical resistance, pyrometer, 416; values of, 398
Electro-chemical series, 139
Electro-slag furnace, 158
Electrode potential, 138
Electrolytic iron, 154
Electrolytic polishing, 47
Electron compounds, 89
Electron microscope, 56
Elongation in tensile test, 9

Embrittlement, caustic, 121; '475', 260; temper, 121, 221
Emissivity, 423
Endurance limit, 25
En specifications, 230
Equi-axed crystals, 63
Equilibrium diagrams, Ag–Cu, 87; Al–Cu, 96; Al–Mg$_2$Si, 97; Al–Si, 338; Cu–Ni, 86; Cu–Sn, 89, 314; Cu–Zn, 304; Fe–Cr, 253; Fe–Cr–C, 253, Fe–Cr–Ni, 252; Fe–Fe$_3$C, 164; Fe–Ni, 253; Fe–W–C, 241; Mg–Sn, 88; Pb–Sb, 83; Sn–Sb, 369; Ternary, 90; interpretation of, 92
Equivalents, Guillet's, 310
Erichsen test for sheets, 33
Errors with pyrometers, 422, 424, 426
Etching reagents, macro, 41; micro, 48
Eutectic colony, 84
Eutectics, 83, 85, 163, 327, 393
Eutectoids, 84, 163, 314, 319
Everdur, 325
Exfoliation, 207
Exhaust valves, steel for, 262
Expansion coefficients, 361, 398
Expansion steels, high, 234
Extra deep drawing steel, 171
Extrusion, 129; die for, 238

Fagotted iron, 154
Failures, investigation of, 43
Fatigue, 24; aluminium alloy, 25; aluminium bronze, 322; of bearings, 370; plastics, 7; residual stress, 30; titanum, 25
Feathery structure in steel, 180
Feeder heads, 64
Fermi level, 286
Feroba, 298
Ferrite, 161
Ferrites, 298
Ferro-manganese, 66, 156
Ferroxcube, 298
Féry radiation pyrometer, 423
Fibre, 117
Fibre reinforcement, 364

Files, 173, 221
Fine-grain steel, 20, 167
Fine grains, 188; thermomechanical treatment, 188; irradiation hardening, 361
Finishing temperature, effect of, 127
Finishing tools, steel for, 236
Firebox and stays, 300
Fire cracking, 120
Fire welding, 394
Firth's FDP and FCB steels, 246
Flakes, 233
Flame hardening, 211
Flow lines, 43
Fluorescent screens, 33
Flux for magnesium, 348; for solders, 393; welding, 395
Forbidden gap, 286
Forging and its effects, 129; steels for, 173; cold, 129
Forks, steel for, 173
FOS, 158
Fracture, brittle, 15; cleavage, 11; ductile, 11; examination of, 39, 43; fatigue, 24; intercrystalline, 12; of pig iron, 152; types of, 115
Fracture toughness, 17; mechanics, 17; aluminium alloys, 345
Frank–Read source, 109
Free-cutting stainless steels, 248
Free-cutting, steel, 173
Freezing point, 59
Fretting, 151
Friction welding, 394, 414
Fry's reagent, 119
Full annealing of steel, 177
Furnace consumable arc, 158; ESR, 159
Furnace parts, steel for, 262
Fusible alloys, 375; solders, 393
FV 520 (B), 250; FV 520 (S), 251

Galvanic series, 139
Gamma loop, 215, 253
Gamma rays, 35
Gases, in metals, 65; in welds, 399

Gassing of copper, 303
Gas solubility effects, 66
Gas thermometer, 415
Gas turbine steels, 236
Gauge length in tensile test, 9, 5·65 $\sqrt{}$(area), 10
Gauges, steel for, 173
Gears, steel for, 26, 173
German silver, 311
Ghosts in steel, 170
Gilding metal, 306
Glacier Du, 372; DQ1, 372; DQ3, 372; DX, 373
Glass ceramics, 269
Glass fibres, 365
Goodman, diagram, 27
GP zones, 98
Grain growth, laws of, 125; in steels, 182, 261
Graphite, atomic structure, 73; micro 277
Gravity die castings, 374
Green rot, 151
Green sand moulds, 69
Grey cast iron, properties of, 234
Griffith's theory, 15
Grinding of micro-sections, 45
Guillet's diagram, 219; equivalents, 310
Gun-metal, 315


Hadfield manganese steel, 217
Hafnium, 361
Hairline cracks, 233
Hammer, forging by, 129
Hammers, steel for, 173
Hardenability, 200
Hardening, dispersion, 100; flame, 211; induction, 211; methods of, 103; of steel, 181, 185, 203; temperatures for, 181
Hard facing, 405
Hard metal, 242
Hardness, measurement of, 2; of quenched steels, 185
Hard zone cracks, 403

Hastelloy, 376
Heat-resisting cast irons, 285
Heat-resisting steels, 260
Heat-treatment, of aluminium bronze, 320; case-hardened steels, 204; for age-hardening, 96; of steel wire, 200; of 60/40 brass, 308
Heterogeneity, chemical, 68
Hiduron, alloys, 313
High-conductivity copper, 290
High-expansion steels, 234
High-frequency electric furnace, 155
High rate forging, 130
High-speed steel, 239; hardening of, 241
High-tensile brasses, 309
High-tensile steel, welding of, 400
Hoes, steel for, 173
Hooke's Law, 2
Hot cracking of welds, 401
Hot cracks, effect of silicon, 414
Hot junction of couple, 418
Hot-working die steels, 237
Hot working of metals, 127, 308; processes, 128
Humphrey's reagent, 42
Huygen's eyepiece, 54
Hycomax, 297
Hydrogen, effect of, on copper, 303
Hydrogen embrittlement, 131, 232, 404, 413

Impingement attack, 145
Impurities, effect on properties, 114, 117, 168, 273
Inclusions, indentification of, 50
Inconel, 263
Indium, 372
Induction hardening, 211
Ingot iron, 155
Ingot structure, 53
Inherent grain size, 167
Inhibitors, 144
Ink print, 42
Inoculants, 60, 61
Inoculated cast iron, 275, 278, 281
Intercrystalline corrosion, MG5, 414
Intercrystalline fracture, 114, 120
Interdendritic porosity, 61, 65
Interdendritic segregation, 68
Intermediate phases and compounds, 89
Internal oxidation, 151
Internal stress, by X-ray, 77
Internal stresses, 120
Interpretation of equilibrium diagram, 84, 92
Interstitial solid solution, 76
Invar, 219
Inverse segregation, 68
Irradiation damage, 361
Irreversible steel, 216
Isotactic, 4
Isothermal, 192
Izod test, 15; relation to fatigue, 28

Jet engine steels, 263
Joists, steel for, 173
Jominy test, 202
Jovignot cupping test, 32

Kaldo, 158
Keys, steel for, 173
Killed steel, 66, 172
Kirkendall effect, 94
Knock-on damage, 361
Knife-edge corrosion, 409

LD process, 116
Ladle degassing, 159
Lamellar tearing, 405
Laps, 69
Latent heat of fusion, 59
Lath martensite, 183
Lattice distortion, 113
Lead, alloys of, 375, 393; corrosion of, 147; in brass, 309; in bronze, 371; in copper, 302; in steel, 134, 173; properties of, 375

# INDEX

Lead–antimony equilibrium diagram, 83
Ledloy steel, 173
Levelling of micro-sections, 46, 47
Limit of proportionality, 6, 10
Limiting creep stress, 23
Liquidus, definition of, 82
Lithium, 302, 413
Lost wax process, 70
Low-expansion steels, 234
Low temperature, effect on properties, 18
Low temperature properties, plastics, 7
Lüder's lines, 119

Machinability, 132, 154, 290, 309, 348
Machining, built up edge, 134
McQuaid–Ehn test, 168
Macro-etching reagents, 41
Macro-sections, preparation of, 40
Magnesium, alloys, 349; flux for, 348; in aluminium, 97, 334, 335; protection of, from corrosion, 350
Magnesium–tin equilibrium diagram, 88
Magnesium–silicide alloy, 333
Magnet alloy, 294
Magnetic particle method, 33
Magnifying power, 55
Magnolia metal, 369
Magnox, 348
Malleability, definition of, 5
Malleable cast iron, 283
Malleable nickel, 376
Mandrel drawing, 132
Manganese bronze, 311
Manganese–copper alloy, 32
Manganese, in brass, 311; in cast iron, 272; in magnesium, 349; in steel, 169, 215
Manganese steel, Hadfields, 217
Manganese sulphide, 169
Maraging, 225; steels, 232
Marfacing, 406
Marforming, 226
Marine propellers, 310

Martempering, 199
Martensite, 183, 195
Martensitic cast iron, 280
Mass effects in steels, 201
Mechanical properties, plastics, 8; relation to structure, 52; tests, 2
Mechanism, of corrosion, 138; of deformation, 106
Melting bronze, 319
Metallic arc welds, 396
Microscope, 52; electron, 56
Micro-sections, preparation of, 45
Micro-structure, appearance of, 50; relation to mechanical properties, 52
Miller indices, 79
Milling cutters, steel for, 173, 237
Miner's hypothesis, 27
Mobium mild steels, 225; superconductors, 292
Moderator, 356
Modification of Al–Si alloys, 338
Modulus, or rupture, 31; Young's, 16
Molybdenum, 268; in high-speed steel, 240; in steel, 223, 230; in stainless and heat-resisting steels, 239, 240, 257, 265
Molybdenum sulphide, 372
Monel metal, 313; corrosion of, 143
Mould shapes, effect on piping, 65
Moulding sands, 69
Mounting small micro-section, 46
Muntz metal, 308
Murakami's reagent, 48
Mushet's steel, 239

Naval brass, 308
Network structure, 165
Neumann bands, 107
Nickel brass, 312
Nickel–chromium steels, 221, 222, 230, 247
Nickel-clad steel, 376
Nickel–copper alloys, 312; equilibrium diagram, 86

Nickel, effects on steels, 216; in aluminium bronze, 322; in brass, 304; in case-hardening steels, 207; in magnets, 297; in stainless and heat-resisting steels, 262
Nickel–iron alloys, 216
Nickel, malleable, 376
Nickel–molybdenum alloy, 376
Nickel–silver, 311
Nickel steel, 9%, 20, 21
Nickel, TD, 102
Nickel, welding, 413
Nicrosilal, 282
Nicuage steel, 229
Ni-Hard, 282
Niobium, properties, 360, 361; welding, 414
Ni-Resist, 280
Nitralloy steels, 209
Nitriding, 209; effect on fatigue, 28
Nitrogen, 157, 399; case-hardening steel by, 209; effects on steel, 171, 209, 399
Nodular iron, 280
Non-destructive tests, 33
Non-metallic inclusions, 50
Non-shrinking tool set, 217, 236
Normalising of steel, 177
Notch bend, 16
Notch brittleness, 16
Nuclear energy, 356
Numerical aperture, 54

Objective in microscope, 54
Octahedral planes, 69, 114, 167
Olsen cupping test, 32
Open-hearth process, 156
Optical pyrometer, 422
Orange peel effect, 25, 308
Order and disorder change, 75
Orientation of crystals, preferred, 80, 118
Orowan, 15
Orton cone, 416
Overheating, over-annealing of steel, 179

Overstressing, 27
Overvoltage, 139, 142
Oxidation of cast iron, by $V_2O_5$5, 150
Oxide films, effect on corrosion, 143, 247, 260
Oxygen, effects in welding, 399; in copper, 301; in steel making, 157
Oxygen-free copper, 301

Paints for preventing corrosion, 148
Passification of zinc, 147
Passivity, 143, 247
Patenting, 200
Pearlite, 161; divorced, 176
Pellini diagram, 18
Peltier effect, 416
Penetrameter, 34
Penetration by non-ferrous metals, 121
Peritectic transformations, 89, 125, 304, 314
Permalloy, 299
Petroforge, 130
Phosphate coats, 149
Phosphates, effects on corrosion, 144
Phosphor-bronze, 315
Phosphorus, in bronze, 217; in cast iron, 273; in copper, 315; in steel, 170
Photoelectric pyrometer, 427
Photography, 56
Pickling inhibitor, 131, 144
Pickling of metals, 131
Pig iron, manufacture of, 152
Pipe in ingots, 64
Pitting by corrosion, 144, 148
Plane notation, 79
Plane, octahedral, 74, 113, 167
Planes in cubic lattice, 78; in HCP lattice, 79
Planes direction, 79
Plastic bronze, 317
Plastic deformation, 104
Plastic strain, effect in ageing, 99
Platinum, 362; iridium, 362; osmium, 362; palladium, 362; rhodium, 362; ruthenium, 362; solders, 393

# INDEX

Plug drawing of tubes, 132
Plumber's solder, 394
Plutonium, 356, 357
Point defects, 93
Pole figure, 81
Poling copper, 301
Polishing, electrolytic, 47; of microsections, 46
Polygonisation, 112
Polytetrafluoroethylene, 372
Porosity in metals, 66, 328, 358
Potentiometer, 420
Pourbaix diagram, 139
Powder metallurgy, 372
Powder X-ray method, 77
Powders, 135
Precipitation hardness, 94
Precision castings, 70
Preece, test, 147
Preferential orientation of crystals, 80, 118
Pressing, 132
Pressure die casting, 374
Pressure welding, 394
Prestal, 366
Proof stress, 8
Propeller, brass for, 310
Properties, and constitution, 92; at high temperature, 21; at low temperature, 18; of fine grained steel, 168
Protection of iron and steel, 146
PTFE, 372
Puddling process, 153
Punches, steel for, 173
Pyrometers, 415; errors in, 422; sheaths for, 419; suction type, 424
Pyroversum pyrometer, 416
Pythagoras, 419

Quench and fracture test, 168
Quench cracks, 189
Quenched and tempered steels, properties of, 174, 206, 208, 230, 232, 238
Quenching media, 185
Quenching of steel, 185

RAE Farnborough, 365
Radiation pyrometers, 422
Radiation shielding, 375
Radiography, 33
Radium, use for tests, 36
Rails, steel for, 173, 217
Rapid cooling, effect on structure, 61, 182
Razor, steel for, 173
Reamers, steel for, 173
Recalescence, 163
Recovery, 112, 122
Recrystallisation of metals, 122, 175
Reducing atmosphere on copper, 303
Reduction of area, 8, 9
Redux bonding, 397
Reeds, 66
Regenerators in furnaces, 157
Relative valency effect, 76
Remanence, 296
Resistance materials, 261, 313, 325
Resolution of structure, 55
Restrainers in pickling, 131
Reversion, 99
RH degassing, 158
Rhodium, 362
Rimming steels, 66
Robertson test, 18
Rock drills, steel for, 217
Rockwell hardness test, 5
Rokes, 68
Rolls, 130
Rotor process, 158
RR alloys, 341
Ruling section, 229
Rupture test, 24
Rust-resisting steels, 245

Sacrificial corrosion, 142
Safety factor, 227
Saws, steel for, 173
Schaeffler diagram, 411
Screwing dies, 173
Seams, 68
Season cracking, 120
Secondary hardening of steel, 242

Seger cone, 416
Segregation in ingots, 40, 68
Selection of micro-sections, 45
Selenium in copper, 411
Semi-conductors, 289
Semi-steel, 278
Sentinel cone, 416
Shafts, steel for, 173, 186, 231
Shallow hardening steels, 201
Shape of particles, 184
Shears, steel for, 173, 239
Sheets, annealing of, 175
Shell moulding, 70
Shells, 68
Shore Scleroscope, 5
Shrinkage, 63
Sigma, 258
Silal, 285
Silchrome steel, 224
Silfos, 394
Silica fibres, 364
Silicon–aluminium alloys, 337
Silicon bronze, 325
Silicon, in bronze, 325; in cast iron, 272, 285; in copper, 291; in steel, 169, 223, 261, 299
Silicon nitride, 268
Silver bearings, 372
Silver bronze, 312
Silver–copper equilibrium diagram, 87
Silver, in copper, 291; in lead, 375
Silver solder, 393
Sinking of tubes, 132
Sintered bearings, 372
Size factor, 76; of crystals, effect on properties, 114
Slag inclusions, 155, 166
Slip bands, 104
Slip, block, 104; direction of, 114
Small angle boundary, 110
Small specimens, mounting of, 46
Smith diagram, 26
'S' monel, 313
Snaps, steel for, 173
Soils, corrosion in, 145
Soldering, 393
Solidification of metals, 59
Solid phase welding, 396

Solid solutions, 76, 85, 116
Solidus, 82
Solubility curve, 96
Solution hardening, 188
Sorbite, 186
Space lattice, 74
Specifications, 2
Speed of loading, 10
Spheroidal graphite, 280
Spheroidising anneal, 181
Spike, 362
Springs, bronze for, 279, 294; faults in, 234; steel for, 173, 234, 245
Stabilised steel, 66
Stacking fault, 93; on fatigue, 30; hardening, 188, 225; on recovery, 123
Stainless iron and steel, 245; welding of, 408
Stainless maraging steel, 233
Stainless precipitation hardens steel, 246, 248
Staybrite, 246
Steel ingots, defects in, 66; pipe in, 64; structure of, 62
Steel sheets, annealing, 175; manufacture of, 132; structure of, 160
Steels, pearlite-reduced, 228; niobium, 228
Stellite, 406
Stereographic projection, 80
Strain, ageing, 171; bands, 105; definition of, 6; on fatigue, 30; twins, 115
Strain ratio, $\bar{r}$, 118
Strain tempering, 226
Stream degassing, 159
Strengthening mechanisms, 227
Stress corrosion, 120, 258, 310, 414
Stress intensity factor $K$, 17
Stress raisers, effect on fatigue, 27
Stress relief, 122; of cast iron, 282
Stress–strain curves, 7
Stretcher strains, 119
Structural alloy steels, 227
Structures, of carbon steels, 160; of ingots, 61; of metals, 61
Sub-critical anneal, 175

# INDEX

Sub-zero properties, 20
Suction pyrometer, 424
Sulfinuz process, 210
Sulphur, in cast iron, 272; in copper, 301; in steel, 169
Sulphurous gases, effect on nickel, 261, 417
Sulphur prints, 40
Superconductors, 292
Supercooling, 59
Superheating cast iron, 278
Super-lattice, 75
Super-plasticity, 366; alloys, 367
Superston alloy, 323
Surface hardening of steel, 203
Swedish wrought iron, 154
Swift test, 33
Switch boxes, iron for, 282
Synthetic moulding sand, 69

Taps, steel for, 237
Tellurium–copper, 277, 411
Tellurium in lead, 375
Temper, brittleness, 221; colours, 176; hardening, 97, 261, 324
Temperatures, effect on properties, 20; for annealing and hardening steel, 176; for forging, 127; measurement of, 415
Tempering, auto, 189; of hardened steel, 186, 187; of high-speed steel, 242; of stainless steel, 253
Tempilstiks, 416
Tensile test, 5
Ternary equilibrium diagrams, 90, 253
Ternary eutectic, 91
Test, bar, 63; fluorescent, 33; for contamination of couple, 422; for intercrystalline corrosion of zinc, 373; for season cracking, 120; for weld decay, 257; McQuaid–Ehn, 168; texture, 81; X-ray, 33
Thermal cycling growth of uranium, 359
Thermal paints, 416
Thermit, 327

Thermocouple pyrometer, 417; errors in, 369
Thermomechanical treatment, 225
Thermoplastics, 384
Thermosetting polymers, 379
Thomas process, 156
Thompson effect, 416
Tin–antimony equilibrium diagram, 369
Tin, corrosion of, 147
Tin-cry, 115
Tin in brass, 308, 310
Tin–magnesium equilibrium diagram, 88
Tinman's solder, 393
Tin sweat, 68
Titanium, alloys, 352
Titanium welding, 414
Tocco hardening, 211
Tool steels, 236; hardening of, 176; high-speed, 240; softening of, 181; tempering of, 186
Tough pitch copper, 301
Transformation civilian, 196; military 196
Transformation hardening, 187
Transformer steel, 299
Transverse test, 31
Triaxial stresses, 13
Troostite, 183
Tropenas converter, 156
TTT curve, 192
Tubes, steel for, 173
Tungsten carbide, 223, 242
Tungsten, in steel, 223, 238, 297
Turbine blades, metal for, 246, 265, 313
Twinned crystals, 115
Twinned martensite, 195
Tyres, steel for, 173

Underannealing of steel, 179, 402
Ultra-high-tensile steel, 232
Undercooling, 59
Unidirection freezing, 362
Uranium, swelling, growth, wrinkling, 329

Vacuum arc furnace, 160
Vacuum degassing, 158
Valency electrons, 71
Valves, steel for, 245, 262
Vanadium in steels, 223, 238, 240
Vermicula iron, 281
Vertical illuminator, 56
Vicker's hardness test, 5
Visco-elastic, 390
Volume change to martensite, 214
Volume changes in steels, 189

Wanner optical pyrometer, 425
Wear, metal to resist, 203, 217, 280
Weldanka, 246, 410
Weld-decay, 256; prevention of, 410
Welding, aluminium bronze, 412; beryllium, 414; brasses, 411; effect on plate, 402; high-tensile steel, 403; light alloys, 413; nickel, 412; niobium, 414; of cast iron, 406; of copper, 410; of stainless steels, 408; silicon bronze, 412; tin bronze, 412; titanium, 414; zirconium, 414
Welds, corrosion of, 409; radiographs of, 35
Whiskers, 365

Whiteheart malleable iron, 283
White metals, 369
Widmanstatten structures, 103, 165, 180, 305, 310, 319
Wien's Law, 426
Wire ropes, steel for, 173
Work hardening, 189
Wrought iron, manufacture of, 153; nature of, 154

X-rays, 33, 77

Y alloy, 341
Yield point, 7, 14, 20
Young's modulus, 6, 363

Zinc, corrosion of, 147
Zinc, die castings, 252
Zinc, hot-working temperature, 128; in brass, 304
Zirconium, corrosion of, 143; in stainless steel, 248; welding, 414
Zircaloy, 360
Zone refining, 86